Finite Difference Computing: Theory and Software Applications

Finite Difference Computing: Theory and Software Applications

Adam Reid

MURPHY & MOORE
www.murphy-moorepublishing.com

Published by Murphy & Moore Publishing
1 Rockefeller Plaza,
New York City, NY 10020, USA
www.murphy-moorepublishing.com

Finite Difference Computing: Theory and Software Applications
Adam Reid

International Standard Book Number: 978-1-63987-227-5 (Hardback)

Cataloging-in-Publication Data

Finite difference computing : theory and software applications / Adam Reid.
    p. cm.
Includes bibliographical references and index.
ISBN 978-1-63987-227-5
1. Finite differences. 2. Finite differences--Data processing. 3. Finite differences--Software.
4. Numerical analysis--Data processing. I. Reid, Adam.
QA431 .F56 2022
515.45--dc23

Contents

Preface

The main aim of this book is to educate learners and enhance their research focus by presenting diverse topics covering this vast field. This is an advanced book which compiles significant studies by distinguished experts in the area of analysis. This book addresses successive solutions to the challenges arising in the area of application, along with it; the book provides scope for future developments.

Finite difference methods (FDM) are a class of numerical techniques which are used for solving differential equations by estimating derivatives with finite differences. It involves discretizing the spatial domain and time interval. The value of the solution at these discrete points is approximated by solving algebraic equations having finite differences and values from adjacent points. Finite difference methods transform ordinary differential equations or partial differential equations, into a system of linear equations that can be solved by matrix algebra techniques. Modern computers can perform these linear algebra computations efficiently, which has led to the widespread use of FDM in modern numerical analysis. It is considered to be one of the most common approaches to the numerical solution of partial differential equations. This book is compiled in such a manner, that it will provide in-depth knowledge about the theory and practice of finite difference computing. Also included herein is a detailed explanation of the various concepts and applications of this method. Students, researchers, experts and all associated with finite-difference methods will benefit alike from this book.

It was a great honour to edit this book, though there were challenges, as it involved a lot of communication and networking between me and the editorial team. However, the end result was this all-inclusive book covering diverse themes in the field.

Finally, it is important to acknowledge the efforts of the contributors for their excellent chapters, through which a wide variety of issues have been addressed. I would also like to thank my colleagues for their valuable feedback during the making of this book.

Adam Reid

1

Diffusion Equations: Methods, Schemes and Applications

The famous *diffusion equation*, also known as the *heat equation*, reads

$$\frac{\partial u}{\partial t} = \alpha \frac{\partial^2 u}{\partial x^2},$$

where $u(x, t)$ is the unknown function to be solved for, x is a coordinate in space, and t is time. The coefficient α is the *diffusion coefficient* and determines how fast u changes in time. A quick short form for the diffusion equation is $u_t = \alpha u_{xx}$.

Compared to the wave equation, $u_{tt} = c^2 u_{xx}$, which looks very similar, the diffusion equation features solutions that are very different from those of the wave equation. Also, the diffusion equation makes quite different demands to the numerical methods.

Typical diffusion problems may experience rapid change in the very beginning, but then the evolution of u becomes slower and slower. The solution is usually very smooth, and after some time, one cannot recognize the initial shape of u. This is in sharp contrast to solutions of the wave equation where the initial shape is preserved in homogeneous media – the solution is then basically a moving initial condition. The standard wave equation $u_{tt} = c^2 u_{xx}$ has solutions that propagate with speed c forever, without changing shape, while the diffusion equation converges to a *stationary solution* $\bar{u}(x)$ as $t \to \infty$. In this limit, $u_t = 0$, and $\bar{u}$ is governed by $\bar{u}''(x) = 0$. This stationary limit of the diffusion equation is called the *Laplace* equation and arises in a very wide range of applications throughout the sciences.

It is possible to solve for $u(x, t)$ using an explicit scheme, as we do in Sect. 1.1, but the time step restrictions soon become much less favorable than for an explicit scheme applied to the wave equation. And of more importance, since the solution u of the diffusion equation is very smooth and changes slowly, small time steps are not convenient and not required by accuracy as the diffusion process converges to a stationary state. Therefore, implicit schemes (as described in Sect. 1.2) are popular, but these require solutions of systems of algebraic equations. We shall use ready-made software for this purpose, but also program some simple iterative methods. The exposition is, as usual in this book, very basic and focuses on the basic ideas and how to implement. More comprehensive mathematical treatments and classical analysis of the methods are found in lots of textbooks. A favorite of ours in this respect is the one by LeVeque [13]. The books by Strikwerda [17] and by Lapidus and Pinder [12] are also highly recommended as additional material on the topic.

1.1 An Explicit Method for the 1D Diffusion Equation

Explicit finite difference methods for the wave equation $u_{tt} = c^2 u_{xx}$ can be used, with small modifications, for solving $u_t = \alpha u_{xx}$ as well. The exposition below assumes that the reader is familiar with the basic ideas of discretization and implementation of A Complete Study of Waves Equations from Chapter 2. Readers not familiar with the Forward Euler, Backward Euler, and Crank-Nicolson (or centered or midpoint) discretization methods in time should consult, e.g., Section 1.1 in [9].

1.1.1 The Initial-Boundary Value Problem for 1D Diffusion

To obtain a unique solution of the diffusion equation, or equivalently, to apply numerical methods, we need initial and boundary conditions. The diffusion equation goes with one initial condition $u(x, 0) = I(x)$, where I is a prescribed function. One boundary condition is required at each point on the boundary, which in 1D means that u must be known, u_x must be known, or some combination of them.

We shall start with the simplest boundary condition: $u = 0$. The complete initial-boundary value diffusion problem in one space dimension can then be specified as

$$\frac{\partial u}{\partial t} = \alpha \frac{\partial^2 u}{\partial x^2} + f, \qquad\qquad x \in (0, L),\ t \in (0, T] \qquad (1.1)$$

$$u(x, 0) = I(x), \qquad\qquad x \in [0, L] \qquad (1.2)$$

$$u(0, t) = 0, \qquad\qquad t > 0, \qquad (1.3)$$

$$u(L, t) = 0, \qquad\qquad t > 0. \qquad (1.4)$$

With only a first-order derivative in time, only one *initial condition* is needed, while the second-order derivative in space leads to a demand for two *boundary conditions*. We have added a source term $f = f(x, t)$, which is convenient when testing implementations.

Diffusion equations like (1.1) have a wide range of applications throughout physical, biological, and financial sciences. One of the most common applications is propagation of heat, where $u(x, t)$ represents the temperature of some substance at point x and time t. Other applications are listed in Sect. 1.8.

1.1.2 Forward Euler Scheme

The first step in the discretization procedure is to replace the domain $[0, L] \times [0, T]$ by a set of mesh points. Here we apply equally spaced mesh points

$$x_i = i \Delta x, \quad i = 0, \ldots, N_x,$$

and

$$t_n = n \Delta t, \quad n = 0, \ldots, N_t.$$

Moreover, u_i^n denotes the mesh function that approximates $u(x_i, t_n)$ for $i = 0, \ldots, N_x$ and $n = 0, \ldots, N_t$. Requiring the PDE (1.1) to be fulfilled at a mesh point (x_i, t_n) leads to the equation

$$\frac{\partial}{\partial t} u(x_i, t_n) = \alpha \frac{\partial^2}{\partial x^2} u(x_i, t_n) + f(x_i, t_n). \tag{1.5}$$

The next step is to replace the derivatives by finite difference approximations. The computationally simplest method arises from using a forward difference in time and a central difference in space:

$$[D_t^+ u = \alpha D_x D_x u + f]_i^n. \tag{1.6}$$

Written out,

$$\frac{u_i^{n+1} - u_i^n}{\Delta t} = \alpha \frac{u_{i+1}^n - 2u_i^n + u_{i-1}^n}{\Delta x^2} + f_i^n. \tag{1.7}$$

We have turned the PDE into algebraic equations, also often called discrete equations. The key property of the equations is that they are algebraic, which makes them easy to solve. As usual, we anticipate that u_i^n is already computed such that u_i^{n+1} is the only unknown in (1.7). Solving with respect to this unknown is easy:

$$u_i^{n+1} = u_i^n + F\left(u_{i+1}^n - 2u_i^n + u_{i-1}^n\right) + \Delta t f_i^n, \tag{1.8}$$

where we have introduced the *mesh Fourier number*:

$$F = \alpha \frac{\Delta t}{\Delta x^2}. \tag{1.9}$$

F is the key parameter in the discrete diffusion equation

Note that F is a *dimensionless* number that lumps the key physical parameter in the problem, α, and the discretization parameters Δx and Δt into a single parameter. Properties of the numerical method are critically dependent upon the value of F (see Sect. 1.3 for details).

The computational algorithm then becomes

1. compute $u_i^0 = I(x_i)$ for $i = 0, \ldots, N_x$
2. for $n = 0, 1, \ldots, N_t$:
   (a) apply (1.8) for all the internal spatial points $i = 1, \ldots, N_x - 1$
   (b) set the boundary values $u_i^{n+1} = 0$ for $i = 0$ and $i = N_x$

The algorithm is compactly and fully specified in Python:

```python
import numpy as np
x = np.linspace(0, L, Nx+1)     # mesh points in space
dx = x[1] - x[0]
t = np.linspace(0, T, Nt+1)     # mesh points in time
dt = t[1] - t[0]
F = a*dt/dx**2
u   = np.zeros(Nx+1)            # unknown u at new time level
u_n = np.zeros(Nx+1)            # u at the previous time level
```

```
# Set initial condition u(x,0) = I(x)
for i in range(0, Nx+1):
    u_n[i] = I(x[i])

for n in range(0, Nt):
    # Compute u at inner mesh points
    for i in range(1, Nx):
        u[i] = u_n[i] + F*(u_n[i-1] - 2*u_n[i] + u_n[i+1]) + \
            dt*f(x[i], t[n])

    # Insert boundary conditions
    u[0] = 0;   u[Nx] = 0

    # Update u_n before next step
    u_n[:]= u
```

Note that we use a for α in the code, motivated by easy visual mapping between the variable name and the mathematical symbol in formulas.

We need to state already now that the shown algorithm does not produce meaningful results unless $F \leq 1/2$. Why is explained in Sect. 3.3.

1.1.3 Implementation

The file diffu1D_u0.py contains a complete function solver_FE_simple for solving the 1D diffusion equation with $u = 0$ on the boundary as specified in the algorithm above:

```
import numpy as np

def solver_FE_simple(I, a, f, L, dt, F, T):
    """
    Simplest expression of the computational algorithm
    using the Forward Euler method and explicit Python loops.
    For this method F <= 0.5 for stability.
    """
    import time;  t0 = time.clock()  # For measuring the CPU time

    Nt = int(round(T/float(dt)))
    t = np.linspace(0, Nt*dt, Nt+1)    # Mesh points in time
    dx = np.sqrt(a*dt/F)
    Nx = int(round(L/dx))
    x = np.linspace(0, L, Nx+1)        # Mesh points in space
    # Make sure dx and dt are compatible with x and t
    dx = x[1] - x[0]
    dt = t[1] - t[0]

    u   = np.zeros(Nx+1)
    u_n = np.zeros(Nx+1)

    # Set initial condition u(x,0) = I(x)
    for i in range(0, Nx+1):
        u_n[i] = I(x[i])
```

```
    for n in range(0, Nt):
        # Compute u at inner mesh points
        for i in range(1, Nx):
            u[i] = u_n[i] + F*(u_n[i-1] - 2*u_n[i] + u_n[i+1]) + \
                   dt*f(x[i], t[n])

        # Insert boundary conditions
        u[0] = 0;  u[Nx] = 0

        # Switch variables before next step
        #u_n[:] = u  # safe, but slow
        u_n, u = u, u_n

    t1 = time.clock()
    return u_n, x, t, t1-t0  # u_n holds latest u
```

A faster alternative is available in the function `solver_FE`, which adds the possibility of solving the finite difference scheme by vectorization. The vectorized version replaces the explicit loop

```
for i in range(1, Nx):
    u[i] = u_n[i] + F*(u_n[i-1] - 2*u_n[i] + u_n[i+1]) \
           + dt*f(x[i], t[n])
```

by arithmetics on displaced slices of the u array:

```
u[1:Nx] = u_n[1:Nx] + F*(u_n[0:Nx-1] - 2*u_n[1:Nx] + u_n[2:Nx+1]) \
          + dt*f(x[1:Nx], t[n])
# or
u[1:-1] = u_n[1:-1] + F*(u_n[0:-2] - 2*u_n[1:-1] + u_n[2:]) \
          + dt*f(x[1:-1], t[n])
```

For example, the vectorized version runs 70 times faster than the scalar version in a case with 100 time steps and a spatial mesh of 10^5 cells.

The `solver_FE` function also features a callback function such that the user can process the solution at each time level. The callback function looks like `user_action(u, x, t, n)`, where u is the array containing the solution at time level n, x holds all the spatial mesh points, while t holds all the temporal mesh points. The `solver_FE` function is very similar to `solver_FE_simple` above:

```
def solver_FE(I, a, f, L, dt, F, T,
              user_action=None, version='scalar'):
    """
    Vectorized implementation of solver_FE_simple.
    """
    import time;  t0 = time.clock()  # for measuring the CPU time
```

```
Nt = int(round(T/float(dt)))
t = np.linspace(0, Nt*dt, Nt+1)    # Mesh points in time
dx = np.sqrt(a*dt/F)
Nx = int(round(L/dx))
x = np.linspace(0, L, Nx+1)         # Mesh points in space
# Make sure dx and dt are compatible with x and t
dx = x[1] - x[0]
dt = t[1] - t[0]

u   = np.zeros(Nx+1)   # solution array
u_n = np.zeros(Nx+1)   # solution at t-dt

# Set initial condition
for i in range(0,Nx+1):
    u_n[i] = I(x[i])

if user_action is not None:
    user_action(u_n, x, t, 0)

for n in range(0, Nt):
    # Update all inner points
    if version == 'scalar':
        for i in range(1, Nx):
            u[i] = u_n[i] +\
                    F*(u_n[i-1] - 2*u_n[i] + u_n[i+1]) +\
                    dt*f(x[i], t[n])

    elif version == 'vectorized':
        u[1:Nx] = u_n[1:Nx] + \
                    F*(u_n[0:Nx-1] - 2*u_n[1:Nx] + u_n[2:Nx+1]) +\
                    dt*f(x[1:Nx], t[n])
    else:
        raise ValueError('version=%s' % version)

    # Insert boundary conditions
    u[0] = 0;  u[Nx] = 0
    if user_action is not None:
        user_action(u, x, t, n+1)

    # Switch variables before next step
    u_n, u = u, u_n

t1 = time.clock()
return t1-t0
```

1.1.4 Verification

Exact solution of discrete equations Before thinking about running the functions
in the previous section, we need to construct a suitable test example for verification.
It appears that a manufactured solution that is linear in time and at most quadratic
in space fulfills the Forward Euler scheme exactly. With the restriction that $u = 0$
for $x = 0, L$, we can try the solution

$$u(x,t) = 5tx(L - x).$$

Inserted in the PDE, it requires a source term

$$f(x,t) = 10\alpha t + 5x(L-x).$$

With the formulas from Appendix A.4 we can easily check that the manufactured u fulfills the scheme:

$$
\begin{aligned}
[D_t^+ u = \alpha D_x D_x u + f]_i^n &= [5x(L-x)D_t^+ t = 5t\alpha D_x D_x (xL - x^2) \\
&\quad + 10\alpha t + 5x(L-x)]_i^n \\
&= [5x(L-x) = 5t\alpha(-2) + 10\alpha t + 5x(L-x)]_i^n,
\end{aligned}
$$

which is a 0=0 expression. The computation of the source term, given any u, is easily automated with sympy:

```
import sympy as sym
x, t, a, L = sym.symbols('x t a L')
u = x*(L-x)*5*t

def pde(u):
    return sym.diff(u, t) - a*sym.diff(u, x, x)

f = sym.simplify(pde(u))
```

Now we can choose any expression for u and automatically get the suitable source term f. However, the manufactured solution u will in general not be exactly reproduced by the scheme: only constant and linear functions are differentiated correctly by a forward difference, while only constant, linear, and quadratic functions are differentiated exactly by a $[D_x D_x u]_i^n$ difference.

The numerical code will need to access the u and f above as Python functions. The exact solution is wanted as a Python function u_exact(x, t), while the source term is wanted as f(x, t). The parameters a and L in u and f above are symbols and must be replaced by float objects in a Python function. This can be done by redefining a and L as float objects and performing substitutions of symbols by numbers in u and f. The appropriate code looks like this:

```
a = 0.5
L = 1.5
u_exact = sym.lambdify(
    [x, t], u.subs('L', L).subs('a', a), modules='numpy')
f = sym.lambdify(
    [x, t], f.subs('L', L).subs('a', a), modules='numpy')
I = lambda x: u_exact(x, 0)
```

Here we also make a function I for the initial condition.

The idea now is that our manufactured solution should be exactly reproduced by the code (to machine precision). For this purpose we make a test function for comparing the exact and numerical solutions at the end of the time interval:

```
def test_solver_FE():
    # Define u_exact, f, I as explained above

    dx = L/3   # 3 cells
    F = 0.5
    dt = F*dx**2

    u, x, t, cpu = solver_FE_simple(
        I=I, a=a, f=f, L=L, dt=dt, F=F, T=2)
    u_e = u_exact(x, t[-1])
    diff = abs(u_e - u).max()
    tol = 1E-14
    assert diff < tol, 'max diff solver_FE_simple: %g' % diff

    u, x, t, cpu = solver_FE(
        I=I, a=a, f=f, L=L, dt=dt, F=F, T=2,
        user_action=None, version='scalar')
    u_e = u_exact(x, t[-1])
    diff = abs(u_e - u).max()
    tol = 1E-14
    assert diff < tol, 'max diff solver_FE, scalar: %g' % diff

    u, x, t, cpu = solver_FE(
        I=I, a=a, f=f, L=L, dt=dt, F=F, T=2,
        user_action=None, version='vectorized')
    u_e = u_exact(x, t[-1])
    diff = abs(u_e - u).max()
    tol = 1E-14
    assert diff < tol, 'max diff solver_FE, vectorized: %g' % diff
```

The critical value $F = 0.5$

We emphasize that the value F=0.5 is critical: the tests above will fail if F has a larger value. This is because the Forward Euler scheme is unstable for $F > 1/2$.

The reader may wonder if $F = 1/2$ is safe or if $F < 1/2$ should be required. Experiments show that $F = 1/2$ works fine for $u_t = \alpha u_{xx}$, so there is no accumulation of rounding errors in this case and hence no need to introduce any safety factor to keep F away from the limiting value 0.5.

Checking convergence rates If our chosen exact solution does not satisfy the discrete equations exactly, we are left with checking the convergence rates, just as we did previously for the wave equation. However, with the Euler scheme here, we have different accuracies in time and space, since we use a second order approximation to the spatial derivative and a first order approximation to the time derivative. Thus, we must expect different convergence rates in time and space. For the numerical error,

$$E = C_t \Delta t^r + C_x \Delta x^p,$$

we should get convergence rates $r = 1$ and $p = 2$ (C_t and C_x are unknown constants). As previously, in Sect. 2.2.3, we simplify matters by introducing a single discretization parameter h:

$$h = \Delta t, \quad \Delta x = K h^{r/p},$$

where K is any constant. This allows us to factor out only *one* discretization parameter h from the formula:

$$E = C_t h + C_x (K h^{r/p})^p = \tilde{C} h^r, \quad \tilde{C} = C_t + C_s K^r .$$

The computed rate r should approach 1 with increasing resolution.

It is tempting, for simplicity, to choose $K = 1$, which gives $\Delta x = h^{r/p}$, expected to be $\sqrt{\Delta t}$. However, we have to control the stability requirement: $F \leq \frac{1}{2}$, which means

$$\frac{\alpha \Delta t}{\Delta x^2} \leq \frac{1}{2} \quad \Rightarrow \quad \Delta x \geq \sqrt{2\alpha} h^{1/2},$$

implying that $K = \sqrt{2\alpha}$ is our choice in experiments where we lie on the stability limit $F = 1/2$.

1.1.5 Numerical Experiments

When a test function like the one above runs silently without errors, we have some evidence for a correct implementation of the numerical method. The next step is to do some experiments with more interesting solutions.

We target a scaled diffusion problem where x/L is a new spatial coordinate and $\alpha t/L^2$ is a new time coordinate. The source term f is omitted, and u is scaled by $\max_{x \in [0,L]} |I(x)|$ (see Section 3.2 in [11] for details). The governing PDE is then

$$\frac{\partial u}{\partial t} = \frac{\partial^2 u}{\partial x^2},$$

in the spatial domain $[0, L]$, with boundary conditions $u(0) = u(1) = 0$. Two initial conditions will be tested: a discontinuous plug,

$$I(x) = \begin{cases} 0, & |x - L/2| > 0.1 \\ 1, & \text{otherwise} \end{cases}$$

and a smooth Gaussian function,

$$I(x) = e^{-\frac{1}{2\sigma^2}(x-L/2)^2} .$$

The functions plug and gaussian in diffu1D_u0.py run the two cases, respectively:

```
def plug(scheme='FE', F=0.5, Nx=50):
    L = 1.
    a = 1.
    T = 0.1
    # Compute dt from Nx and F
    dx = L/Nx;   dt = F/a*dx**2
```

```
    def I(x):
        """Plug profile as initial condition."""
        if abs(x-L/2.0) > 0.1:
            return 0
        else:
            return 1

    cpu = viz(I, a, L, dt, F, T,
              umin=-0.1, umax=1.1,
              scheme=scheme, animate=True, framefiles=True)
    print 'CPU time:', cpu

def gaussian(scheme='FE', F=0.5, Nx=50, sigma=0.05):
    L = 1.
    a = 1.
    T = 0.1
    # Compute dt from Nx and F
    dx = L/Nx;  dt = F/a*dx**2

    def I(x):
        """Gaussian profile as initial condition."""
        return exp(-0.5*((x-L/2.0)**2)/sigma**2)

    u, cpu = viz(I, a, L, dt, F, T,
                 umin=-0.1, umax=1.1,
                 scheme=scheme, animate=True, framefiles=True)
    print 'CPU time:', cpu
```

These functions make use of the function `viz` for running the solver and visualizing the solution using a callback function with plotting:

```
def viz(I, a, L, dt, F, T, umin, umax,
        scheme='FE', animate=True, framefiles=True):

    def plot_u(u, x, t, n):
        plt.plot(x, u, 'r-', axis=[0, L, umin, umax],
                 title='t=%f' % t[n])
        if framefiles:
            plt.savefig('tmp_frame%04d.png' % n)
        if t[n] == 0:
            time.sleep(2)
        elif not framefiles:
            # It takes time to write files so pause is needed
            # for screen only animation
            time.sleep(0.2)

    user_action = plot_u if animate else lambda u,x,t,n: None

    cpu = eval('solver_'+scheme)(I, a, L, dt, F, T,
                                 user_action=user_action)
    return cpu
```

Notice that this `viz` function stores all the solutions in a list `solutions` in the callback function. Modern computers have hardly any problem with storing a lot

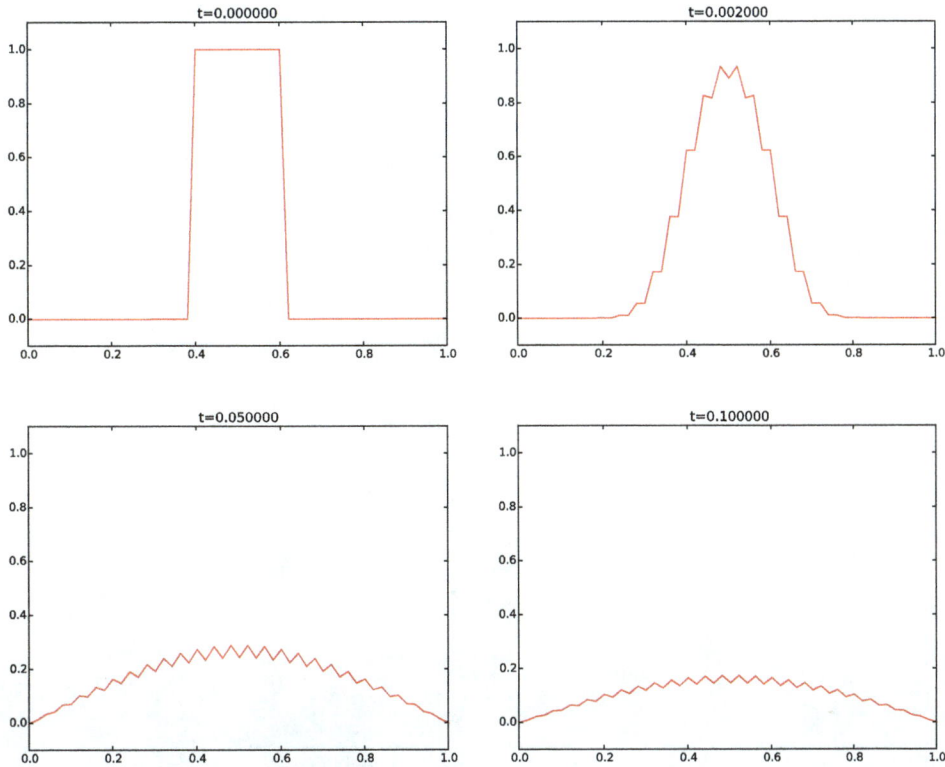

Fig. 1.1 Forward Euler scheme for $F = 0.5$

of such solutions for moderate values of N_x in 1D problems, but for 2D and 3D problems, this technique cannot be used and solutions must be stored in files.

Our experiments employ a time step $\Delta t = 0.0002$ and simulate for $t \in [0, 0.1]$. First we try the highest value of F: $F = 0.5$. This resolution corresponds to $N_x = 50$. A possible terminal command is

```
┌─────────┐
│Terminal │
└─────────┘
Terminal> python -c 'from diffu1D_u0 import gaussian
          gaussian("solver_FE", F=0.5, dt=0.0002)'
```

The $u(x, t)$ curve as a function of x is shown in Fig. 1.1 at four time levels.

Movie 1 https://raw.githubusercontent.com/hplgit/fdm-book/master/doc/pub/book/html/mov-diffu/diffu1D_u0_FE_plug/movie.ogg

We see that the curves have saw-tooth waves in the beginning of the simulation. This non-physical noise is smoothed out with time, but solutions of the diffusion equations are known to be smooth, and this numerical solution is definitely not smooth. Lowering F helps: $F \leq 0.25$ gives a smooth solution, see Fig. 1.2 (and a movie[1]).

---

[1] http://tinyurl.com/gokgkov/mov-diffu/diffu1D_u0_FE_plug_F025/movie.ogg

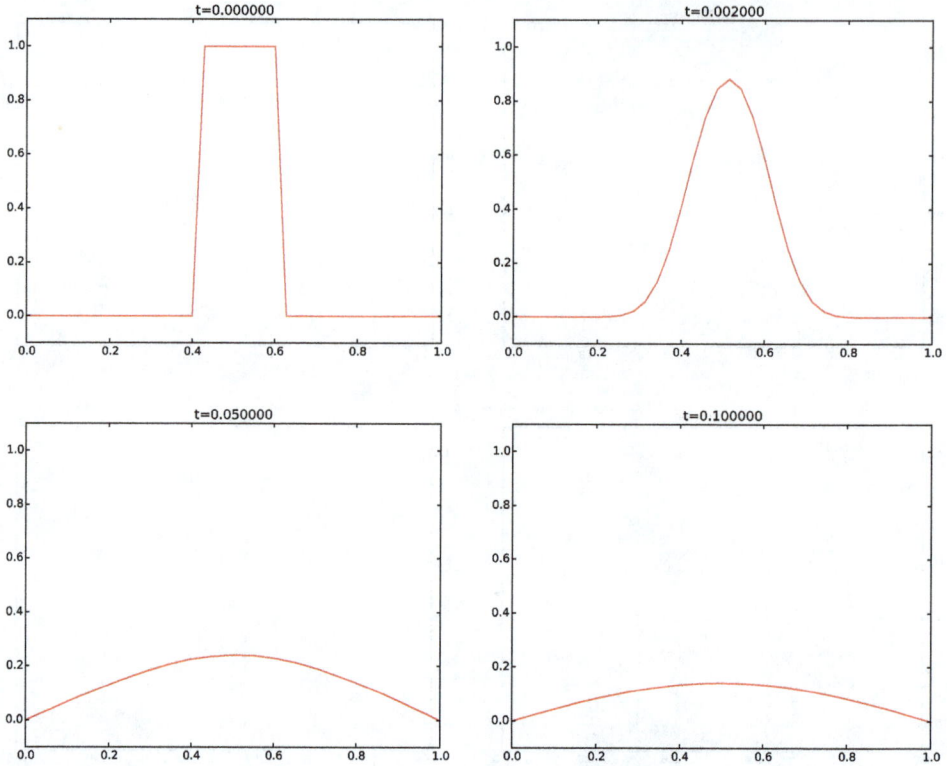

Fig. 1.2 Forward Euler scheme for $F = 0.25$

Increasing F slightly beyond the limit 0.5, to $F = 0.51$, gives growing, non-physical instabilities, as seen in Fig. 1.3.

Instead of a discontinuous initial condition we now try the smooth Gaussian function for $I(x)$. A simulation for $F = 0.5$ is shown in Fig. 1.4. Now the numerical solution is smooth for all times, and this is true for any $F \leq 0.5$.

Experiments with these two choices of $I(x)$ reveal some important observations:

- The Forward Euler scheme leads to growing solutions if $F > \frac{1}{2}$.
- $I(x)$ as a discontinuous plug leads to a saw tooth-like noise for $F = \frac{1}{2}$, which is absent for $F \leq \frac{1}{4}$.
- The smooth Gaussian initial function leads to a smooth solution for all relevant F values ($F \leq \frac{1}{2}$).

1.2   Implicit Methods for the 1D Diffusion Equation

Simulations with the Forward Euler scheme show that the time step restriction, $F \leq \frac{1}{2}$, which means $\Delta t \leq \Delta x^2/(2\alpha)$, may be relevant in the beginning of the diffusion process, when the solution changes quite fast, but as time increases, the process slows down, and a small Δt may be inconvenient. With *implicit schemes*, which lead to coupled systems of linear equations to be solved at each time level,

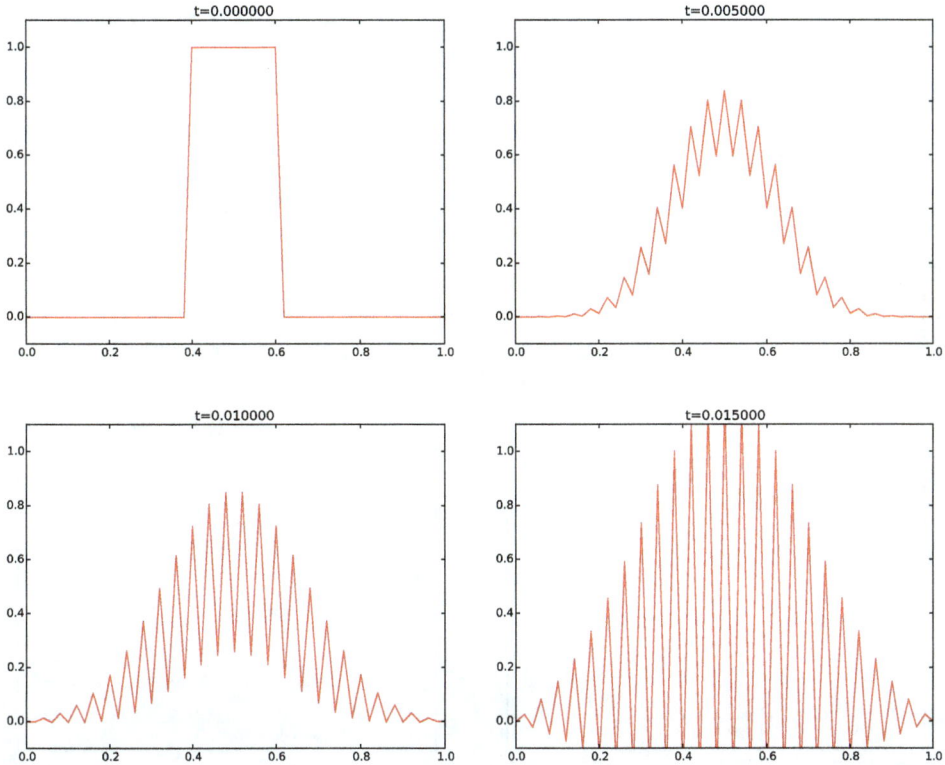

Fig. 1.3 Forward Euler scheme for $F = 0.51$

any size of Δt is possible (but the accuracy decreases with increasing Δt). The Backward Euler scheme, derived and implemented below, is the simplest implicit scheme for the diffusion equation.

1.2.1 Backward Euler Scheme

In (1.5), we now apply a backward difference in time, but the same central difference in space:

$$[D_t^- u = D_x D_x u + f]_i^n, \tag{3.10}$$

which written out reads

$$\frac{u_i^n - u_i^{n-1}}{\Delta t} = \alpha \frac{u_{i+1}^n - 2u_i^n + u_{i-1}^n}{\Delta x^2} + f_i^n. \tag{3.11}$$

Now we assume u_i^{n-1} is already computed, but that all quantities at the "new" time level n are unknown. This time it is not possible to solve with respect to u_i^n because this value couples to its neighbors in space, u_{i-1}^n and u_{i+1}^n, which are also unknown. Let us examine this fact for the case when $N_x = 3$. Equation (3.11) written for

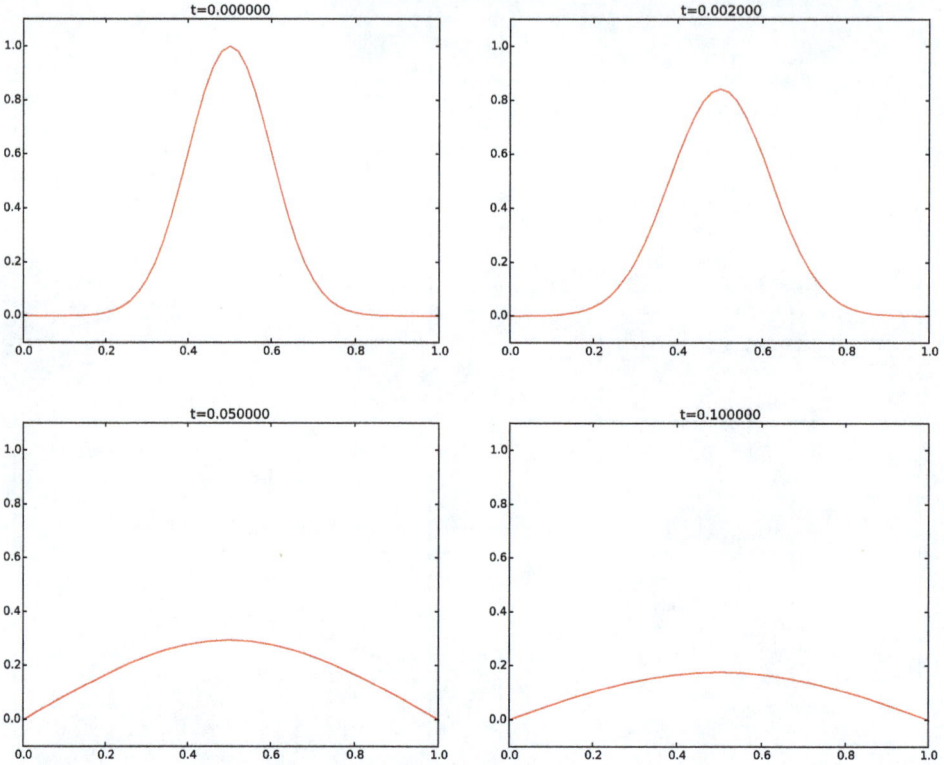

Fig. 1.4 Forward Euler scheme for $F = 0.5$

$i = 1, \ldots, Nx - 1 = 1, 2$ becomes

$$\frac{u_1^n - u_1^{n-1}}{\Delta t} = \alpha \frac{u_2^n - 2u_1^n + u_0^n}{\Delta x^2} + f_1^n \tag{3.12}$$

$$\frac{u_2^n - u_2^{n-1}}{\Delta t} = \alpha \frac{u_3^n - 2u_2^n + u_1^n}{\Delta x^2} + f_2^n. \tag{3.13}$$

The boundary values u_0^n and u_3^n are known as zero. Collecting the unknown new values u_1^n and u_2^n on the left-hand side and multiplying by Δt gives

$$(1 + 2F) u_1^n - F u_2^n = u_1^{n-1} + \Delta t f_1^n, \tag{3.14}$$

$$-F u_1^n + (1 + 2F) u_2^n = u_2^{n-1} + \Delta t f_2^n. \tag{3.15}$$

This is a coupled 2×2 system of algebraic equations for the unknowns u_1^n and u_2^n. The equivalent matrix form is

$$\begin{pmatrix} 1 + 2F & -F \\ -F & 1 + 2F \end{pmatrix} \begin{pmatrix} u_1^n \\ u_2^n \end{pmatrix} = \begin{pmatrix} u_1^{n-1} + \Delta t f_1^n \\ u_2^{n-1} + \Delta t f_2^n \end{pmatrix}.$$

Terminology: implicit vs. explicit methods

Discretization methods that lead to a coupled system of equations for the unknown function at a new time level are said to be *implicit methods*. The counterpart, *explicit methods*, refers to discretization methods where there is a simple explicit formula for the values of the unknown function at each of the spatial mesh points at the new time level. From an implementational point of view, implicit methods are more comprehensive to code since they require the solution of coupled equations, i.e., a matrix system, at each time level. With explicit methods we have a closed-form formula for the value of the unknown at each mesh point.

Very often explicit schemes have a restriction on the size of the time step that can be relaxed by using implicit schemes. In fact, implicit schemes are frequently unconditionally stable, so the size of the time step is governed by accuracy and not by stability. This is the great advantage of implicit schemes.

In the general case, (1.11) gives rise to a coupled $(N_x - 1) \times (N_x - 1)$ system of algebraic equations for all the unknown u_i^n at the interior spatial points $i = 1, \ldots, N_x - 1$. Collecting the unknowns on the left-hand side, (1.11) can be written

$$- F u_{i-1}^n + (1 + 2F) u_i^n - F u_{i+1}^n = u_i^{n-1}, \tag{3.16}$$

for $i = 1, \ldots, N_x - 1$. One can either view these equations as a system where the u_i^n values at the internal mesh points, $i = 1, \ldots, N_x - 1$, are unknown, or we may append the boundary values u_0^n and $u_{N_x}^n$ to the system. In the latter case, all u_i^n for $i = 0, \ldots, N_x$ are considered unknown, and we must add the boundary equations to the $N_x - 1$ equations in (1.16):

$$u_0^n = 0, \tag{3.17}$$
$$u_{N_x}^n = 0. \tag{3.18}$$

A coupled system of algebraic equations can be written on matrix form, and this is important if we want to call up ready-made software for solving the system. The equations (1.16) and (1.17)–(1.18) correspond to the matrix equation

$$AU = b$$

where $U = (u_0^n, \ldots, u_{N_x}^n)$, and the matrix A has the following structure:

$$A = \begin{pmatrix}
A_{0,0} & A_{0,1} & 0 & \cdots & & \cdots & \cdots & \cdots & & \cdots & & 0 \\
A_{1,0} & A_{1,1} & A_{1,2} & \ddots & & & & & & & & \vdots \\
0 & A_{2,1} & A_{2,2} & A_{2,3} & \ddots & & & & & & & \vdots \\
\vdots & \ddots & & \ddots & \ddots & 0 & & & & & & \vdots \\
\vdots & & \ddots & \ddots & \ddots & \ddots & \ddots & & & & & \vdots \\
\vdots & & & 0 & A_{i,i-1} & A_{i,i} & A_{i,i+1} & \ddots & & & & \vdots \\
\vdots & & & & \ddots & \ddots & \ddots & \ddots & & 0 & & \\
\vdots & & & & & \ddots & \ddots & \ddots & A_{N_x-1,N_x} & & \\
0 & \cdots & \cdots & \cdots & & \cdots & 0 & A_{N_x,N_x-1} & A_{N_x,N_x}
\end{pmatrix}.$$

$$\tag{1.19}$$

The nonzero elements are given by

$$A_{i,i-1} = -F \tag{1.20}$$
$$A_{i,i} = 1 + 2F \tag{1.21}$$
$$A_{i,i+1} = -F \tag{1.22}$$

in the equations for internal points, $i = 1, \ldots, N_x - 1$. The first and last equation correspond to the boundary condition, where we know the solution, and therefore we must have

$$A_{0,0} = 1, \tag{1.23}$$
$$A_{0,1} = 0, \tag{1.24}$$
$$A_{N_x,N_x-1} = 0, \tag{1.25}$$
$$A_{N_x,N_x} = 1. \tag{1.26}$$

The right-hand side b is written as

$$b = \begin{pmatrix} b_0 \\ b_1 \\ \vdots \\ b_i \\ \vdots \\ b_{N_x} \end{pmatrix} \tag{1.27}$$

with

$$b_0 = 0, \tag{1.28}$$
$$b_i = u_i^{n-1}, \quad i = 1, \ldots, N_x - 1, \tag{1.29}$$
$$b_{N_x} = 0. \tag{1.30}$$

We observe that the matrix A contains quantities that do not change in time. Therefore, A can be formed once and for all before we enter the recursive formulas for the time evolution. The right-hand side b, however, must be updated at each time step. This leads to the following computational algorithm, here sketched with Python code:

```
x = np.linspace(0, L, Nx+1)    # mesh points in space
dx = x[1] - x[0]
t = np.linspace(0, T, N+1)     # mesh points in time
u   = np.zeros(Nx+1)           # unknown u at new time level
u_n = np.zeros(Nx+1)           # u at the previous time level
```

```
# Data structures for the linear system
A = np.zeros((Nx+1, Nx+1))
b = np.zeros(Nx+1)

for i in range(1, Nx):
    A[i,i-1] = -F
    A[i,i+1] = -F
    A[i,i]   = 1 + 2*F
A[0,0] = A[Nx,Nx] = 1

# Set initial condition u(x,0) = I(x)
for i in range(0, Nx+1):
    u_n[i] = I(x[i])

import scipy.linalg

for n in range(0, Nt):
    # Compute b and solve linear system
    for i in range(1, Nx):
        b[i] = -u_n[i]
    b[0] = b[Nx] = 0
    u[:] = scipy.linalg.solve(A, b)

    # Update u_n before next step
    u_n[:] = u
```

Regarding verification, the same considerations apply as for the Forward Euler method (Sect. 3.1.4).

1.2.2   Sparse Matrix Implementation

We have seen from (1.19) that the matrix A is tridiagonal. The code segment above used a full, dense matrix representation of A, which stores a lot of values we know are zero beforehand, and worse, the solution algorithm computes with all these zeros. With $N_x + 1$ unknowns, the work by the solution algorithm is $\frac{1}{3}(N_x + 1)^3$ and the storage requirements $(N_x + 1)^2$. By utilizing the fact that A is tridiagonal and employing corresponding software tools that work with the three diagonals, the work and storage demands can be proportional to N_x only. This leads to a dramatic improvement: with $N_x = 200$, which is a realistic resolution, the code runs about 40,000 times faster and reduces the storage to just 1.5 %! It is no doubt that we should take advantage of the fact that A is tridiagonal.

The key idea is to apply a data structure for a tridiagonal or sparse matrix. The scipy.sparse package has relevant utilities. For example, we can store only the nonzero diagonals of a matrix. The package also has linear system solvers that operate on sparse matrix data structures. The code below illustrates how we can store only the main diagonal and the upper and lower diagonals.

```
# Representation of sparse matrix and right-hand side
main  = np.zeros(Nx+1)
lower = np.zeros(Nx)
upper = np.zeros(Nx)
b     = np.zeros(Nx+1)

# Precompute sparse matrix
main[:] = 1 + 2*F
lower[:] = -F
upper[:] = -F
# Insert boundary conditions
main[0] = 1
main[Nx] = 1

A = scipy.sparse.diags(
    diagonals=[main, lower, upper],
    offsets=[0, -1, 1], shape=(Nx+1, Nx+1),
    format='csr')
print A.todense()  # Check that A is correct

# Set initial condition
for i in range(0,Nx+1):
    u_n[i] = I(x[i])

for n in range(0, Nt):
    b = u_n
    b[0] = b[-1] = 0.0  # boundary conditions
    u[:] = scipy.sparse.linalg.spsolve(A, b)
    u_n[:] = u
```

The `scipy.sparse.linalg.spsolve` function utilizes the sparse storage structure of A and performs, in this case, a very efficient Gaussian elimination solve.

The program `diffu1D_u0.py` contains a function `solver_BE`, which implements the Backward Euler scheme sketched above. As mentioned in Sect. 1.1.2, the functions `plug` and `gaussian` run the case with $I(x)$ as a discontinuous plug or a smooth Gaussian function. All experiments point to two characteristic features of the Backward Euler scheme: 1) it is always stable, and 2) it always gives a smooth, decaying solution.

1.2.3 Crank-Nicolson Scheme

The idea in the Crank-Nicolson scheme is to apply centered differences in space and time, combined with an average in time. We demand the PDE to be fulfilled at the spatial mesh points, but midway between the points in the time mesh:

$$\frac{\partial}{\partial t}u\left(x_i, t_{n+\frac{1}{2}}\right) = \alpha\frac{\partial^2}{\partial x^2}u\left(x_i, t_{n+\frac{1}{2}}\right) + f\left(x_i, t_{n+\frac{1}{2}}\right),$$

for $i = 1, \ldots, N_x - 1$ and $n = 0, \ldots, N_t - 1$.

With centered differences in space and time, we get

$$[D_t u = \alpha D_x D_x u + f]_i^{n+\frac{1}{2}} .$$

On the right-hand side we get an expression

$$\frac{1}{\Delta x^2} \left(u_{i-1}^{n+\frac{1}{2}} - 2u_i^{n+\frac{1}{2}} + u_{i+1}^{n+\frac{1}{2}} \right) + f_i^{n+\frac{1}{2}} .$$

This expression is problematic since $u_i^{n+\frac{1}{2}}$ is not one of the unknowns we compute. A possibility is to replace $u_i^{n+\frac{1}{2}}$ by an arithmetic average:

$$u_i^{n+\frac{1}{2}} \approx \frac{1}{2} \left(u_i^n + u_i^{n+1} \right) .$$

In the compact notation, we can use the arithmetic average notation $\overline{u}^t$:

$$[D_t u = \alpha D_x D_x \overline{u}^t + f]_i^{n+\frac{1}{2}} .$$

We can also use an average for $f_i^{n+\frac{1}{2}}$:

$$[D_t u = \alpha D_x D_x \overline{u}^t + \overline{f}^t]_i^{n+\frac{1}{2}} .$$

After writing out the differences and average, multiplying by Δt, and collecting all unknown terms on the left-hand side, we get

$$u_i^{n+1} - \frac{1}{2} F \left(u_{i-1}^{n+1} - 2u_i^{n+1} + u_{i+1}^{n+1} \right) = u_i^n + \frac{1}{2} F \left(u_{i-1}^n - 2u_i^n + u_{i+1}^n \right)$$
$$+ \frac{1}{2} f_i^{n+1} + \frac{1}{2} f_i^n . \qquad (1.31)$$

Also here, as in the Backward Euler scheme, the new unknowns u_{i-1}^{n+1}, u_i^{n+1}, and u_{i+1}^{n+1} are coupled in a linear system $AU = b$, where A has the same structure as in (1.19), but with slightly different entries:

$$A_{i,i-1} = -\frac{1}{2} F \qquad (1.32)$$

$$A_{i,i} = 1 + F \qquad (1.33)$$

$$A_{i,i+1} = -\frac{1}{2} F \qquad (1.34)$$

in the equations for internal points, $i = 1, \ldots, N_x - 1$. The equations for the boundary points correspond to

$$A_{0,0} = 1, \qquad (1.35)$$

$$A_{0,1} = 0, \qquad (1.36)$$

$$A_{N_x, N_x-1} = 0, \qquad (1.37)$$

$$A_{N_x, N_x} = 1 . \qquad (1.38)$$

The right-hand side b has entries

$$b_0 = 0, \tag{1.39}$$

$$b_i = u_i^{n-1} + \frac{1}{2}\left(f_i^n + f_i^{n+1}\right), \quad i = 1,\ldots,N_x - 1, \tag{1.40}$$

$$b_{N_x} = 0. \tag{1.41}$$

When verifying some implementation of the Crank-Nicolson scheme by convergence rate testing, one should note that the scheme is second order accurate in both space and time. The numerical error then reads

$$E = C_t \Delta t^r + C_x \Delta x^r,$$

where $r = 2$ (C_t and C_x are unknown constants, as before). When introducing a single discretization parameter, we may now simply choose

$$h = \Delta x = \Delta t,$$

which gives

$$E = C_t h^r + C_x h^r = (C_t + C_x)h^r,$$

where r should approach 2 as resolution is increased in the convergence rate computations.

1.2.4   The Unifying θ Rule

For the equation

$$\frac{\partial u}{\partial t} = G(u),$$

where $G(u)$ is some spatial differential operator, the θ-rule looks like

$$\frac{u_i^{n+1} - u_i^n}{\Delta t} = \theta G(u_i^{n+1}) + (1 - \theta)G(u_i^n).$$

The important feature of this time discretization scheme is that we can implement one formula and then generate a family of well-known and widely used schemes:

- $\theta = 0$ gives the Forward Euler scheme in time
- $\theta = 1$ gives the Backward Euler scheme in time
- $\theta = \frac{1}{2}$ gives the Crank-Nicolson scheme in time

In the compact difference notation, we write the θ rule as

$$[\overline{D}_t u = \alpha D_x D_x u]^{n+\theta}.$$

We have that $t_{n+\theta} = \theta t_{n+1} + (1 - \theta)t_n$.

Applied to the 1D diffusion problem, the θ-rule gives

$$\frac{u_i^{n+1} - u_i^n}{\Delta t} = \alpha \left(\theta \frac{u_{i+1}^{n+1} - 2u_i^{n+1} + u_{i-1}^{n+1}}{\Delta x^2} + (1 - \theta) \frac{u_{i+1}^n - 2u_i^n + u_{i-1}^n}{\Delta x^2} \right)$$
$$+ \theta f_i^{n+1} + (1 - \theta) f_i^n \,.$$

This scheme also leads to a matrix system with entries

$$A_{i,i-1} = -F\theta, \quad A_{i,i} = 1 + 2F\theta, \quad A_{i,i+1} = -F\theta,$$

while right-hand side entry b_i is

$$b_i = u_i^n + F(1 - \theta) \frac{u_{i+1}^n - 2u_i^n + u_{i-1}^n}{\Delta x^2} + \Delta t \theta f_i^{n+1} + \Delta t (1 - \theta) f_i^n \,.$$

The corresponding entries for the boundary points are as in the Backward Euler and Crank-Nicolson schemes listed earlier.

Note that convergence rate testing with implementations of the theta rule must adjust the error expression according to which of the underlying schemes is actually being run. That is, if $\theta = 0$ (i.e., Forward Euler) or $\theta = 1$ (i.e., Backward Euler), there should be first order convergence, whereas with $\theta = 0.5$ (i.e., Crank-Nicolson), one should get second order convergence (as outlined in previous sections).

1.2.5 Experiments

We can repeat the experiments from Sect. 1.1.5 to see if the Backward Euler or Crank-Nicolson schemes have problems with sawtooth-like noise when starting with a discontinuous initial condition. We can also verify that we can have $F > \frac{1}{2}$, which allows larger time steps than in the Forward Euler method.

The Backward Euler scheme always produces smooth solutions for any F. Figure 3.5 shows one example. Note that the mathematical discontinuity at $t = 0$ leads to a linear variation on a mesh, but the approximation to a jump becomes better as N_x increases. In our simulation, we specify Δt and F, and set N_x to $L/\sqrt{\alpha \Delta t / F}$. Since $N_x \sim \sqrt{F}$, the discontinuity looks sharper in the Crank-Nicolson simulations with larger F.

The Crank-Nicolson method produces smooth solutions for small F, $F \le \frac{1}{2}$, but small noise gets more and more evident as F increases. Figures 1.6 and 1.7 demonstrate the effect for $F = 3$ and $F = 10$, respectively. Section 1.3 explains why such noise occur.

1.2.6 The Laplace and Poisson Equation

The Laplace equation, $\nabla^2 u = 0$, and the Poisson equation, $-\nabla^2 u = f$, occur in numerous applications throughout science and engineering. In 1D these equations read $u''(x) = 0$ and $-u''(x) = f(x)$, respectively. We can solve 1D variants of the

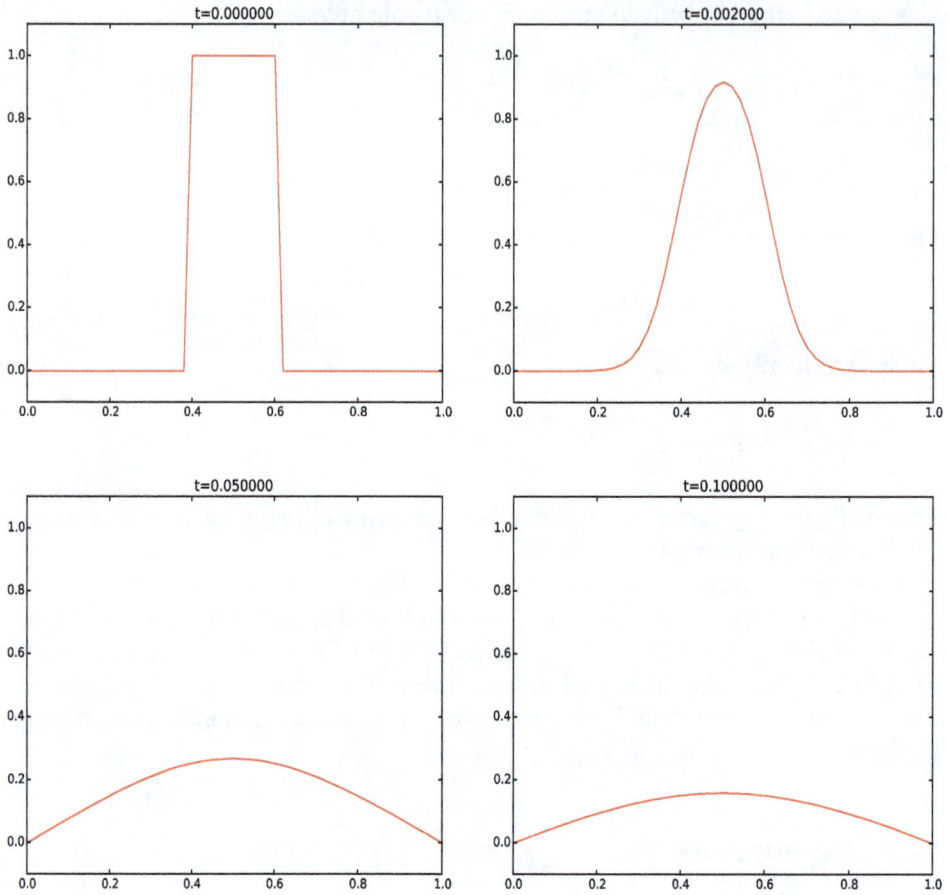

Fig. 1.5 Backward Euler scheme for $F = 0.5$

Laplace equations with the listed software, because we can interpret $u_{xx} = 0$ as the limiting solution of $u_t = \alpha u_{xx}$ when u reaches a steady state limit where $u_t \to 0$. Similarly, Poisson's equation $-u_{xx} = f$ arises from solving $u_t = u_{xx} + f$ and letting $t \to \infty$ so $u_t \to 0$.

Technically in a program, we can simulate $t \to \infty$ by just taking one large time step: $\Delta t \to \infty$. In the limit, the Backward Euler scheme gives

$$-\frac{u_{i+1}^{n+1} - 2u_i^{n+1} + u_{i-1}^{n+1}}{\Delta x^2} = f_i^{n+1},$$

which is nothing but the discretization $[-D_x D_x u = f]_i^{n+1} = 0$ of $-u_{xx} = f$.

The result above means that the Backward Euler scheme can solve the limit equation directly and hence produce a solution of the 1D Laplace equation. With the Forward Euler scheme we must do the time stepping since $\Delta t > \Delta x^2/\alpha$ is illegal and leads to instability. We may interpret this time stepping as solving the equation system from $-u_{xx} = f$ by iterating on a pseudo time variable.

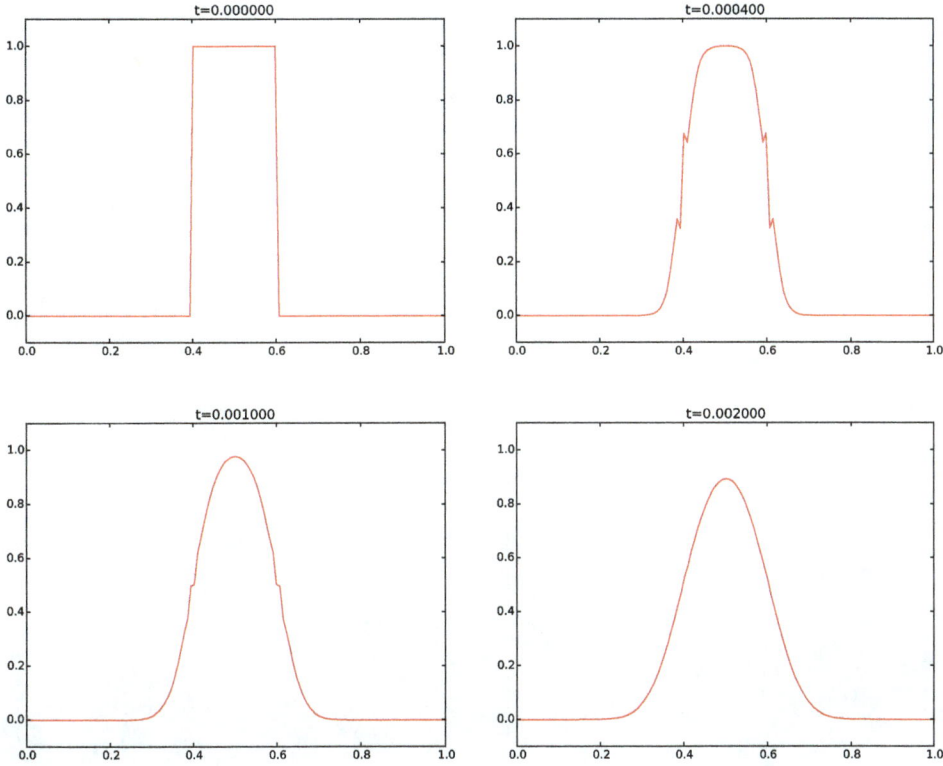

Fig. 1.6 Crank-Nicolson scheme for $F = 3$

1.3 Analysis of Schemes for the Diffusion Equation

The numerical experiments in Sect. 1.1.5 and 1.2.5 reveal that there are some numerical problems with the Forward Euler and Crank-Nicolson schemes: sawtooth-like noise is sometimes present in solutions that are, from a mathematical point of view, expected to be smooth. This section presents a mathematical analysis that explains the observed behavior and arrives at criteria for obtaining numerical solutions that reproduce the qualitative properties of the exact solutions. In short, we shall explain what is observed in Fig. 1.1–3.7.

1.3.1 Properties of the Solution

A particular characteristic of diffusive processes, governed by an equation like

$$u_t = \alpha u_{xx}, \tag{1.42}$$

is that the initial shape $u(x, 0) = I(x)$ spreads out in space with time, along with a decaying amplitude. Three different examples will illustrate the spreading of u in space and the decay in time.

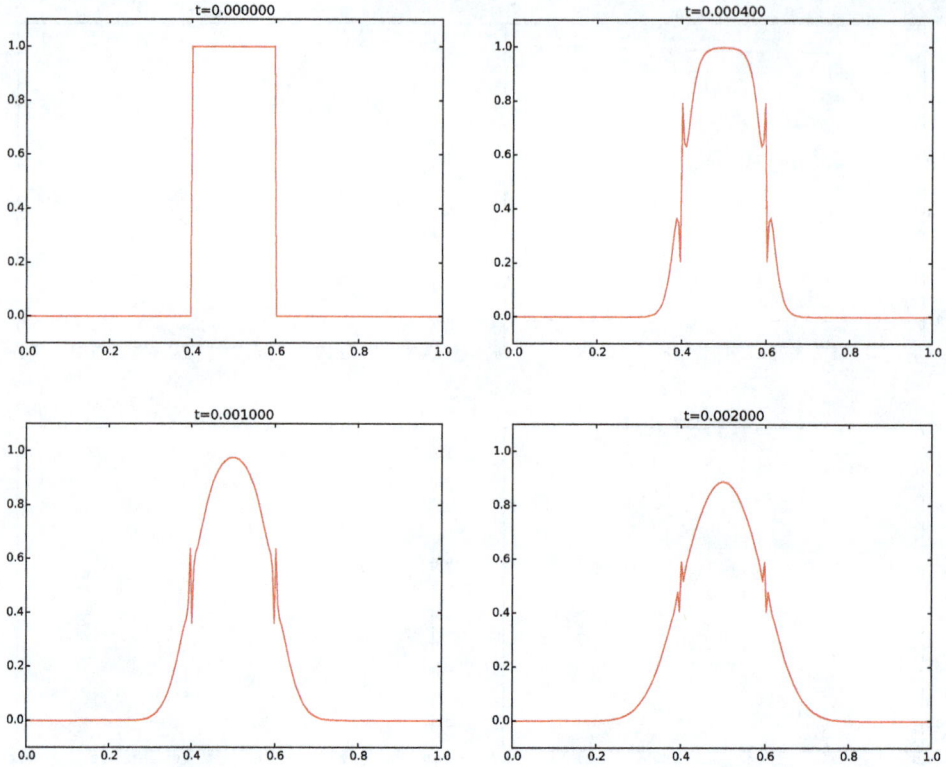

Fig. 1.7 Crank-Nicolson scheme for $F = 10$

Similarity solution The diffusion equation (1.42) admits solutions that depend on $\eta = (x - c)/\sqrt{4\alpha t}$ for a given value of c. One particular solution is

$$u(x, t) = a \operatorname{erf}(\eta) + b, \tag{1.43}$$

where

$$\operatorname{erf}(\eta) = \frac{2}{\sqrt{\pi}} \int_0^\eta e^{-\zeta^2} d\zeta, \tag{1.44}$$

is the *error function*, and a and b are arbitrary constants. The error function lies in $(-1, 1)$, is odd around $\eta = 0$, and goes relatively quickly to ± 1:

$$\lim_{\eta \to -\infty} \operatorname{erf}(\eta) = -1,$$

$$\lim_{\eta \to \infty} \operatorname{erf}(\eta) = 1,$$

$$\operatorname{erf}(\eta) = -\operatorname{erf}(-\eta),$$

$$\operatorname{erf}(0) = 0,$$

$$\operatorname{erf}(2) = 0.99532227,$$

$$\operatorname{erf}(3) = 0.99997791 .$$

As $t \to 0$, the error function approaches a step function centered at $x = c$. For a diffusion problem posed on the unit interval $[0, 1]$, we may choose the step at $x = 1/2$ (meaning $c = 1/2$), $a = -1/2$, $b = 1/2$. Then

$$u(x,t) = \frac{1}{2}\left(1 - \text{erf}\left(\frac{x - \frac{1}{2}}{\sqrt{4\alpha t}}\right)\right) = \frac{1}{2}\text{erfc}\left(\frac{x - \frac{1}{2}}{\sqrt{4\alpha t}}\right), \tag{1.45}$$

where we have introduced the *complementary error function* $\text{erfc}(\eta) = 1 - \text{erf}(\eta)$. The solution (1.45) implies the boundary conditions

$$u(0,t) = \frac{1}{2}\left(1 - \text{erf}\left(\frac{-1/2}{\sqrt{4\alpha t}}\right)\right), \tag{1.46}$$

$$u(1,t) = \frac{1}{2}\left(1 - \text{erf}\left(\frac{1/2}{\sqrt{4\alpha t}}\right)\right). \tag{1.47}$$

For small enough t, $u(0,t) \approx 1$ and $u(1,t) \approx 0$, but as $t \to \infty$, $u(x,t) \to 1/2$ on $[0, 1]$.

Solution for a Gaussian pulse The standard diffusion equation $u_t = \alpha u_{xx}$ admits a Gaussian function as solution:

$$u(x,t) = \frac{1}{\sqrt{4\pi\alpha t}}\exp\left(-\frac{(x-c)^2}{4\alpha t}\right). \tag{1.48}$$

At $t = 0$ this is a Dirac delta function, so for computational purposes one must start to view the solution at some time $t = t_\epsilon > 0$. Replacing t by $t_\epsilon + t$ in (1.48) makes it easy to operate with a (new) t that starts at $t = 0$ with an initial condition with a finite width. The important feature of (1.48) is that the standard deviation σ of a sharp initial Gaussian pulse increases in time according to $\sigma = \sqrt{2\alpha t}$, making the pulse diffuse and flatten out.

Solution for a sine component Also, (1.42) admits a solution of the form

$$u(x,t) = Q e^{-at} \sin(kx). \tag{1.49}$$

The parameters Q and k can be freely chosen, while inserting (1.49) in (1.42) gives the constraint

$$a = -\alpha k^2.$$

A very important feature is that the initial shape $I(x) = Q\sin(kx)$ undergoes a damping $\exp(-\alpha k^2 t)$, meaning that rapid oscillations in space, corresponding to large k, are very much faster dampened than slow oscillations in space, corresponding to small k. This feature leads to a smoothing of the initial condition with time. (In fact, one can use a few steps of the diffusion equation as a method for removing noise in signal processing.) To judge how good a numerical method is, we may look at its ability to smoothen or dampen the solution in the same way as the PDE does.

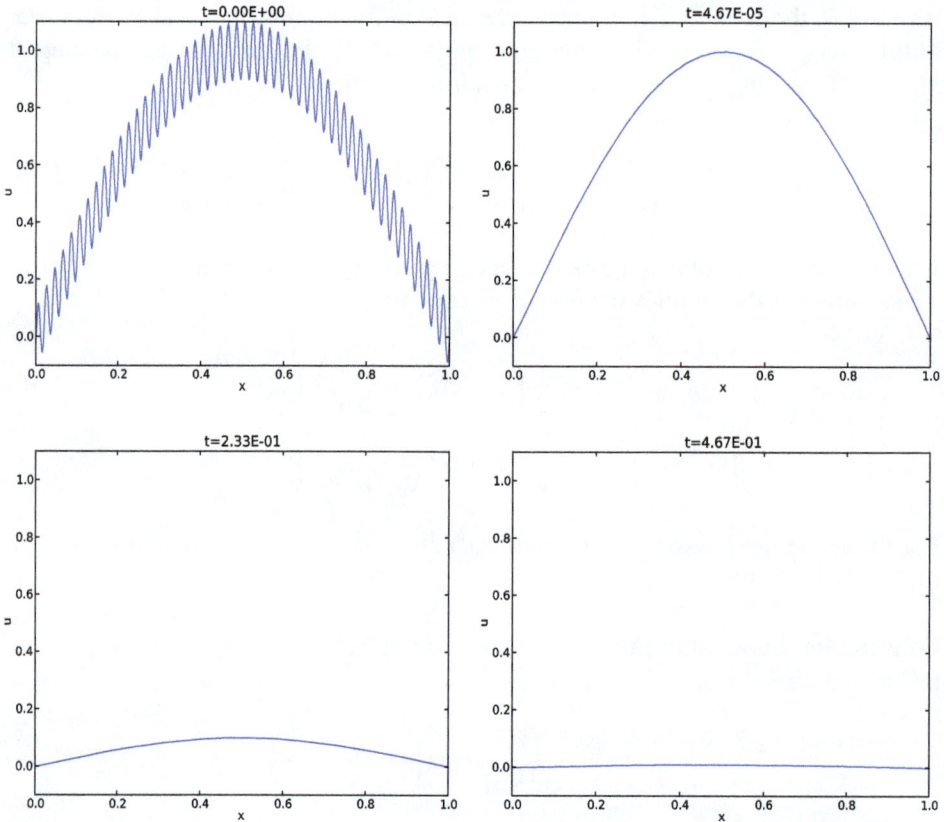

Fig. 1.8 Evolution of the solution of a diffusion problem: initial condition (*upper left*), 1/100 reduction of the small waves (*upper right*), 1/10 reduction of the long wave (*lower left*), and 1/100 reduction of the long wave (*lower right*)

The following example illustrates the damping properties of (1.49). We consider the specific problem

$$
\begin{aligned}
u_t &= u_{xx}, & x \in (0,1), \ t \in (0,T], \\
u(0,t) &= u(1,t) = 0, & t \in (0,T], \\
u(x,0) &= \sin(\pi x) + 0.1 \sin(100\pi x).
\end{aligned}
$$

The initial condition has been chosen such that adding two solutions like (1.49) constructs an analytical solution to the problem:

$$
u(x,t) = e^{-\pi^2 t} \sin(\pi x) + 0.1 e^{-\pi^2 10^4 t} \sin(100\pi x). \tag{1.50}
$$

Figure 3.8 illustrates the rapid damping of rapid oscillations $\sin(100\pi x)$ and the very much slower damping of the slowly varying $\sin(\pi x)$ term. After about $t = 0.5 \cdot 10^{-4}$ the rapid oscillations do not have a visible amplitude, while we have to wait until $t \sim 0.5$ before the amplitude of the long wave $\sin(\pi x)$ becomes very small.

1.3.2 Analysis of Discrete Equations

A counterpart to (1.49) is the complex representation of the same function:

$$u(x,t) = Q e^{-at} e^{ikx},$$

where $i = \sqrt{-1}$ is the imaginary unit. We can add such functions, often referred to as wave components, to make a Fourier representation of a general solution of the diffusion equation:

$$u(x,t) \approx \sum_{k \in K} b_k e^{-\alpha k^2 t} e^{ikx}, \qquad (1.51)$$

where K is a set of an infinite number of k values needed to construct the solution. In practice, however, the series is truncated and K is a finite set of k values needed to build a good approximate solution. Note that (1.50) is a special case of (1.51) where $K = \{\pi, 100\pi\}$, $b_\pi = 1$, and $b_{100\pi} = 0.1$.

The amplitudes b_k of the individual Fourier waves must be determined from the initial condition. At $t = 0$ we have $u \approx \sum_k b_k \exp(ikx)$ and find K and b_k such that

$$I(x) \approx \sum_{k \in K} b_k e^{ikx} . \qquad (1.52)$$

(The relevant formulas for b_k come from Fourier analysis, or equivalently, a least-squares method for approximating $I(x)$ in a function space with basis $\exp(ikx)$.)

Much insight about the behavior of numerical methods can be obtained by investigating how a wave component $\exp(-\alpha k^2 t) \exp(ikx)$ is treated by the numerical scheme. It appears that such wave components are also solutions of the schemes, but the damping factor $\exp(-\alpha k^2 t)$ varies among the schemes. To ease the forthcoming algebra, we write the damping factor as A^n. The exact amplification factor corresponding to A is $A_e = \exp(-\alpha k^2 \Delta t)$.

1.3.3 Analysis of the Finite Difference Schemes

We have seen that a general solution of the diffusion equation can be built as a linear combination of basic components

$$e^{-\alpha k^2 t} e^{ikx} .$$

A fundamental question is whether such components are also solutions of the finite difference schemes. This is indeed the case, but the amplitude $\exp(-\alpha k^2 t)$ might be modified (which also happens when solving the ODE counterpart $u' = -\alpha u$). We therefore look for numerical solutions of the form

$$u_q^n = A^n e^{ikq\Delta x} = A^n e^{ikx}, \qquad (1.53)$$

where the amplification factor A must be determined by inserting the component into an actual scheme. Note that A^n means A raised to the power of n, n being the index in the time mesh, while the superscript n in u_q^n just denotes u at time t_n.

Stability The exact amplification factor is $A_e = \exp(-\alpha^2 k^2 \Delta t)$. We should there-
fore require $|A| < 1$ to have a decaying numerical solution as well. If $-1 \le A < 0$,
A^n will change sign from time level to time level, and we get stable, non-physical
oscillations in the numerical solutions that are not present in the exact solution.

Accuracy To determine how accurately a finite difference scheme treats one wave
component (1.53), we see that the basic deviation from the exact solution is reflected
in how well A^n approximates A_e^n, or how well A approximates A_e. We can plot A_e
and the various expressions for A, and we can make Taylor expansions of A/A_e to
see the error more analytically.

Truncation error As an alternative to examining the accuracy of the damping of
a wave component, we can perform a general truncation error analysis as explained
in Appendix B. Such results are more general, but less detailed than what we get
from the wave component analysis. The truncation error can almost always be
computed and represents the error in the numerical model when the exact solution
is substituted into the equations. In particular, the truncation error analysis tells
the order of the scheme, which is of fundamental importance when verifying codes
based on empirical estimation of convergence rates.

1.3.4   Analysis of the Forward Euler Scheme

The Forward Euler finite difference scheme for $u_t = \alpha u_{xx}$ can be written as

$$[D_t^+ u = \alpha D_x D_x u]_q^n .$$

Inserting a wave component (1.53) in the scheme demands calculating the terms

$$e^{ikq\Delta x}[D_t^+ A]^n = e^{ikq\Delta x} A^n \frac{A-1}{\Delta t},$$

and

$$A^n D_x D_x [e^{ikx}]_q = A^n \left(-e^{ikq\Delta x} \frac{4}{\Delta x^2} \sin^2\left(\frac{k\Delta x}{2}\right)\right).$$

Inserting these terms in the discrete equation and dividing by $A^n e^{ikq\Delta x}$ leads to

$$\frac{A-1}{\Delta t} = -\alpha \frac{4}{\Delta x^2} \sin^2\left(\frac{k\Delta x}{2}\right),$$

and consequently

$$A = 1 - 4F \sin^2 p \tag{1.54}$$

where

$$F = \frac{\alpha \Delta t}{\Delta x^2} \tag{1.55}$$

is the *numerical Fourier number*, and $p = k\Delta x/2$. The complete numerical solution is then

$$u_q^n = \left(1 - 4F \sin^2 p\right)^n e^{ikq\Delta x} . \tag{1.56}$$

Stability We easily see that $A \leq 1$. However, the A can be less than -1, which will lead to growth of a numerical wave component. The criterion $A \geq -1$ implies

$$4F \sin^2(p/2) \leq 2.$$

The worst case is when $\sin^2(p/2) = 1$, so a sufficient criterion for stability is

$$F \leq \frac{1}{2}, \tag{1.57}$$

or expressed as a condition on Δt:

$$\Delta t \leq \frac{\Delta x^2}{2\alpha}. \tag{1.58}$$

Note that halving the spatial mesh size, $\Delta x \to \frac{1}{2}\Delta x$, requires Δt to be reduced by a factor of $1/4$. The method hence becomes very expensive for fine spatial meshes.

Accuracy Since A is expressed in terms of F and the parameter we now call $p = k\Delta x/2$, we should also express A_e by F and p. The exponent in A_e is $-\alpha k^2 \Delta t$, which equals $-Fk^2\Delta x^2 = -F4p^2$. Consequently,

$$A_e = \exp(-\alpha k^2 \Delta t) = \exp(-4Fp^2).$$

All our A expressions as well as A_e are now functions of the two dimensionless parameters F and p.

Computing the Taylor series expansion of A/A_e in terms of F can easily be done with aid of sympy:

```
def A_exact(F, p):
    return exp(-4*F*p**2)

def A_FE(F, p):
    return 1 - 4*F*sin(p)**2

from sympy import *
F, p = symbols('F p')
A_err_FE = A_FE(F, p)/A_exact(F, p)
print A_err_FE.series(F, 0, 6)
```

The result is

$$\frac{A}{A_e} = 1 - 4F \sin^2 p + 2Fp^2 - 16F^2 p^2 \sin^2 p + 8F^2 p^4 + \cdots$$

Recalling that $F = \alpha\Delta t/\Delta x^2$, $p = k\Delta x/2$, and that $\sin^2 p \leq 1$, we realize that the dominating terms in A/A_e are at most

$$1 - 4\alpha \frac{\Delta t}{\Delta x^2} + \alpha\Delta t - 4\alpha^2 \Delta t^2 + \alpha^2 \Delta t^2 \Delta x^2 + \cdots.$$

Truncation error We follow the theory explained in Appendix B. The recipe is to set up the scheme in operator notation and use formulas from Appendix B.2.4 to derive an expression for the residual. The details are documented in Appendix B.6.1. We end up with a truncation error

$$R_i^n = \mathcal{O}(\Delta t) + \mathcal{O}(\Delta x^2).$$

Although this is not the true error $u_e(x_i, t_n) - u_i^n$, it indicates that the true error is of the form

$$E = C_t \Delta t + C_x \Delta x^2$$

for two unknown constants C_t and C_x.

1.3.5 Analysis of the Backward Euler Scheme

Discretizing $u_t = \alpha u_{xx}$ by a Backward Euler scheme,

$$[D_t^- u = \alpha D_x D_x u]_q^n,$$

and inserting a wave component (1.53), leads to calculations similar to those arising from the Forward Euler scheme, but since

$$e^{ikq\Delta x}[D_t^- A]^n = A^n e^{ikq\Delta x} \frac{1 - A^{-1}}{\Delta t},$$

we get

$$\frac{1 - A^{-1}}{\Delta t} = -\alpha \frac{4}{\Delta x^2} \sin^2 \left(\frac{k\Delta x}{2} \right),$$

and then

$$A = \left(1 + 4F \sin^2 p \right)^{-1}. \tag{1.59}$$

The complete numerical solution can be written

$$u_q^n = \left(1 + 4F \sin^2 p \right)^{-n} e^{ikq\Delta x}. \tag{1.60}$$

Stability We see from (1.59) that $0 < A < 1$, which means that all numerical wave components are stable and non-oscillatory for any $\Delta t > 0$.

Truncation error The derivation of the truncation error for the Backward Euler scheme is almost identical to that for the Forward Euler scheme. We end up with

$$R_i^n = \mathcal{O}(\Delta t) + \mathcal{O}(\Delta x^2).$$

1.3.6   Analysis of the Crank-Nicolson Scheme

The Crank-Nicolson scheme can be written as

$$[D_t u = \alpha D_x D_x \bar{u}^x]_q^{n+\frac{1}{2}},$$

or

$$[D_t u]_q^{n+\frac{1}{2}} = \frac{1}{2}\alpha \left([D_x D_x u]_q^n + [D_x D_x u]_q^{n+1} \right).$$

Inserting (1.53) in the time derivative approximation leads to

$$[D_t A^n e^{ikq\Delta x}]^{n+\frac{1}{2}} = A^{n+\frac{1}{2}} e^{ikq\Delta x} \frac{A^{\frac{1}{2}} - A^{-\frac{1}{2}}}{\Delta t} = A^n e^{ikq\Delta x} \frac{A-1}{\Delta t}.$$

Inserting (1.53) in the other terms and dividing by $A^n e^{ikq\Delta x}$ gives the relation

$$\frac{A-1}{\Delta t} = -\frac{1}{2}\alpha \frac{4}{\Delta x^2} \sin^2\left(\frac{k\Delta x}{2}\right)(1+A),$$

and after some more algebra,

$$A = \frac{1 - 2F\sin^2 p}{1 + 2F\sin^2 p}. \tag{1.61}$$

The exact numerical solution is hence

$$u_q^n = \left(\frac{1 - 2F\sin^2 p}{1 + 2F\sin^2 p}\right)^n e^{ikq\Delta x}. \tag{1.62}$$

Stability The criteria $A > -1$ and $A < 1$ are fulfilled for any $\Delta t > 0$. Therefore, the solution cannot grow, but it will oscillate if $1 - 2F\sin^p < 0$. To avoid such non-physical oscillations, we must demand $F \le \frac{1}{2}$.

Truncation error The truncation error is derived in Appendix B.6.1:

$$R_i^{n+\frac{1}{2}} = \mathcal{O}(\Delta x^2) + \mathcal{O}(\Delta t^2).$$

1.3.7   Analysis of the Leapfrog Scheme

An attractive feature of the Forward Euler scheme is the explicit time stepping and no need for solving linear systems. However, the accuracy in time is only $\mathcal{O}(\Delta t)$. We can get an explicit *second-order* scheme in time by using the Leapfrog method:

$$[D_{2t} u = \alpha D_x D x u + f]_q^n.$$

Written out,

$$u_q^{n+1} = u_q^{n-1} + \frac{2\alpha\Delta t}{\Delta x^2}\left(u_{q+1}^n - 2u_q^n + u_{q-1}^n\right) + f(x_q, t_n).$$

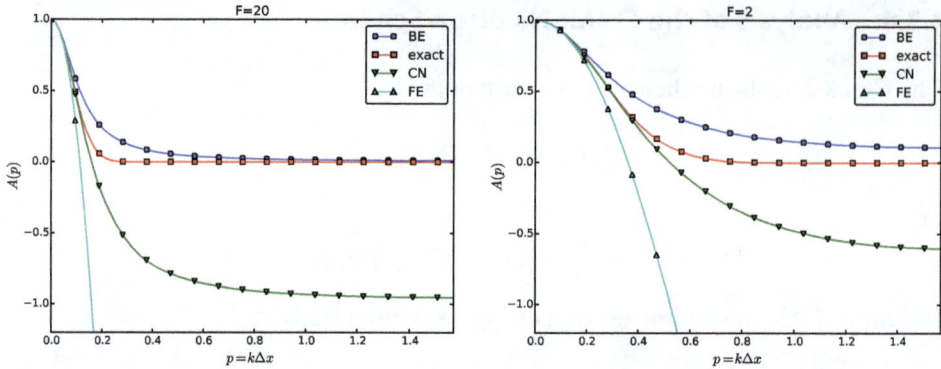

Fig. 1.9 Amplification factors for large time steps

We need some formula for the first step, u_q^1, but for that we can use a Forward Euler step.

Unfortunately, the Leapfrog scheme is always unstable for the diffusion equation. To see this, we insert a wave component $A^n e^{ikx}$ and get

$$\frac{A - A^{-1}}{\Delta t} = -\alpha \frac{4}{\Delta x^2} \sin^2 p,$$

or

$$A^2 + 4F \sin^2 p \, A - 1 = 0,$$

which has roots

$$A = -2F \sin^2 p \pm \sqrt{4F^2 \sin^4 p + 1}.$$

Both roots have $|A| > 1$ so the amplitude always grows, which is not in accordance with the physics of the problem. However, for a PDE with a first-order derivative in space, instead of a second-order one, the Leapfrog scheme performs very well. Details are provided in Sect. 4.1.3.

1.3.8 Summary of Accuracy of Amplification Factors

We can plot the various amplification factors against $p = k\Delta x/2$ for different choices of the F parameter. Figures 1.9, 1.10, and 1.11 show how long and small waves are damped by the various schemes compared to the exact damping. As long as all schemes are stable, the amplification factor is positive, except for Crank-Nicolson when $F > 0.5$.

The effect of negative amplification factors is that A^n changes sign from one time level to the next, thereby giving rise to oscillations in time in an animation of the solution. We see from Fig. 1.9 that for $F = 20$, waves with $p \geq \pi/4$ undergo a damping close to -1, which means that the amplitude does not decay and that the wave component jumps up and down (flips amplitude) in time. For $F = 2$ we have a damping of a factor of 0.5 from one time level to the next, which is very much smaller than the exact damping. Short waves will therefore fail to be effectively

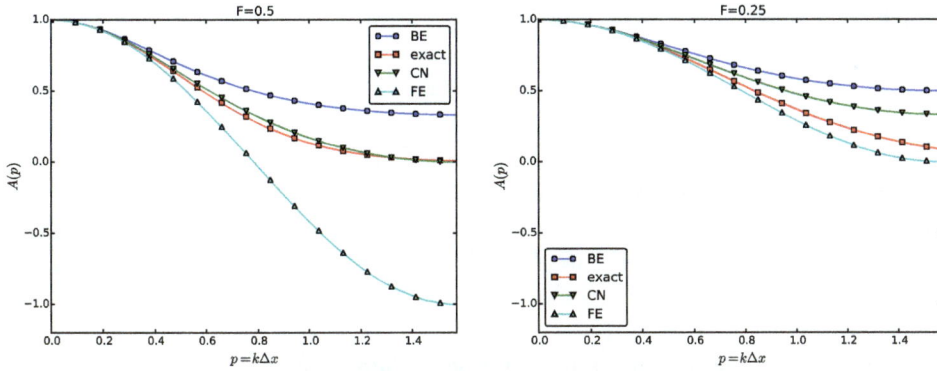

Fig. 1.10 Amplification factors for time steps around the Forward Euler stability limit

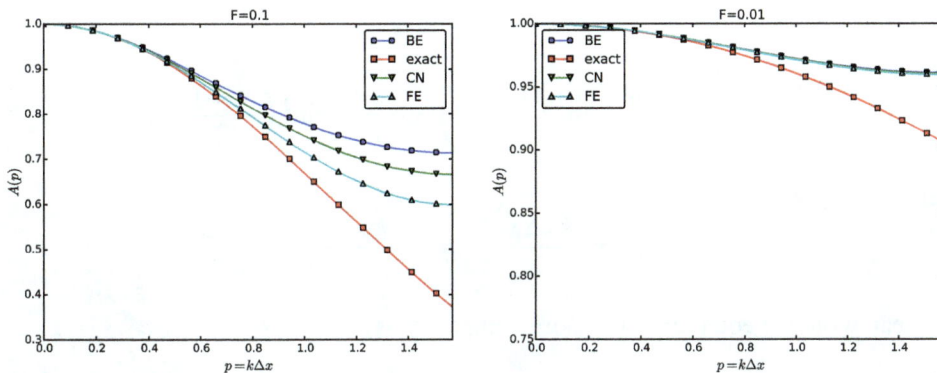

Fig. 1.11 Amplification factors for small time steps

dampened. These waves will manifest themselves as high frequency oscillatory noise in the solution.

A value $p = \pi/4$ corresponds to four mesh points per wave length of e^{ikx}, while $p = \pi/2$ implies only two points per wave length, which is the smallest number of points we can have to represent the wave on the mesh.

To demonstrate the oscillatory behavior of the Crank-Nicolson scheme, we choose an initial condition that leads to short waves with significant amplitude. A discontinuous $I(x)$ will in particular serve this purpose: Figures 1.6 and 1.7 correspond to $F = 3$ and $F = 10$, respectively, and we see how short waves pollute the overall solution.

1.3.9 Analysis of the 2D Diffusion Equation

Diffusion in several dimensions is treated later, but it is appropriate to include the analysis here. We first consider the 2D diffusion equation

$$u_t = \alpha(u_{xx} + u_{yy}),$$

which has Fourier component solutions of the form

$$u(x, y, t) = Ae^{-\alpha k^2 t} e^{i(k_x x + k_y y)},$$

and the schemes have discrete versions of this Fourier component:

$$u_{q,r}^n = A\xi^n e^{i(k_x q \Delta x + k_y r \Delta y)}.$$

The Forward Euler scheme For the Forward Euler discretization,

$$[D_t^+ u = \alpha(D_x D_x u + D_y D_y u)]_{q,r}^n,$$

we get

$$\frac{\xi - 1}{\Delta t} = -\alpha \frac{4}{\Delta x^2} \sin^2 \left(\frac{k_x \Delta x}{2} \right) - \alpha \frac{4}{\Delta y^2} \sin^2 \left(\frac{k_y \Delta y}{2} \right).$$

Introducing

$$p_x = \frac{k_x \Delta x}{2}, \quad p_y = \frac{k_y \Delta y}{2},$$

we can write the equation for ξ more compactly as

$$\frac{\xi - 1}{\Delta t} = -\alpha \frac{4}{\Delta x^2} \sin^2 p_x - \alpha \frac{4}{\Delta y^2} \sin^2 p_y,$$

and solve for ξ:

$$\xi = 1 - 4F_x \sin^2 p_x - 4F_y \sin^2 p_y. \tag{1.63}$$

The complete numerical solution for a wave component is

$$u_{q,r}^n = A(1 - 4F_x \sin^2 p_x - 4F_y \sin^2 p_y)^n e^{i(k_x q \Delta x + k_y r \Delta y)}. \tag{1.64}$$

For stability we demand $-1 \leq \xi \leq 1$, and $-1 \leq \xi$ is the critical limit, since clearly $\xi \leq 1$, and the worst case happens when the sines are at their maximum. The stability criterion becomes

$$F_x + F_y \leq \frac{1}{2}. \tag{1.65}$$

For the special, yet common, case $\Delta x = \Delta y = h$, the stability criterion can be written as

$$\Delta t \leq \frac{h^2}{2d\alpha},$$

where d is the number of space dimensions: $d = 1, 2, 3$.

The Backward Euler scheme The Backward Euler method,

$$[D_t^- u = \alpha(D_x D_x u + D_y D_y u)]_{q,r}^n,$$

results in

$$1 - \xi^{-1} = -4F_x \sin^2 p_x - 4F_y \sin^2 p_y,$$

and

$$\xi = (1 + 4F_x \sin^2 p_x + 4F_y \sin^2 p_y)^{-1},$$

which is always in $(0, 1]$. The solution for a wave component becomes

$$u_{q,r}^n = A(1 + 4F_x \sin^2 p_x + 4F_y \sin^2 p_y)^{-n} e^{i(k_x q \Delta x + k_y r \Delta y)}. \qquad (1.66)$$

The Crank-Nicolson scheme With a Crank-Nicolson discretization,

$$[D_t u]_{q,r}^{n+\frac{1}{2}} = \frac{1}{2}[\alpha(D_x D_x u + D_y D_y u)]_{q,r}^{n+1} + \frac{1}{2}[\alpha(D_x D_x u + D_y D_y u)]_{q,r}^n,$$

we have, after some algebra,

$$\xi = \frac{1 - 2(F_x \sin^2 p_x + F_x \sin^2 p_y)}{1 + 2(F_x \sin^2 p_x + F_x \sin^2 p_y)}.$$

The fraction on the right-hand side is always less than 1, so stability in the sense of non-growing wave components is guaranteed for all physical and numerical parameters. However, the fraction can become negative and result in non-physical oscillations. This phenomenon happens when

$$F_x \sin^2 p_x + F_x \sin^2 p_y > \frac{1}{2}.$$

A criterion against non-physical oscillations is therefore

$$F_x + F_y \leq \frac{1}{2},$$

which is the same limit as the stability criterion for the Forward Euler scheme.

The exact discrete solution is

$$u_{q,r}^n = A \left(\frac{1 - 2(F_x \sin^2 p_x + F_x \sin^2 p_y)}{1 + 2(F_x \sin^2 p_x + F_x \sin^2 p_y)} \right)^n e^{i(k_x q \Delta x + k_y r \Delta y)}. \qquad (1.67)$$

1.3.10   Explanation of Numerical Artifacts

The behavior of the solution generated by Forward Euler discretization in time (and centered differences in space) is summarized at the end of Sect. 1.1.5. Can we, from the analysis above, explain the behavior?

We may start by looking at Fig. 1.3 where $F = 0.51$. The figure shows that the solution is unstable and grows in time. The stability limit for such growth is $F = 0.5$ and since the F in this simulation is slightly larger, growth is unavoidable.

Figure 1.1 has unexpected features: we would expect the solution of the diffusion equation to be smooth, but the graphs in Fig. 1.1 contain non-smooth noise. Turning to Fig. 1.4, which has a quite similar initial condition, we see that the curves are indeed smooth. The problem with the results in Fig. 1.1 is that the initial condition is discontinuous. To represent it, we need a significant amplitude on the shortest waves in the mesh. However, for $F = 0.5$, the shortest wave ($p = \pi/2$) gives the amplitude in the numerical solution as $(1-4F)^n$, which oscillates between negative and positive values at subsequent time levels for $F > \frac{1}{4}$. Since the shortest waves have visible amplitudes in the solution profile, the oscillations becomes visible. The smooth initial condition in Fig. 1.4, on the other hand, leads to very small amplitudes of the shortest waves. That these waves then oscillate in a non-physical way for $F = 0.5$ is not a visible effect. The oscillations in time in the amplitude $(1 - 4F)^n$ disappear for $F \leq \frac{1}{4}$, and that is why also the discontinuous initial condition always leads to smooth solutions in Fig. 1.2, where $F = \frac{1}{4}$.

Turning the attention to the Backward Euler scheme and the experiments in Fig. 1.5, we see that even the discontinuous initial condition gives smooth solutions for $F = 0.5$ (and in fact all other F values). From the exact expression of the numerical amplitude, $(1 + 4F \sin^2 p)^{-1}$, we realize that this factor can never flip between positive and negative values, and no instabilities can occur. The conclusion is that the Backward Euler scheme always produces smooth solutions. Also, the Backward Euler scheme guarantees that the solution cannot grow in time (unless we add a source term to the PDE, but that is meant to represent a physically relevant growth).

Finally, we have some small, strange artifacts when simulating the development of the initial plug profile with the Crank-Nicolson scheme, see Fig. 1.7, where $F = 3$. The Crank-Nicolson scheme cannot give growing amplitudes, but it may give oscillating amplitudes in time. The critical factor is $1 - 2F \sin^2 p$, which for the shortest waves ($p = \pi/2$) indicates a stability limit $F = 0.5$. With the discontinuous initial condition, we have enough amplitude on the shortest waves so their wrong behavior is visible, and this is what we see as small instabilities in Fig. 1.7. The only remedy is to lower the F value.

1.4   Exercises

Exercise 1.1: Explore symmetry in a 1D problem

This exercise simulates the exact solution (1.48). Suppose for simplicity that $c = 0$.

a) Formulate an initial-boundary value problem that has (1.48) as solution in the domain $[-L, L]$. Use the exact solution (1.48) as Dirichlet condition at the boundaries. Simulate the diffusion of the Gaussian peak. Observe that the solution is symmetric around $x = 0$.

b) Show from (1.48) that $u_x(c, t) = 0$. Since the solution is symmetric around $x = c = 0$, we can solve the numerical problem in half of the domain, using a *symmetry boundary condition* $u_x = 0$ at $x = 0$. Set up the initial-boundary value problem in this case. Simulate the diffusion problem in $[0, L]$ and compare with the solution in a).

Filename: `diffu_symmetric_gaussian`.

Exercise 1.2: Investigate approximation errors from a $u_x = 0$ boundary condition

We consider the problem solved in Exercise 1.1 part b). The boundary condition $u_x(0, t) = 0$ can be implemented in two ways: 1) by a standard symmetric finite difference $[D_{2x}u]_i^n = 0$, or 2) by a one-sided difference $[D^+u = 0]_i^n = 0$. Investigate the effect of these two conditions on the convergence rate in space.

Hint If you use a Forward Euler scheme, choose a discretization parameter $h = \Delta t = \Delta x^2$ and assume the error goes like $E \sim h^r$. The error in the scheme is $\mathcal{O}(\Delta t, \Delta x^2)$ so one should expect that the estimated r approaches 1. The question is if a one-sided difference approximation to $u_x(0, t) = 0$ destroys this convergence rate.

Filename: `diffu_onesided_fd`.

Exercise 1.3: Experiment with open boundary conditions in 1D

We address diffusion of a Gaussian function as in Exercise 1.1, in the domain $[0, L]$, but now we shall explore different types of boundary conditions on $x = L$. In real-life problems we do not know the exact solution on $x = L$ and must use something simpler.

a) Imagine that we want to solve the problem numerically on $[0, L]$, with a symmetry boundary condition $u_x = 0$ at $x = 0$, but we do not know the exact solution and cannot of that reason assign a correct Dirichlet condition at $x = L$. One idea is to simply set $u(L, t) = 0$ since this will be an accurate approximation before the diffused pulse reaches $x = L$ and even thereafter it might be a satisfactory condition if the exact u has a small value. Let u_e be the exact solution and let u be the solution of $u_t = \alpha u_{xx}$ with an initial Gaussian pulse and the boundary conditions $u_x(0, t) = u(L, t) = 0$. Derive a diffusion problem for the error $e = u_e - u$. Solve this problem numerically using an exact Dirichlet condition at $x = L$. Animate the evolution of the error and make a curve plot of the error measure

$$E(t) = \sqrt{\frac{\int_0^L e^2 dx}{\int_0^L u dx}}.$$

Is this a suitable error measure for the present problem?

b) Instead of using $u(L, t) = 0$ as approximate boundary condition for letting the diffused Gaussian pulse move out of our finite domain, one may try $u_x(L, t) = 0$ since the solution for large t is quite flat. Argue that this condition gives a completely wrong asymptotic solution as $t \to 0$. To do this, integrate the diffusion equation from 0 to L, integrate u_{xx} by parts (or use Gauss' divergence theorem in 1D) to arrive at the important property

$$\frac{d}{dt}\int_0^L u(x,t)dx = 0,$$

implying that $\int_0^L u\,dx$ must be constant in time, and therefore

$$\int_0^L u(x,t)dx = \int_0^L I(x)dx\,.$$

The integral of the initial pulse is 1.

c) Another idea for an artificial boundary condition at $x = L$ is to use a cooling law

$$-\alpha u_x = q(u - u_S), \tag{1.68}$$

where q is an unknown heat transfer coefficient and u_S is the surrounding temperature in the medium outside of $[0, L]$. (Note that arguing that u_S is approximately $u(L,t)$ gives the $u_x = 0$ condition from the previous subexercise that is qualitatively wrong for large t.) Develop a diffusion problem for the error in the solution using (1.68) as boundary condition. Assume one can take $u_S = 0$ "outside the domain" since $u_e \to 0$ as $x \to \infty$. Find a function $q = q(t)$ such that the exact solution obeys the condition (1.68). Test some constant values of q and animate how the corresponding error function behaves. Also compute $E(t)$ curves as defined above.

Filename: `diffu_open_BC`.

Exercise 1.4: Simulate a diffused Gaussian peak in 2D/3D

a) Generalize (1.48) to multi dimensions by assuming that one-dimensional solutions can be multiplied to solve $u_t = \alpha\nabla^2 u$. Set $c = 0$ such that the peak of the Gaussian is at the origin.

b) One can from the exact solution show that $u_x = 0$ on $x = 0$, $u_y = 0$ on $y = 0$, and $u_z = 0$ on $z = 0$. The approximately correct condition $u = 0$ can be set on the remaining boundaries (say $x = L$, $y = L$, $z = L$), cf. Exercise 1.3. Simulate a 2D case and make an animation of the diffused Gaussian peak.

c) The formulation in b) makes use of symmetry of the solution such that we can solve the problem in the first quadrant (2D) or octant (3D) only. To check that the symmetry assumption is correct, formulate the problem without symmetry in a domain $[-L, L]\times[L, L]$ in 2D. Use $u = 0$ as approximately correct boundary condition. Simulate the same case as in b), but in a four times as large domain. Make an animation and compare it with the one in b).

Filename: `diffu_symmetric_gaussian_2D`.

Exercise 1.5: Examine stability of a diffusion model with a source term

Consider a diffusion equation with a linear u term:

$$u_t = \alpha u_{xx} + \beta u \,.$$

a) Derive in detail the Forward Euler, Backward Euler, and Crank-Nicolson schemes for this type of diffusion model. Thereafter, formulate a θ-rule to summarize the three schemes.

b) Assume a solution like (1.49) and find the relation between a, k, α, and β.

Hint Insert (1.49) in the PDE problem.

c) Calculate the stability of the Forward Euler scheme. Design numerical experiments to confirm the results.

Hint Insert the discrete counterpart to (1.49) in the numerical scheme. Run experiments at the stability limit and slightly above.

d) Repeat c) for the Backward Euler scheme.

e) Repeat c) for the Crank-Nicolson scheme.

f) How does the extra term bu impact the accuracy of the three schemes?

Hint For analysis of the accuracy, compare the numerical and exact amplification factors, in graphs and/or by Taylor series expansion.

Filename: `diffu_stability_uterm`.

1.5 Diffusion in Heterogeneous Media

Diffusion in heterogeneous media normally implies a non-constant diffusion coefficient $\alpha = \alpha(x)$. A 1D diffusion model with such a variable diffusion coefficient reads

$$\frac{\partial u}{\partial t} = \frac{\partial}{\partial x}\left(\alpha(x)\frac{\partial u}{\partial x}\right) + f(x,t), \qquad x \in (0, L),\ t \in (0, T], \qquad (1.69)$$

$$u(x, 0) = I(x), \qquad\qquad x \in [0, L], \qquad (1.70)$$

$$u(0, t) = U_0, \qquad\qquad t > 0, \qquad (1.71)$$

$$u(L, t) = U_L, \qquad\qquad t > 0. \qquad (1.72)$$

A short form of the diffusion equation with variable coefficients is $u_t = (\alpha u_x)_x + f$.

1.5.1 Discretization

We can discretize (1.69) by a θ-rule in time and centered differences in space:

$$[D_t u]_i^{n+\frac{1}{2}} = \theta[D_x(\overline{\alpha}^x D_x u) + f]_i^{n+1} + (1 - \theta)[D_x(\overline{\alpha}^x D_x u) + f]_i^n \,.$$

Written out, this becomes

$$\frac{u_i^{n+1} - u_i^n}{\Delta t} = \theta \frac{1}{\Delta x^2} \left(\alpha_{i+\frac{1}{2}}(u_{i+1}^{n+1} - u_i^{n+1}) - \alpha_{i-\frac{1}{2}}(u_i^{n+1} - u_{i-1}^{n+1}) \right)$$
$$+ (1-\theta)\frac{1}{\Delta x^2} \left(\alpha_{i+\frac{1}{2}}(u_{i+1}^n - u_i^n) - \alpha_{i-\frac{1}{2}}(u_i^n - u_{i-1}^n) \right)$$
$$+ \theta f_i^{n+1} + (1-\theta) f_i^n,$$

where, e.g., an arithmetic mean can to be used for $\alpha_{i+\frac{1}{2}}$:

$$\alpha_{i+\frac{1}{2}} = \frac{1}{2}(\alpha_i + \alpha_{i+1}).$$

1.5.2 Implementation

Suitable code for solving the discrete equations is very similar to what we created
for a constant α. Since the Fourier number has no meaning for varying α, we
introduce a related parameter $D = \Delta t / \Delta x^2$.

```
def solver_theta(I, a, L, Nx, D, T, theta=0.5, u_L=1, u_R=0,
                 user_action=None):
    x = linspace(0, L, Nx+1)    # mesh points in space
    dx = x[1] - x[0]
    dt = D*dx**2
    Nt = int(round(T/float(dt)))
    t = linspace(0, T, Nt+1)    # mesh points in time

    u   = zeros(Nx+1)    # solution array at t[n+1]
    u_n = zeros(Nx+1)    # solution at t[n]

    Dl = 0.5*D*theta
    Dr = 0.5*D*(1-theta)

    # Representation of sparse matrix and right-hand side
    diagonal = zeros(Nx+1)
    lower    = zeros(Nx)
    upper    = zeros(Nx)
    b        = zeros(Nx+1)

    # Precompute sparse matrix (scipy format)
    diagonal[1:-1] = 1 + Dl*(a[2:] + 2*a[1:-1] + a[:-2])
    lower[:-1] = -Dl*(a[1:-1] + a[:-2])
    upper[1:]  = -Dl*(a[2:] + a[1:-1])
    # Insert boundary conditions
    diagonal[0] = 1
    upper[0] = 0
    diagonal[Nx] = 1
    lower[-1] = 0

    A = scipy.sparse.diags(
        diagonals=[diagonal, lower, upper],
        offsets=[0, -1, 1],
        shape=(Nx+1, Nx+1),
        format='csr')
```

```
# Set initial condition
for i in range(0,Nx+1):
    u_n[i] = I(x[i])

if user_action is not None:
    user_action(u_n, x, t, 0)

# Time loop
for n in range(0, Nt):
    b[1:-1] = u_n[1:-1] + Dr*(
        (a[2:] + a[1:-1])*(u_n[2:] - u_n[1:-1]) -
        (a[1:-1] + a[0:-2])*(u_n[1:-1] - u_n[:-2]))
    # Boundary conditions
    b[0]  = u_L(t[n+1])
    b[-1] = u_R(t[n+1])
    # Solve
    u[:] = scipy.sparse.linalg.spsolve(A, b)

    if user_action is not None:
        user_action(u, x, t, n+1)

    # Switch variables before next step
    u_n, u = u, u_n
```

The code is found in the file diffu1D_vc.py.

1.5.3 Stationary Solution

As $t \to \infty$, the solution of the problem (1.69)–(1.72) will approach a stationary limit where $\partial u/\partial t = 0$. The governing equation is then

$$\frac{d}{dx}\left(\alpha \frac{du}{dx}\right) = 0, \qquad (1.73)$$

with boundary conditions $u(0) = U_0$ and $u(L) = U_L$. It is possible to obtain an exact solution of (1.73) for any α. Integrating twice and applying the boundary conditions to determine the integration constants gives

$$u(x) = U_0 + (U_L - U_0)\frac{\int_0^x (\alpha(\xi))^{-1}d\xi}{\int_0^L (\alpha(\xi))^{-1}d\xi}. \qquad (1.74)$$

1.5.4 Piecewise Constant Medium

Consider a medium built of M layers. The layer boundaries are denoted $b_0, \ldots, b_M$, where $b_0 = 0$ and $b_M = L$. If the layers potentially have different material properties, but these properties are constant within each layer, we can express α as a *piecewise constant function* according to

$$
\alpha(x) = \begin{cases} \alpha_0, & b_0 \le x < b_1, \\ \vdots \\ \alpha_i, & b_i \le x < b_{i+1}, \\ \vdots \\ \alpha_{M-1}, & b_{M-1} \le x \le b_M. \end{cases} \tag{1.75}
$$

The exact solution (1.74) in case of such a piecewise constant α function is easy to derive. Assume that x is in the m-th layer: $x \in [b_m, b_{m+1}]$. In the integral $\int_0^x (a(\xi))^{-1} d\xi$ we must integrate through the first $m - 1$ layers and then add the contribution from the remaining part $x - b_m$ into the m-th layer:

$$
u(x) = U_0 + (U_L - U_0) \frac{\sum_{j=0}^{m-1} (b_{j+1} - b_j)/\alpha(b_j) + (x - b_m)/\alpha(b_m)}{\sum_{j=0}^{M-1} (b_{j+1} - b_j)/\alpha(b_j)}. \tag{1.76}
$$

Remark It may sound strange to have a discontinuous α in a differential equation where one is to differentiate, but a discontinuous α is compensated by a discontinuous u_x such that αu_x is continuous and therefore can be differentiated as $(\alpha u_x)_x$.

1.5.5 Implementation of Diffusion in a Piecewise Constant Medium

Programming with piecewise function definitions quickly becomes cumbersome as the most naive approach is to test for which interval x lies, and then start evaluating a formula like (1.76). In Python, vectorized expressions may help to speed up the computations. The convenience classes `PiecewiseConstant` and `IntegratedPiecewiseConstant` in the `Heaviside` module were made to simplify programming with functions like (1.75) and expressions like (1.76). These utilities not only represent piecewise constant functions, but also *smoothed* versions of them where the discontinuities can be smoothed out in a controlled fashion.

The `PiecewiseConstant` class is created by sending in the domain as a 2-tuple or 2-list and a `data` object describing the boundaries $b_0, \ldots, b_M$ and the corresponding function values $\alpha_0, \ldots, \alpha_{M-1}$. More precisely, `data` is a nested list, where `data[i][0]` holds b_i and `data[i][1]` holds the corresponding value α_i, for $i = 0, \ldots, M - 1$. Given b_i and α_i in arrays b and a, it is easy to fill out the nested list `data`.

In our application, we want to represent α and $1/\alpha$ as piecewise constant functions, in addition to the $u(x)$ function which involves the integrals of $1/\alpha$. A class creating the functions we need and a method for evaluating u, can take the form

```
class SerialLayers:
    """

    b: coordinates of boundaries of layers, b[0] is left boundary
    and b[-1] is right boundary of the domain [0,L].
    a: values of the functions in each layer (len(a) = len(b)-1).
    U_0: u(x) value at left boundary x=0=b[0].
    U_L: u(x) value at right boundary x=L=b[0].
    """
```

```
    def __init__(self, a, b, U_0, U_L, eps=0):
        self.a, self.b = np.asarray(a), np.asarray(b)
        self.eps = eps  # smoothing parameter for smoothed a
        self.U_0, self.U_L = U_0, U_L

        a_data = [[bi, ai] for bi, ai in zip(self.b, self.a)]
        domain = [b[0], b[-1]]
        self.a_func = PiecewiseConstant(domain, a_data, eps)

        # inv_a = 1/a is needed in formulas
        inv_a_data = [[bi, 1./ai] for bi, ai in zip(self.b, self.a)]
        self.inv_a_func = \
            PiecewiseConstant(domain, inv_a_data, eps)
        self.integral_of_inv_a_func = \
            IntegratedPiecewiseConstant(domain, inv_a_data, eps)
        # Denominator in the exact formula is constant
        self.inv_a_0L = self.integral_of_inv_a_func(b[-1])

    def __call__(self, x):
        solution = self.U_0 + (self.U_L-self.U_0)*\
                    self.integral_of_inv_a_func(x)/self.inv_a_0L
        return solution
```

A visualization method is also convenient to have. Below we plot $u(x)$ along with $\alpha(x)$ (which works well as long as $\max \alpha(x)$ is of the same size as $\max u = \max(U_0, U_L)$).

```
class SerialLayers:
    ...

    def plot(self):
        x, y_a = self.a_func.plot()
        x = np.asarray(x); y_a = np.asarray(y_a)
        y_u = self.u_exact(x)
        import matplotlib.pyplot as plt
        plt.figure()
        plt.plot(x, y_u, 'b')
        plt.hold('on')  # Matlab style
        plt.plot(x, y_a, 'r')
        ymin = -0.1
        ymax = 1.2*max(y_u.max(), y_a.max())
        plt.axis([x[0], x[-1], ymin, ymax])
        plt.legend(['solution $u$', 'coefficient $a$'], loc='upper left')
        if self.eps > 0:
            plt.title('Smoothing eps: %s' % self.eps)
        plt.savefig('tmp.pdf')
        plt.savefig('tmp.png')
        plt.show()
```

Figure 1.12 shows the case where

```
b = [0, 0.25, 0.5, 1]    # material boundaries
a = [0.2, 0.4, 4]        # material values
U_0 = 0.5;  U_L = 5      # boundary conditions
```

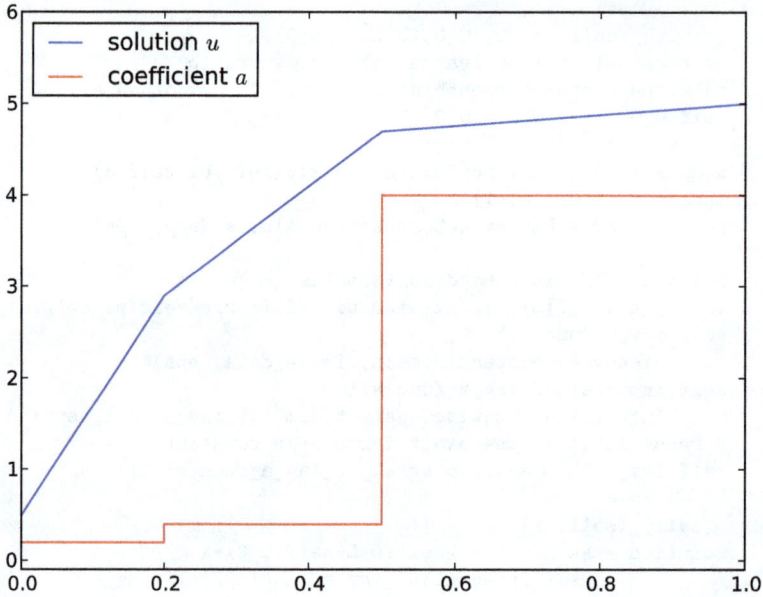

Fig. 1.12 Solution of the stationary diffusion equation corresponding to a piecewise constant diffusion coefficient

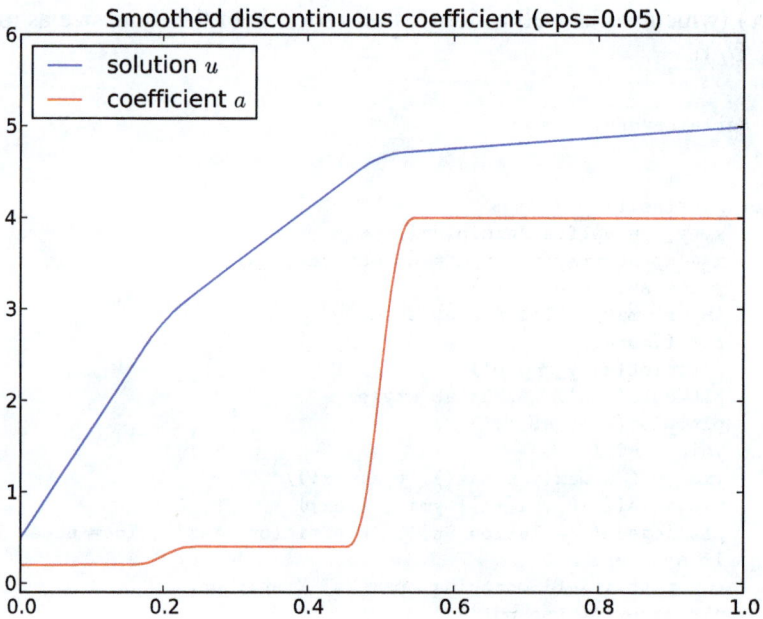

Fig. 1.13 Solution of the stationary diffusion equation corresponding to a *smoothed* piecewise constant diffusion coefficient

By adding the `eps` parameter to the constructor of the `SerialLayers` class, we can experiment with smoothed versions of α and see the (small) impact on u. Figure 1.13 shows the result.

1.5.6   Axi-Symmetric Diffusion

Suppose we have a diffusion process taking place in a straight tube with radius R. We assume axi-symmetry such that u is just a function of r and t, with r being the radial distance from the center axis of the tube to a point. With such axi-symmetry it is advantageous to introduce *cylindrical coordinates* r, θ, and z, where z is in the direction of the tube and (r, θ) are polar coordinates in a cross section. Axi-symmetry means that all quantities are independent of θ. From the relations $x = \cos\theta$, $y = \sin\theta$, and $z = z$, between Cartesian and cylindrical coordinates, one can (with some effort) derive the diffusion equation in cylindrical coordinates, which with axi-symmetry takes the form

$$\frac{\partial u}{\partial t} = \frac{1}{r}\frac{\partial}{\partial r}\left(r\alpha(r,z)\frac{\partial u}{\partial r}\right) + \frac{\partial}{\partial z}\left(\alpha(r,z)\frac{\partial u}{\partial z}\right) + f(r,z,t).$$

Let us assume that u does not change along the tube axis so it suffices to compute variations in a cross section. Then $\partial u/\partial z = 0$ and we have a 1D diffusion equation in the radial coordinate r and time t. In particular, we shall address the initial-boundary value problem

$$\frac{\partial u}{\partial t} = \frac{1}{r}\frac{\partial}{\partial r}\left(r\alpha(r)\frac{\partial u}{\partial r}\right) + f(t), \qquad r \in (0,R),\ t \in (0,T], \qquad (1.77)$$

$$\frac{\partial u}{\partial r}(0,t) = 0, \qquad\qquad t \in (0,T], \qquad (1.78)$$

$$u(R,t) = 0, \qquad\qquad t \in (0,T], \qquad (1.79)$$

$$u(r,0) = I(r), \qquad\qquad r \in [0,R]. \qquad (1.80)$$

The condition (1.78) is a necessary symmetry condition at $r = 0$, while (1.79) could be any Dirichlet or Neumann condition (or Robin condition in case of cooling or heating).

The finite difference approximation will need the discretized version of the PDE for $r = 0$ (just as we use the PDE at the boundary when implementing Neumann conditions). However, discretizing the PDE at $r = 0$ poses a problem because of the $1/r$ factor. We therefore need to work out the PDE for discretization at $r = 0$ with care. Let us, for the case of constant α, expand the spatial derivative term to

$$\alpha\frac{\partial^2 u}{\partial r^2} + \alpha\frac{1}{r}\frac{\partial u}{\partial r}.$$

The last term faces a difficulty at $r = 0$, since it becomes a 0/0 expression caused by the symmetry condition at $r = 0$. However, L'Hosptial's rule can be used:

$$\lim_{r\to 0}\frac{1}{r}\frac{\partial u}{\partial r} = \frac{\partial^2 u}{\partial r^2}.$$

The PDE at $r = 0$ therefore becomes

$$\frac{\partial u}{\partial t} = 2\alpha\frac{\partial^2 u}{\partial r^2} + f(t). \qquad (1.81)$$

For a variable coefficient $\alpha(r)$ the expanded spatial derivative term reads

$$\alpha(r)\frac{\partial^2 u}{\partial r^2} + \frac{1}{r}(\alpha(r) + r\alpha'(r))\frac{\partial u}{\partial r} \,.$$

We are interested in this expression for $r = 0$. A necessary condition for u to be axi-symmetric is that all input data, including α, must also be axi-symmetric, implying that $\alpha'(0) = 0$ (the second term vanishes anyway because of $r = 0$). The limit of interest is

$$\lim_{r \to 0} \frac{1}{r}\alpha(r)\frac{\partial u}{\partial r} = \alpha(0)\frac{\partial^2 u}{\partial r^2} \,.$$

The PDE at $r = 0$ now looks like

$$\frac{\partial u}{\partial t} = 2\alpha(0)\frac{\partial^2 u}{\partial r^2} + f(t), \tag{1.82}$$

so there is no essential difference between the constant coefficient and variable coefficient cases.

The second-order derivative in (1.81) and (1.82) is discretized in the usual way.

$$2\alpha\frac{\partial^2}{\partial r^2}u(r_0, t_n) \approx [2\alpha D_r D_r u]_0^n = 2\alpha\frac{u_1^n - 2u_0^n + u_{-1}^n}{\Delta r^2} \,.$$

The fictitious value u_{-1}^n can be eliminated using the discrete symmetry condition

$$[D_{2r}u = 0]_0^n \quad \Rightarrow \quad u_{-1}^n = u_1^n,$$

which then gives the modified approximation to the term with the second-order derivative of u in r at $r = 0$:

$$4\alpha\frac{u_1^n - u_0^n}{\Delta r^2} \,. \tag{1.83}$$

The discretization of the term with the second-order derivative in r at any internal mesh point is straightforward:

$$\left[\frac{1}{r}\frac{\partial}{\partial r}\left(r\alpha\frac{\partial u}{\partial r}\right)\right]_i^n \approx [r^{-1}D_r(r\alpha D_r u)]_i^n$$

$$= \frac{1}{r_i}\frac{1}{\Delta r^2}\left(r_{i+\frac{1}{2}}\alpha_{i+\frac{1}{2}}(u_{i+1}^n - u_i^n) - r_{i-\frac{1}{2}}\alpha_{i-\frac{1}{2}}(u_i^n - u_{i-1}^n)\right) \,.$$

To complete the discretization, we need a scheme in time, but that can be done as before and does not interfere with the discretization in space.

1.5.7 Spherically-Symmetric Diffusion

Discretization in spherical coordinates Let us now pose the problem from Sect. 1.5.6 in spherical coordinates, where u only depends on the radial coordinate

r and time t. That is, we have spherical symmetry. For simplicity we restrict the diffusion coefficient α to be a constant. The PDE reads

$$\frac{\partial u}{\partial t} = \frac{\alpha}{r^\gamma} \frac{\partial}{\partial r} \left(r^\gamma \frac{\partial u}{\partial r} \right) + f(t), \tag{1.84}$$

for $r \in (0, R)$ and $t \in (0, T]$. The parameter γ is 2 for spherically-symmetric problems and 1 for axi-symmetric problems. The boundary and initial conditions have the same mathematical form as in (1.77)–(1.80).

Since the PDE in spherical coordinates has the same form as the PDE in Sect. 1.5.6, just with the γ parameter being different, we can use the same discretization approach. At the origin $r = 0$ we get problems with the term

$$\frac{\gamma}{r} \frac{\partial u}{\partial t},$$

but L'Hosptial's rule shows that this term equals $\gamma \partial^2 u / \partial r^2$, and the PDE at $r = 0$ becomes

$$\frac{\partial u}{\partial t} = (\gamma + 1)\alpha \frac{\partial^2 u}{\partial r^2} + f(t). \tag{1.85}$$

The associated discrete form is then

$$\left[D_t u = \frac{1}{2}(\gamma + 1)\alpha D_r D_r \overline{u}^t + \overline{f}^t \right]_i^{n+\frac{1}{2}}, \tag{1.86}$$

for a Crank-Nicolson scheme.

Discretization in Cartesian coordinates The spherically-symmetric spatial derivative can be transformed to the Cartesian counterpart by introducing

$$v(r, t) = ru(r, t).$$

Inserting $u = v/r$ in

$$\frac{1}{r^2} \frac{\partial}{\partial r} \left(\alpha(r) r^2 \frac{\partial u}{\partial r} \right),$$

yields

$$r \left(\frac{d\alpha}{dr} \frac{\partial v}{\partial r} + \alpha \frac{\partial^2 v}{\partial r^2} \right) - \frac{d\alpha}{dr} v.$$

The two terms in the parenthesis can be combined to

$$r \frac{\partial}{\partial r} \left(\alpha \frac{\partial v}{\partial r} \right).$$

The PDE for v takes the form

$$\frac{\partial v}{\partial t} = \frac{\partial}{\partial r} \left(\alpha \frac{\partial v}{\partial r} \right) - \frac{1}{r} \frac{d\alpha}{dr} v + rf(r, t), \quad r \in (0, R), \ t \in (0, T]. \tag{1.87}$$

For α constant we immediately realize that we can reuse a solver in Cartesian co-ordinates to compute v. With variable α, a "reaction" term v/r needs to be added to the Cartesian solver. The boundary condition $\partial u/\partial r = 0$ at $r = 0$, implied by symmetry, forces $v(0, t) = 0$, because

$$\frac{\partial u}{\partial r} = \frac{1}{r^2} \left(r \frac{\partial v}{\partial r} - v \right) = 0, \quad r = 0.$$

1.6 Diffusion in 2D

We now address diffusion in two space dimensions:

$$\frac{\partial u}{\partial t} = \alpha \left(\frac{\partial^2 u}{\partial x^2} + \frac{\partial^2 u}{\partial x^2} \right) + f(x, y), \tag{1.88}$$

in a domain

$$(x, y) \in (0, L_x) \times (0, L_y), \quad t \in (0, T],$$

with $u = 0$ on the boundary and $u(x, y, 0) = I(x, y)$ as initial condition.

1.6.1 Discretization

For generality, it is natural to use a θ-rule for the time discretization. Standard, second-order accurate finite differences are used for the spatial derivatives. We sample the PDE at a space-time point $(i, j, n + \frac{1}{2})$ and apply the difference approx-imations:

$$[D_t u]^{n+\frac{1}{2}} = \theta [\alpha (D_x D_x u + D_y D_y u) + f]^{n+1}$$
$$+ (1 - \theta)[\alpha (D_x D_x u + D_y D_y u) + f]^n . \tag{1.89}$$

Written out,

$$\frac{u_{i,j}^{n+1} - u_{i,j}^n}{\Delta t}$$
$$= \theta \left(\alpha \left(\frac{u_{i-1,j}^{n+1} - 2u_{i,j}^{n+1} + u_{i+1,j}^{n+1}}{\Delta x^2} + \frac{u_{i,j-1}^{n+1} - 2u_{i,j}^{n+1} + u_{i,j+1}^{n+1}}{\Delta y^2} \right) + f_{i,j}^{n+1} \right)$$
$$+ (1 - \theta) \left(\alpha \left(\frac{u_{i-1,j}^n - 2u_{i,j}^n + u_{i+1,j}^n}{\Delta x^2} + \frac{u_{i,j-1}^n - 2u_{i,j}^n + u_{i,j+1}^n}{\Delta y^2} \right) + f_{i,j}^n \right) . \tag{1.90}$$

We collect the unknowns on the left-hand side

$$u_{i,j}^{n+1} - \theta \left(F_x(u_{i-1,j}^{n+1} - 2u_{i,j}^{n+1} + u_{i+1,j}^{n+1}) + F_y(u_{i,j-1}^{n+1} - 2u_{i,j}^{n+1} + u_{i,j+1}^{n+1}) \right)$$
$$= (1 - \theta) \left(F_x(u_{i-1,j}^n - 2u_{i,j}^n + u_{i+1,j}^n) + F_y(u_{i,j-1}^n - 2u_{i,j}^n + u_{i,j+1}^n) \right)$$
$$+ \theta \Delta t f_{i,j}^{n+1} + (1 - \theta) \Delta t f_{i,j}^n + u_{i,j}^n, \tag{1.91}$$

```
(0,2): 8          (1,2): 9              (2,2): 10            (3,2): 11




(0,1): 4          (1,1): 5             (2,1): 6            (3,1): 7




(0,0): 0          (1,0): 1             (2,0): 2            (3,0): 3
```

Fig. 1.14 3x2 2D mesh

where

$$F_x = \frac{\alpha \Delta t}{\Delta x^2}, \quad F_y = \frac{\alpha \Delta t}{\Delta y^2},$$

are the Fourier numbers in x and y direction, respectively.

1.6.2 Numbering of Mesh Points Versus Equations and Unknowns

The equations (1.91) are coupled at the new time level $n+1$. That is, we must solve a system of (linear) algebraic equations, which we will write as $Ac = b$, where A is the coefficient matrix, c is the vector of unknowns, and b is the right-hand side.

Let us examine the equations in $Ac = b$ on a mesh with $N_x = 3$ and $N_y = 2$ cells in the respective spatial directions. The spatial mesh is depicted in Fig. 1.14. The equations at the boundary just implement the boundary condition $u = 0$:

$$u_{0,0}^{n+1} = u_{1,0}^{n+1} = u_{2,0}^{n+1} = u_{3,0}^{n+1} = u_{0,1}^{n+1}$$
$$= u_{3,1}^{n+1} = u_{0,2}^{n+1} = u_{1,2}^{n+1} = u_{2,2}^{n+1} = u_{3,2}^{n+1} = 0.$$

We are left with two interior points, with $i = 1$, $j = 1$ and $i = 2$, $j = 1$. The corresponding equations are

$$u_{i,j}^{n+1} - \theta \left(F_x(u_{i-1,j}^{n+1} - 2u_{i,j}^{n+1} + u_{i+1,j}^{n+1}) + F_y(u_{i,j-1}^{n+1} - 2u_{i,j}^{n+1} + u_{i,j+1}^{n+1}) \right)$$
$$= (1 - \theta) \left(F_x(u_{i-1,j}^{n} - 2u_{i,j}^{n} + u_{i+1,j}^{n}) + F_y(u_{i,j-1}^{n} - 2u_{i,j}^{n} + u_{i,j+1}^{n}) \right)$$
$$+ \theta \Delta t f_{i,j}^{n+1} + (1 - \theta) \Delta t f_{i,j}^{n} + u_{i,j}^{n} .$$

There are in total 12 unknowns $u_{i,j}^{n+1}$ for $i = 0, 1, 2, 3$ and $j = 0, 1, 2$. To solve the equations, we need to form a matrix system $Ac = b$. In that system, the solution vector c can only have one index. Thus, we need a numbering of the unknowns with one index, not two as used in the mesh. We introduce a mapping $m(i, j)$ from a mesh point with indices (i, j) to the corresponding unknown p in the equation system:

$$p = m(i, j) = j(N_x + 1) + i .$$

When i and j run through their values, we see the following mapping to p:

$$(0, 0) \to 0, \ (0, 1) \to 1, \ (0, 2) \to 2, \ (0, 3) \to 3,$$
$$(1, 0) \to 4, \ (1, 1) \to 5, \ (1, 2) \to 6, \ (1, 3) \to 7,$$
$$(2, 0) \to 8, \ (2, 1) \to 9, \ (2, 2) \to 10, \ (2, 3) \to 11 .$$

That is, we number the points along the x axis, starting with $y = 0$, and then progress one "horizontal" mesh line at a time. In Fig. 3.14 you can see that the (i, j) and the corresponding single index (p) are listed for each mesh point.

We could equally well have numbered the equations in other ways, e.g., let the j index be the fastest varying index: $p = m(i, j) = i(N_y + 1) + j$.

Let us form the coefficient matrix A, or more precisely, insert a matrix element (according Python's convention with zero as base index) for each of the nonzero elements in A (the indices run through the values of p, i.e., $p = 0, \dots, 11$):

$$
\begin{pmatrix}
(0,0) & 0 & 0 & 0 & 0 & 0 & 0 & 0 & 0 & 0 & 0 & 0 \\
0 & (1,1) & 0 & 0 & 0 & 0 & 0 & 0 & 0 & 0 & 0 & 0 \\
0 & 0 & (2,2) & 0 & 0 & 0 & 0 & 0 & 0 & 0 & 0 & 0 \\
0 & 0 & 0 & (3,3) & 0 & 0 & 0 & 0 & 0 & 0 & 0 & 0 \\
0 & 0 & 0 & 0 & (4,4) & 0 & 0 & 0 & 0 & 0 & 0 & 0 \\
0 & (5,1) & 0 & 0 & (5,4) & (5,5) & (5,6) & 0 & 0 & (5,9) & 0 & 0 \\
0 & 0 & (6,2) & 0 & 0 & (6,5) & (6,6) & (6,7) & 0 & 0 & (6,10) & 0 \\
0 & 0 & 0 & 0 & 0 & 0 & 0 & (7,7) & 0 & 0 & 0 & 0 \\
0 & 0 & 0 & 0 & 0 & 0 & 0 & 0 & (8,8) & 0 & 0 & 0 \\
0 & 0 & 0 & 0 & 0 & 0 & 0 & 0 & 0 & (9,9) & 0 & 0 \\
0 & 0 & 0 & 0 & 0 & 0 & 0 & 0 & 0 & 0 & (10,10) & 0 \\
0 & 0 & 0 & 0 & 0 & 0 & 0 & 0 & 0 & 0 & 0 & (11,11)
\end{pmatrix} .
$$

Here is a more compact visualization of the coefficient matrix where we insert dots for zeros and bullets for non-zero elements:

$$
\begin{pmatrix}
\bullet & \cdot & \cdot & \cdot & \cdot & \cdot & \cdot & \cdot & \cdot & \cdot & \cdot & \cdot \\
\cdot & \bullet & \cdot & \cdot & \cdot & \cdot & \cdot & \cdot & \cdot & \cdot & \cdot & \cdot \\
\cdot & \cdot & \bullet & \cdot & \cdot & \cdot & \cdot & \cdot & \cdot & \cdot & \cdot & \cdot \\
\cdot & \cdot & \cdot & \bullet & \cdot & \cdot & \cdot & \cdot & \cdot & \cdot & \cdot & \cdot \\
\cdot & \cdot & \cdot & \cdot & \bullet & \cdot & \cdot & \cdot & \cdot & \cdot & \cdot & \cdot \\
\cdot & \bullet & \cdot & \cdot & \bullet & \bullet & \bullet & \cdot & \cdot & \bullet & \cdot & \cdot \\
\cdot & \cdot & \bullet & \cdot & \cdot & \bullet & \bullet & \bullet & \cdot & \cdot & \bullet & \cdot \\
\cdot & \cdot & \cdot & \cdot & \cdot & \cdot & \bullet & \cdot & \cdot & \cdot & \cdot & \cdot \\
\cdot & \cdot & \cdot & \cdot & \cdot & \cdot & \cdot & \bullet & \cdot & \cdot & \cdot & \cdot \\
\cdot & \cdot & \cdot & \cdot & \cdot & \cdot & \cdot & \cdot & \bullet & \cdot & \cdot & \cdot \\
\cdot & \cdot & \cdot & \cdot & \cdot & \cdot & \cdot & \cdot & \cdot & \bullet & \cdot & \cdot \\
\cdot & \cdot & \cdot & \cdot & \cdot & \cdot & \cdot & \cdot & \cdot & \cdot & \cdot & \bullet
\end{pmatrix}
$$

It is clearly seen that most of the elements are zero. This is a general feature of coefficient matrices arising from discretizing PDEs by finite difference methods. We say that the matrix is *sparse*.

Let $A_{p,q}$ be the value of element (p, q) in the coefficient matrix A, where p and q now correspond to the numbering of the unknowns in the equation system. We have $A_{p,q} = 1$ for $p = q = 0, 1, 2, 3, 4, 7, 8, 9, 10, 11$, corresponding to all the known boundary values. Let p be $m(i, j)$, i.e., the single index corresponding to mesh point (i, j). Then we have

$$A_{m(i,j),m(i,j)} = A_{p,p} = 1 + \theta(F_x + F_y), \tag{1.92}$$

$$A_{p,m(i-1,j)} = A_{p,p-1} = -\theta F_x, \tag{1.93}$$

$$A_{p,m(i+1,j)} = A_{p,p+1} = -\theta F_x, \tag{1.94}$$

$$A_{p,m(i,j-1)} = A_{p,p-(N_x+1)} = -\theta F_y, \tag{1.95}$$

$$A_{p,m(i,j+1)} = A_{p,p+(N_x+1)} = -\theta F_y, \tag{1.96}$$

for the equations associated with the two interior mesh points. At these interior points, the single index p takes on the specific values $p = 5, 6$, corresponding to the values $(1, 1)$ and $(1, 2)$ of the pair (i, j).

The above values for $A_{p,q}$ can be inserted in the matrix:

$$
\begin{pmatrix}
1 & 0 & 0 & 0 & 0 & 0 & 0 & 0 & 0 & 0 & 0 & 0 \\
0 & 1 & 0 & 0 & 0 & 0 & 0 & 0 & 0 & 0 & 0 & 0 \\
0 & 0 & 1 & 0 & 0 & 0 & 0 & 0 & 0 & 0 & 0 & 0 \\
0 & 0 & 0 & 1 & 0 & 0 & 0 & 0 & 0 & 0 & 0 & 0 \\
0 & 0 & 0 & 0 & 1 & 0 & 0 & 0 & 0 & 0 & 0 & 0 \\
0 & -\theta F_y & 0 & 0 & -\theta F_x & 1 + 2\theta F_x & -\theta F_x & 0 & 0 & -\theta F_y & 0 & 0 \\
0 & 0 & -\theta F_y & 0 & 0 & -\theta F_x & 1 + 2\theta F_x & -\theta F_x & 0 & 0 & -\theta F_y & 0 \\
0 & 0 & 0 & 0 & 0 & 0 & 0 & 1 & 0 & 0 & 0 & 0 \\
0 & 0 & 0 & 0 & 0 & 0 & 0 & 0 & 1 & 0 & 0 & 0 \\
0 & 0 & 0 & 0 & 0 & 0 & 0 & 0 & 0 & 1 & 0 & 0 \\
0 & 0 & 0 & 0 & 0 & 0 & 0 & 0 & 0 & 0 & 1 & 0 \\
0 & 0 & 0 & 0 & 0 & 0 & 0 & 0 & 0 & 0 & 0 & 1
\end{pmatrix}
$$

The corresponding right-hand side vector in the equation system has the entries b_p, where p numbers the equations. We have

$$b_0 = b_1 = b_2 = b_3 = b_4 = b_7 = b_8 = b_9 = b_{10} = b_{11} = 0,$$

for the boundary values. For the equations associated with the interior points, we get for $p = 5, 6$, corresponding to $i = 1, 2$ and $j = 1$:

$$b_p = u_{i,j}^n + (1 - \theta)\left(F_x(u_{i-1,j}^n - 2u_{i,j}^n + u_{i+1,j}^n) + F_y(u_{i,j-1}^n - 2u_{i,j}^n + u_{i,j+1}^n)\right)$$
$$+ \theta \Delta t f_{i,j}^{n+1} + (1 - \theta)\Delta t f_{i,j}^n .$$

Recall that $p = m(i, j) = j(N_x + 1) + j$ in this expression.

We can, as an alternative, leave the boundary mesh points out of the matrix system. For a mesh with $N_x = 3$ and $N_y = 2$ there are only two internal mesh points whose unknowns will enter the matrix system. We must now number the unknowns at the interior points:

$$p = (j - 1)(N_x - 1) + i,$$

for $i = 1, \ldots, N_x - 1$, $j = 1, \ldots, N_y - 1$.

We can continue with illustrating a bit larger mesh, $N_x = 4$ and $N_y = 3$, see Fig. 3.15. The corresponding coefficient matrix with dots for zeros and bullets for non-zeroes looks as follows (values at boundary points are included in the equation system):

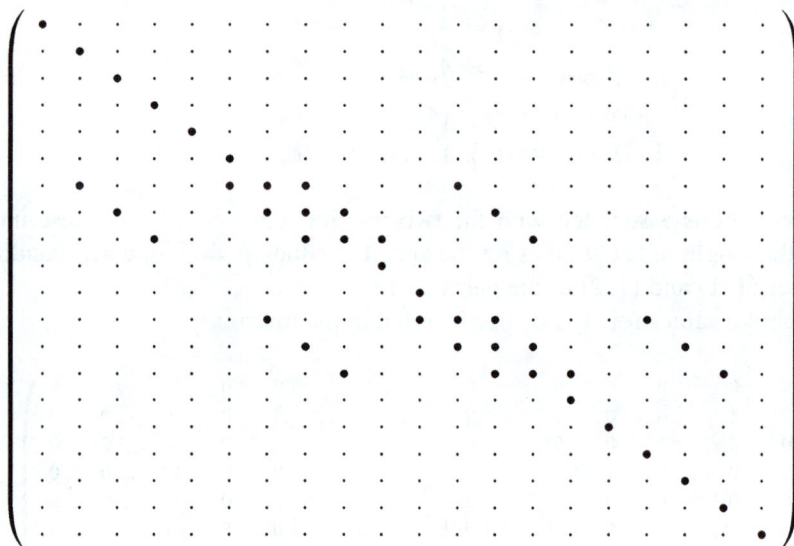

The coefficient matrix is banded

Besides being sparse, we observe that the coefficient matrix is *banded*: it has five distinct bands. We have the diagonal $A_{i,i}$, the subdiagonal $A_{i-1,j}$, the superdiagonal $A_{i,i+1}$, a lower diagonal $A_{i,i-(Nx+1)}$, and an upper diagonal $A_{i,i+(Nx+1)}$.

(0,3): 15	(1,3): 16	(2,3): 17	(3,3): 18	(4,3): 19
(0,2): 10	(1,2): 11	(2,2): 12	(3,2): 13	(4,2): 14
(0,1): 5	(1,1): 6	(2,1): 7	(3,1): 8	(4,1): 9
(0,0): 0	(1,0): 1	(2,0): 2	(3,0): 3	(4,0): 4

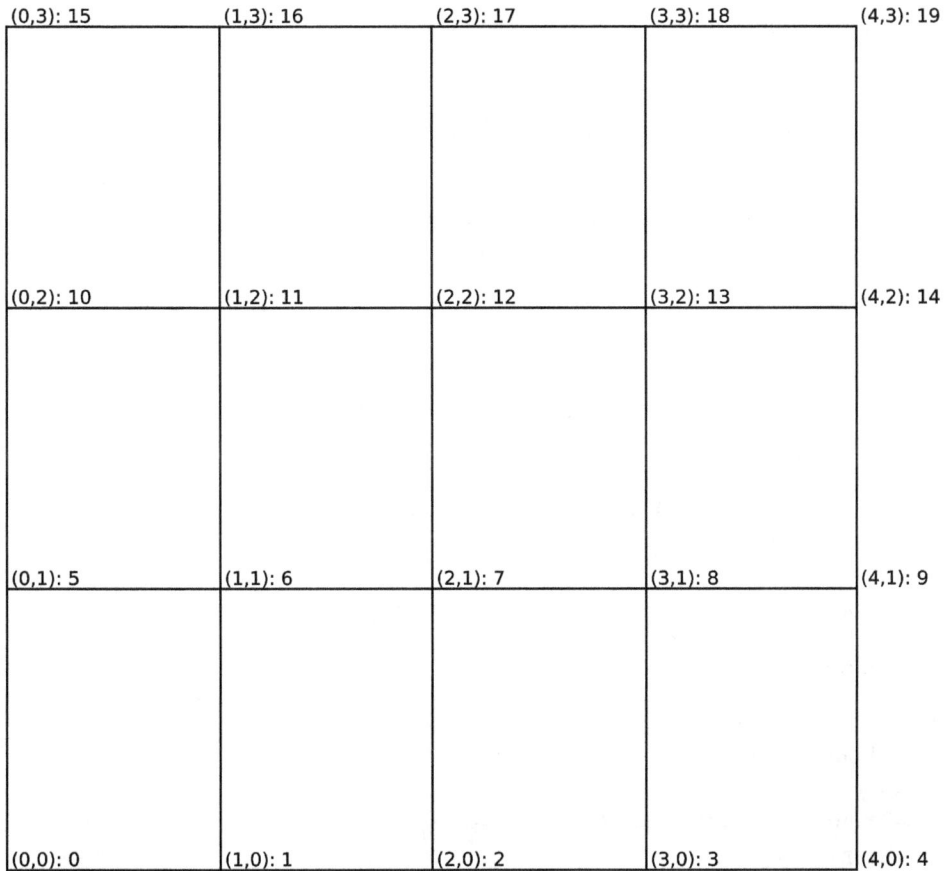

Fig. 3.15 4x3 2D mesh

The other matrix entries are known to be zero. With $N_x + 1 = N_y + 1 = N$, only a fraction $5N^{-2}$ of the matrix entries are nonzero, so the matrix is clearly very sparse for relevant N values. The more we can compute with the nonzeros only, the faster the solution methods will potentially be.

1.6.3 Algorithm for Setting Up the Coefficient Matrix

We looked at a specific mesh in the previous section, formulated the equations, and saw what the corresponding coefficient matrix and right-hand side are. Now our aim is to set up a general algorithm, for any choice of N_x and N_y, that produces the coefficient matrix and the right-hand side vector. We start with a zero matrix and vector, run through each mesh point, and fill in the values depending on whether the mesh point is an interior point or on the boundary.

- for $i = 0, \ldots, N_x$
    - for $j = 0, \ldots, N_y$
        * $p = j(N_x + 1) + i$

* if point (i, j) is on the boundary:
  · $A_{p,p} = 1, b_p = 0$
* else:
  · fill $A_{p,m(i-1,j)}$, $A_{p,m(i+1,j)}$, $A_{p,m(i,j)}$, $A_{p,m(i,j-1)}$, $A_{p,m(i,j+1)}$, and b_p

To ease the test on whether (i, j) is on the boundary or not, we can split the loops a bit, starting with the boundary line $j = 0$, then treat the interior lines $1 \leq j < N_y$, and finally treat the boundary line $j = N_y$:

* for $i = 0, \ldots, N_x$
  - boundary $j = 0$: $p = j(N_x + 1) + i$, $A_{p,p} = 1$
* for $j = 0, \ldots, N_y$
  - boundary $i = 0$: $p = j(N_x + 1) + i$, $A_{p,p} = 1$
  - for $i = 1, \ldots, N_x - 1$
    * interior point $p = j(N_x + 1) + i$
    * fill $A_{p,m(i-1,j)}$, $A_{p,m(i+1,j)}$, $A_{p,m(i,j)}$, $A_{p,m(i,j-1)}$, $A_{p,m(i,j+1)}$, and b_p
  - boundary $i = N_x$: $p = j(N_x + 1) + i$, $A_{p,p} = 1$
* for $i = 0, \ldots, N_x$
  - boundary $j = N_y$: $p = j(N_x + 1) + i$, $A_{p,p} = 1$

The right-hand side is set up as follows.

* for $i = 0, \ldots, N_x$
  - boundary $j = 0$: $p = j(N_x + 1) + i$, $b_p = 0$
* for $j = 0, \ldots, N_y$
  - boundary $i = 0$: $p = j(N_x + 1) + i$, $b_p = 0$
  - for $i = 1, \ldots, N_x - 1$
    * interior point $p = j(N_x + 1) + i$
    * fill b_p
  - boundary $i = N_x$: $p = j(N_x + 1) + i$, $b_p = 0$
* for $i = 0, \ldots, N_x$
  - boundary $j = N_y$: $p = j(N_x + 1) + i$, $b_p = 0$

1.6.4 Implementation with a Dense Coefficient Matrix

The goal now is to map the algorithms in the previous section to Python code. One should, for computational efficiency reasons, take advantage of the fact that the coefficient matrix is sparse and/or banded, i.e., take advantage of all the zeros. However, we first demonstrate how to fill an $N \times N$ dense square matrix, where N is the number of unknowns, here $N = (N_x + 1)(N_y + 1)$. The dense matrix is much easier to understand than the sparse matrix case.

```
import numpy as np

def solver_dense(
    I, a, f, Lx, Ly, Nx, Ny, dt, T, theta=0.5, user_action=None):
    """
    Solve u_t = a*(u_xx + u_yy) + f, u(x,y,0)=I(x,y), with u=0
    on the boundary, on [0,Lx]x[0,Ly]x[0,T], with time step dt,
    using the theta-scheme.
    """
```

```
x = np.linspace(0, Lx, Nx+1)        # mesh points in x dir
y = np.linspace(0, Ly, Ny+1)        # mesh points in y dir
dx = x[1] - x[0]
dy = y[1] - y[0]

dt = float(dt)                      # avoid integer division
Nt = int(round(T/float(dt)))
t = np.linspace(0, Nt*dt, Nt+1)   # mesh points in time

# Mesh Fourier numbers in each direction
Fx = a*dt/dx**2
Fy = a*dt/dy**2
```

The $u_{i,j}^{n+1}$ and $u_{i,j}^n$ mesh functions are represented by their spatial values at the mesh points:

```
u   = np.zeros((Nx+1, Ny+1))        # unknown u at new time level
u_n = np.zeros((Nx+1, Ny+1))        # u at the previous time level
```

It is a good habit (for extensions) to introduce index sets for all mesh points:

```
Ix = range(0, Nx+1)
Iy = range(0, Ny+1)
It = range(0, Nt+1)
```

The initial condition is easy to fill in:

```
# Load initial condition into u_n
for i in Ix:
    for j in Iy:
        u_n[i,j] = I(x[i], y[j])
```

The memory for the coefficient matrix and right-hand side vector is allocated by

```
N = (Nx+1)*(Ny+1)  # no of unknowns
A = np.zeros((N, N))
b = np.zeros(N)
```

The filling of A goes like this:

```
m = lambda i, j: j*(Nx+1) + i

# Equations corresponding to j=0, i=0,1,... (u known)
j = 0
for i in Ix:
    p = m(i,j);  A[p, p] = 1
```

```
# Loop over all internal mesh points in y diretion
# and all mesh points in x direction
for j in Iy[1:-1]:
    i = 0;  p = m(i,j);  A[p, p] = 1   # Boundary
    for i in Ix[1:-1]:                  # Interior points
        p = m(i,j)
        A[p, m(i,j-1)] = - theta*Fy
        A[p, m(i-1,j)] = - theta*Fx
        A[p, p]        = 1 + 2*theta*(Fx+Fy)
        A[p, m(i+1,j)] = - theta*Fx
        A[p, m(i,j+1)] = - theta*Fy
    i = Nx;  p = m(i,j);  A[p, p] = 1  # Boundary
# Equations corresponding to j=Ny, i=0,1,... (u known)
j = Ny
for i in Ix:
    p = m(i,j);  A[p, p] = 1
```

Since A is independent of time, it can be filled once and for all before the time loop.
The right-hand side vector must be filled at each time level inside the time loop:

```
import scipy.linalg

for n in It[0:-1]:
    # Compute b
    j = 0
    for i in Ix:
        p = m(i,j);  b[p] = 0              # Boundary
    for j in Iy[1:-1]:
        i = 0; p = m(i,j); b[p] = 0   # Boundary
        for i in Ix[1:-1]:                # Interior points
            p = m(i,j)
            b[p] = u_n[i,j] + \
               (1-theta)*(
               Fx*(u_n[i+1,j] - 2*u_n[i,j] + u_n[i-1,j]) +\
               Fy*(u_n[i,j+1] - 2*u_n[i,j] + u_n[i,j-1]))\
                  + theta*dt*f(i*dx,j*dy,(n+1)*dt) + \
               (1-theta)*dt*f(i*dx,j*dy,n*dt)
        i = Nx; p = m(i,j); b[p] = 0  # Boundary
    j = Ny
    for i in Ix:
        p = m(i,j);  b[p] = 0              # Boundary

    # Solve matrix system A*c = b
    c = scipy.linalg.solve(A, b)

    # Fill u with vector c
    for i in Ix:
        for j in Iy:
            u[i,j] = c[m(i,j)]

    # Update u_n before next step
    u_n, u = u, u_n
```

We use solve from scipy.linalg and not from numpy.linalg. The difference
is stated below.

`scipy.linalg` **versus** `numpy.linalg`

Quote from the SciPy documentation[2]:

> `scipy.linalg` contains all the functions in `numpy.linalg` plus some other more advanced ones not contained in `numpy.linalg`.
>
> Another advantage of using `scipy.linalg` over numpy.linalg is that it is always compiled with BLAS/LAPACK support, while for NumPy this is optional. Therefore, the SciPy version might be faster depending on how NumPy was installed.
>
> Therefore, unless you don't want to add SciPy as a dependency to your NumPy program, use `scipy.linalg` instead of `numpy.linalg`.

The code shown above is available in the `solver_dense` function in the file `diffu2D_u0.py`, differing only in the boundary conditions, which in the code can be an arbitrary function along each side of the domain.

We do not bother to look at vectorized versions of filling A since a dense matrix is just used of pedagogical reasons for the very first implementation. Vectorization will be treated when A has a sparse matrix representation, as in Sect. 3.6.7.

How to debug the computation of A and b

A good starting point for debugging the filling of A and b is to choose a very coarse mesh, say $N_x = N_y = 2$, where there is just one internal mesh point, compute the equations by hand, and print out A and b for comparison in the code. If wrong elements in A or b occur, print out each assignment to elements in A and b inside the loops and compare with what you expect.

To let the user store, analyze, or visualize the solution at each time level, we include a callback function, named `user_action`, to be called before the time loop and in each pass in that loop. The function has the signature

```
user_action(u, x, xv, y, yv, t, n)
```

where u is a two-dimensional array holding the solution at time level n and time t[n]. The x and y coordinates of the mesh points are given by the arrays x and y, respectively. The arrays xv and yv are vectorized representations of the mesh points such that vectorized function evaluations can be invoked. The xv and yv arrays are defined by

```
xv = x[:,np.newaxis]
yv = y[np.newaxis,:]
```

One can then evaluate, e.g., $f(x, y, t)$ at all internal mesh points at time level n by first evaluating f at all points,

```
f_a = f(xv, yv, t[n])
```

---

[2] http://docs.scipy.org/doc/scipy/reference/tutorial/linalg.html

and then use slices to extract a view of the values at the internal mesh points: f_a[1:-1,1:-1]. The next section features an example on writing a user_action callback function.

1.6.5 Verification: Exact Numerical Solution

A good test example to start with is one that preserves the solution $u = 0$, i.e., $f = 0$ and $I(x, y) = 0$. This trivial solution can uncover some bugs.

The first real test example is based on having an exact solution of the discrete equations. This solution is linear in time and quadratic in space:

$$u(x, y, t) = 5tx(L_x - x)y(y - L_y).$$

Inserting this manufactured solution in the PDE shows that the source term f must be

$$f(x, y, t) = 5x(L_x - x)y(y - L_y) + 10\alpha t(x(L_x - x) + y(y - L_y)).$$

We can use the user_action function to compare the numerical solution with the exact solution at each time level. A suitable helper function for checking the solution goes like this:

```
def quadratic(theta, Nx, Ny):

    def u_exact(x, y, t):
        return 5*t*x*(Lx-x)*y*(Ly-y)
    def I(x, y):
        return u_exact(x, y, 0)
    def f(x, y, t):
        return 5*x*(Lx-x)*y*(Ly-y) + 10*a*t*(y*(Ly-y)+x*(Lx-x))

    # Use rectangle to detect errors in switching i and j in scheme
    Lx = 0.75
    Ly = 1.5
    a = 3.5
    dt = 0.5
    T = 2

    def assert_no_error(u, x, xv, y, yv, t, n):
        """Assert zero error at all mesh points."""
        u_e = u_exact(xv, yv, t[n])
        diff = abs(u - u_e).max()
        tol = 1E-12
        msg = 'diff=%g, step %d, time=%g' % (diff, n, t[n])
        print msg
        assert diff < tol, msg

    solver_dense(
        I, a, f, Lx, Ly, Nx, Ny,
        dt, T, theta, user_action=assert_no_error)
```

A true test function for checking the quadratic solution for several different meshes and θ values can take the form

```
def test_quadratic():
    # For each of the three schemes (theta = 1, 0.5, 0), a series of
    # meshes are tested (Nx > Ny and Nx < Ny)
    for theta in [1, 0.5, 0]:
        for Nx in range(2, 6, 2):
            for Ny in range(2, 6, 2):
                print 'testing for %dx%d mesh' % (Nx, Ny)
                quadratic(theta, Nx, Ny)
```

1.6.6 Verification: Convergence Rates

For 2D verification with convergence rate computations, the expressions and computations just build naturally on what we saw for 1D diffusion. Truncation error analysis and other forms of error analysis point to a numerical error formula like

$$E = C_t \Delta t^p + C_x \Delta x^2 + C_y \Delta y^2,$$

where p, C_t, C_x, and C_y are constants. Often, the analysis of a Crank-Nicolson method can show that $p = 2$, while the Forward and Backward Euler schemes have $p = 1$.

When checking the error formula empirically, we need to reduce it to a form $E = Ch^r$ with a single discretization parameter h and some rate r to be estimated. For the Backward Euler method, where $p = 1$, we can introduce a single discretization parameter according to

$$h = \Delta x^2 = \Delta y^2, \quad h = K^{-1}\Delta t,$$

where K is a constant. The error formula then becomes

$$E = C_t K h + C_x h + C_y h = \tilde{C}h, \quad \tilde{C} = C_t K + C_x + C_y.$$

The simplest choice is obviously $K = 1$. With the Forward Euler method, however, stability requires $\Delta t = hK \le h/(4\alpha)$, so $K \le 1/(4\alpha)$.

For the Crank-Nicolson method, $p = 2$, and we can simply choose

$$h = \Delta x = \Delta y = \Delta t,$$

since there is no restriction on Δt in terms of Δx and Δy.

A frequently used error measure is the ℓ^2 norm of the error mesh point values. Section 2.2.3 and the formula (2.26) shows the error measure for a 1D time-dependent problem. The extension to the current 2D problem reads

$$E = \left(\Delta t \Delta x \Delta y \sum_{n=0}^{N_t} \sum_{i=0}^{N_x} \sum_{j=0}^{N_y} (u_e(x_i, y_j, t_n) - u_{i,j}^n)^2 \right)^{\frac{1}{2}}.$$

One attractive manufactured solution is

$$u_e = e^{-pt} \sin(k_x x) \sin(k_y y), \quad k_x = \frac{\pi}{L_x}, k_y = \frac{\pi}{L_y},$$

where p can be arbitrary. The required source term is

$$f = (\alpha(k_x^2 + k_y^2) - p)u_e\,.$$

The function `convergence_rates` in `diffu2D_u0.py` implements a convergence rate test. Two potential difficulties are important to be aware of:

1. The error formula is assumed to be correct when $h \to 0$, so for coarse meshes the estimated rate r may be somewhat away from the expected value. Fine meshes may lead to prohibitively long execution times.
2. Choosing $p = \alpha(k_x^2 + k_y^2)$ in the manufactured solution above seems attractive ($f = 0$), but leads to a slower approach to the asymptotic range where the error formula is valid (i.e., r fluctuates and needs finer meshes to stabilize).

1.6.7 Implementation with a Sparse Coefficient Matrix

We used a sparse matrix implementation in Sect. 1.2.2 for a 1D problem with a tridiagonal matrix. The present matrix, arising from a 2D problem, has five diagonals, but we can use the same sparse matrix data structure `scipy.sparse.diags`.

Understanding the diagonals Let us look closer at the diagonals in the example with a 4×3 mesh as depicted in Fig. 1.15 and its associated matrix visualized by dots for zeros and bullets for nonzeros. From the example mesh, we may generalize to an $N_x \times N_y$ mesh.

$$
\begin{aligned}
0 &= m(0,0)\\
1 &= m(1,0)\\
2 &= m(2,0)\\
3 &= m(3,0)\\
N_x &= m(N_x,0)\\
N_x + 1 &= m(0,1)\\
(N_x + 1) + 1 &= m(1,1)\\
(N_x + 1) + 2 &= m(2,1)\\
(N_x + 1) + 3 &= m(3,1)\\
(N_x + 1) + N_x &= m(N_x,1)\\
2(N_x + 1) &= m(0,2)\\
2(N_x + 1) + 1 &= m(1,2)\\
2(N_x + 1) + 2 &= m(2,2)\\
2(N_x + 1) + 3 &= m(3,2)\\
2(N_x + 1) + N_x &= m(N_x,2)\\
N_y(N_x + 1) &= m(0,N_y)\\
N_y(N_x + 1) + 1 &= m(1,N_y)\\
N_y(N_x + 1) + 2 &= m(2,N_y)\\
N_y(N_x + 1) + 3 &= m(3,N_y)\\
N_y(N_x + 1) + N_x &= m(N_x,N_y)
\end{aligned}
$$

The main diagonal has $N = (N_x + 1)(N_y + 1)$ elements, while the sub- and super-diagonals have $N - 1$ elements. By looking at the matrix above, we realize that the lower diagonal starts in row $N_x + 1$ and goes to row N, so its length is $N - (N_x + 1)$. Similarly, the upper diagonal starts at row 0 and lasts to row $N - (N_x + 1)$, so it has the same length. Based on this information, we declare the diagonals by

```
main    = np.zeros(N)            # diagonal
lower   = np.zeros(N-1)          # subdiagonal
upper   = np.zeros(N-1)          # superdiagonal
lower2  = np.zeros(N-(Nx+1))     # lower diagonal
upper2  = np.zeros(N-(Nx+1))     # upper diagonal
b       = np.zeros(N)            # right-hand side
```

Filling the diagonals We run through all mesh points and fill in elements on the various diagonals. The line of mesh points corresponding to $j = 0$ are all on the boundary, and only the main diagonal gets a contribution:

```
m = lambda i, j: j*(Nx+1) + i
j = 0; main[m(0,j):m(Nx+1,j)] = 1  # j=0 boundary line
```

Then we run through all interior $j = $ const lines of mesh points. The first and the last point on each line, $i = 0$ and $i = N_x$, correspond to boundary points:

```
for j in Iy[1:-1]:              # Interior mesh lines j=1,...,Ny-1
    i = 0;    main[m(i,j)] = 1
    i = Nx;   main[m(i,j)] = 1  # Boundary
```

For the interior mesh points $i = 1, \ldots, N_x - 1$ on a mesh line $y = $ const we can start with the main diagonal. The entries to be filled go from $i = 1$ to $i = N_x - 1$ so the relevant slice in the main vector is m(1,j):m(Nx,j):

```
main[m(1,j):m(Nx,j)] = 1 + 2*theta*(Fx+Fy)
```

The upper array for the superdiagonal has its index 0 corresponding to row 0 in the matrix, and the array entries to be set go from $m(1, j)$ to $m(N_x - 1, j)$:

```
upper[m(1,j):m(Nx,j)] = - theta*Fx
```

The subdiagonal (lower array), however, has its index 0 corresponding to row 1, so there is an offset of 1 in indices compared to the matrix. The first nonzero occurs (interior point) at a mesh line $j = $ const corresponding to matrix row $m(1, j)$, and the corresponding array index in lower is then $m(1, j)$. To fill the entries from $m(1, j)$ to $m(N_x - 1, j)$ we set the following slice in lower:

```
lower_offset = 1
lower[m(1,j)-lower_offset:m(Nx,j)-lower_offset] = - theta*Fx
```

For the upper diagonal, its index 0 corresponds to matrix row 0, so there is no offset and we can set the entries correspondingly to upper:

```
upper2[m(1,j):m(Nx,j)] = - theta*Fy
```

The `lower2` diagonal, however, has its first index 0 corresponding to row $N_x + 1$, so here we need to subtract the offset $N_x + 1$:

```
lower2_offset = Nx+1
lower2[m(1,j)-lower2_offset:m(Nx,j)-lower2_offset] = - theta*Fy
```

We can now summarize the above code lines for setting the entries in the sparse matrix representation of the coefficient matrix:

```
lower_offset = 1
lower2_offset = Nx+1
m = lambda i, j: j*(Nx+1) + i

j = 0; main[m(0,j):m(Nx+1,j)] = 1  # j=0 boundary line
for j in Iy[1:-1]:                 # Interior mesh lines j=1,...,Ny-1
    i = 0;   main[m(i,j)] = 1  # Boundary
    i = Nx;  main[m(i,j)] = 1  # Boundary
    # Interior i points: i=1,...,N_x-1
    lower2[m(1,j)-lower2_offset:m(Nx,j)-lower2_offset] = - theta*Fy
    lower[m(1,j)-lower_offset:m(Nx,j)-lower_offset] = - theta*Fx
    main[m(1,j):m(Nx,j)] = 1 + 2*theta*(Fx+Fy)
    upper[m(1,j):m(Nx,j)] = - theta*Fx
    upper2[m(1,j):m(Nx,j)] = - theta*Fy
j = Ny; main[m(0,j):m(Nx+1,j)] = 1  # Boundary line
```

The next task is to create the sparse matrix from these diagonals:

```
import scipy.sparse

A = scipy.sparse.diags(
    diagonals=[main, lower, upper, lower2, upper2],
    offsets=[0, -lower_offset, lower_offset,
             -lower2_offset, lower2_offset],
    shape=(N, N), format='csr')
```

Filling the right-hand side; scalar version Setting the entries in the right-hand side is easier, since there are no offsets in the array to take into account. The right-hand side is in fact similar to the one previously shown, when we used a dense matrix representation (the right-hand side vector is, of course, independent of what type of representation we use for the coefficient matrix). The complete time loop goes as follows.

```
import scipy.sparse.linalg

for n in It[0:-1]:
    # Compute b
    j = 0
    for i in Ix:
        p = m(i,j);  b[p] = 0                              # Boundary
```

```
        for j in Iy[1:-1]:
            i = 0;  p = m(i,j);  b[p] = 0              # Boundary
            for i in Ix[1:-1]:
                p = m(i,j)                             # Interior
                b[p] = u_n[i,j] + \
                    (1-theta)*(
                    Fx*(u_n[i+1,j] - 2*u_n[i,j] + u_n[i-1,j]) +\
                    Fy*(u_n[i,j+1] - 2*u_n[i,j] + u_n[i,j-1]))\
                        + theta*dt*f(i*dx,j*dy,(n+1)*dt) + \
                    (1-theta)*dt*f(i*dx,j*dy,n*dt)
            i = Nx;  p = m(i,j);  b[p] = 0             # Boundary
        j = Ny
        for i in Ix:
            p = m(i,j);  b[p] = 0                      # Boundary

        # Solve matrix system A*c = b
        c = scipy.sparse.linalg.spsolve(A, b)

        # Fill u with vector c
        for i in Ix:
            for j in Iy:
                u[i,j] = c[m(i,j)]

        # Update u_n before next step
        u_n, u = u, u_n
```

Filling the right-hand side; vectorized version. Since we use a sparse matrix and try to speed up the computations, we should examine the loops and see if some can be easily removed by vectorization. In the filling of A we have already used vectorized expressions at each $j = $ const line of mesh points. We can very easily do the same in the code above and remove the need for loops over the i index:

```
for n in It[0:-1]:
    # Compute b, vectorized version

    # Precompute f in array so we can make slices
    f_a_np1 = f(xv, yv, t[n+1])
    f_a_n   = f(xv, yv, t[n])

    j = 0; b[m(0,j):m(Nx+1,j)] = 0      # Boundary
    for j in Iy[1:-1]:
        i = 0;  p = m(i,j);  b[p] = 0 # Boundary
        i = Nx;  p = m(i,j);  b[p] = 0 # Boundary
        imin = Ix[1]
        imax = Ix[-1]  # for slice, max i index is Ix[-1]-1
        b[m(imin,j):m(imax,j)] = u_n[imin:imax,j] + \
            (1-theta)*(Fx*(
          u_n[imin+1:imax+1,j] -
        2*u_n[imin:imax,j] + \
          u_n[imin-1:imax-1,j]) +
                            Fy*(
          u_n[imin:imax,j+1] -
        2*u_n[imin:imax,j] +
          u_n[imin:imax,j-1])) + \
            theta*dt*f_a_np1[imin:imax,j] + \
          (1-theta)*dt*f_a_n[imin:imax,j]
    j = Ny;  b[m(0,j):m(Nx+1,j)] = 0 # Boundary
```

```
# Solve matrix system A*c = b
c = scipy.sparse.linalg.spsolve(A, b)

# Fill u with vector c
u[:,:] = c.reshape(Ny+1,Nx+1).T

# Update u_n before next step
u_n, u = u, u_n
```

The most tricky part of this code snippet is the loading of values from the one-dimensional array c into the two-dimensional array u. With our numbering of unknowns from left to right along "horizontal" mesh lines, the correct reordering of the one-dimensional array c as a two-dimensional array requires first a reshaping to an (Ny+1,Nx+1) two-dimensional array and then taking the transpose. The result is an (Nx+1,Ny+1) array compatible with u both in size and appearance of the function values.

The spsolve function in scipy.sparse.linalg is an efficient version of Gaussian elimination suited for matrices described by diagonals. The algorithm is known as *sparse Gaussian elimination*, and spsolve calls up a well-tested C code called SuperLU[3].

The complete code utilizing spsolve is found in the solver_sparse function in the file diffu2D_u0.py.

Verification   We can easily extend the function quadratic from Sect. 1.6.5 to include a test of the solver_sparse function as well.

```
def quadratic(theta, Nx, Ny):
    ...
    t, cpu = solver_sparse(
        I, a, f, Lx, Ly, Nx, Ny,
        dt, T, theta, user_action=assert_no_error)
```

1.6.8   The Jacobi Iterative Method

So far we have created a matrix and right-hand side of a linear system $Ac = b$ and solved the system for c by calling an exact algorithm based on Gaussian elimination. A much simpler implementation, which requires no memory for the coefficient matrix A, arises if we solve the system by *iterative* methods. These methods are only approximate, and the core algorithm is repeated many times until the solution is considered to be converged.

Numerical scheme and linear system   To illustrate the idea of the Jacobi method, we simplify the numerical scheme to the Backward Euler case, $\theta = 1$, so there are

---

[3] http://crd-legacy.lbl.gov/~xiaoye/SuperLU/

fewer terms to write:

$$u_{i,j}^{n+1} - \left(F_x \left(u_{i-1,j}^{n+1} - 2u_{i,j}^{n+1} + u_{i+1,j}^{n+1} \right) + F_y \left(u_{i,j-1}^{n+1} - 2u_{i,j}^{n+1} + u_{i,j+1}^{n+1} \right) \right)$$
$$= u_{i,j}^n + \Delta t f_{i,j}^{n+1}.$$

(1.97)

The idea of the *Jacobi* iterative method is to introduce an iteration, here with index r, where we in each iteration treat $u_{i,j}^{n+1}$ as unknown, but use values from the previous iteration for the other unknowns $u_{i\pm1,j\pm1}^{n+1}$.

Iterations Let $u_{i,j}^{n+1,r}$ be the approximation to $u_{i,j}^{n+1}$ in iteration r, for all relevant i and j indices. We first solve with respect to $u_{i,j}^{n+1}$ to get the equation to solve:

$$u_{i,j}^{n+1} = (1 + 2F_x + 2F_y)^{-1} \left(F_x \left(u_{i-1,j}^{n+1} + u_{i+1,j}^{n+1} \right) + F_y \left(u_{i,j-1}^{n+1} + u_{i,j+1}^{n+1} \right) \right)$$
$$+ u_{i,j}^n + \Delta t f_{i,j}^{n+1}.$$

(1.98)

The iteration is introduced by using iteration index r, for computed values, on the right-hand side and $r + 1$ (unknown in this iteration) on the left-hand side:

$$u_{i,j}^{n+1,r+1} = (1 + 2F_x + 2F_y)^{-1} \left(F_x \left(u_{i-1,j}^{n+1,r} + u_{i+1,j}^{n+1,r} \right) + F_y \left(u_{i,j-1}^{n+1,r} + u_{i,j+1}^{n+1,r} \right) \right)$$
$$+ u_{i,j}^n + \Delta t f_{i,j}^{n+1}.$$

(1.99)

Initial guess We start the iteration with the computed values at the previous time level:

$$u_{i,j}^{n+1,0} = u_{i,j}^n, \quad i = 0, \ldots, N_x, \ j = 0, \ldots, N_y.$$

(1.100)

Relaxation A common technique in iterative methods is to introduce a *relaxation*, which means that the new approximation is a weighted mean of the approximation as suggested by the algorithm and the previous approximation. Naming the quantity on the left-hand side of (1.99) as $u_{i,j}^{n+1,*}$, a new approximation based on relaxation reads

$$u^{n+1,r+1} = \omega u_{i,j}^{n+1,*} + (1 - \omega) u_{i,j}^{n+1,r}.$$

(1.101)

Under-relaxation means $\omega < 1$, while over-relaxation has $\omega > 1$.

Stopping criteria The iteration can be stopped when the change from one iteration to the next is sufficiently small ($\leq \epsilon$), using either an infinity norm,

$$\max_{i,j} \left| u_{i,j}^{n+1,r+1} - u_{i,j}^{n+1,r} \right| \leq \epsilon,$$

(1.102)

or an L^2 norm,

$$\left(\Delta x \Delta y \sum_{i,j} (u_{i,j}^{n+1,r+1} - u_{i,j}^{n+1,r})^2 \right)^{\frac{1}{2}} \leq \epsilon.$$

(1.103)

Another widely used criterion measures how well the equations are solved by looking at the residual (essentially $b - Ac^{r+1}$ if c^{r+1} is the approximation to the solution in iteration $r + 1$). The residual, defined in terms of the finite difference stencil, is

$$
R_{i,j} = u_{i,j}^{n+1,r+1} - \left(F_x \left(u_{i-1,j}^{n+1,r+1} - 2u_{i,j}^{n+1,r+1} + u_{i+1,j}^{n+1,r+1} \right) \right.
$$

$$
\left. + F_y \left(u_{i,j-1}^{n+1,r+1} - 2u_{i,j}^{n+1,r+1} + u_{i,j+1}^{n+1,r+1} \right) \right)
$$

$$
- u_{i,j}^n - \Delta t f_{i,j}^{n+1} .
\tag{1.104}
$$

One can then iterate until the norm of the mesh function $R_{i,j}$ is less than some tolerance:

$$
\left(\Delta x \Delta y \sum_{i,j} R_{i,j}^2 \right)^{\frac{1}{2}} \le \epsilon .
\tag{1.105}
$$

Code-friendly notation To make the mathematics as close as possible to what we will write in a computer program, we may introduce some new notation: $u_{i,j}$ is a short notation for $u_{i,j}^{n+1,r+1}$, $u_{i,j}^-$ is a short notation for $u_{i,j}^{n+1,r}$, and $u_{i,j}^{(s)}$ denotes $u_{i,j}^{n+1-s}$. That is, $u_{i,j}$ is the unknown, $u_{i,j}^-$ is its most recently computed approximation, and s counts time levels backwards in time. The Jacobi method (1.99) takes the following form with the new notation:

$$
u_{i,j}^* = (1 + 2F_x + 2F_y)^{-1} \left(\left(F_x(u_{i-1,j}^- + u_{i+1,j}^-) + F_y(u_{i,j-1}^- + u_{i,j+1}^-) \right) \right.
$$

$$
\left. + u_{i,j}^{(1)} + \Delta t f_{i,j}^{n+1} \right) .
\tag{1.106}
$$

Generalization of the scheme We can also quite easily introduce the θ rule for discretization in time and write up the Jacobi iteration in that case as well:

$$
u_{i,j}^* = (1 + 2\theta(F_x + F_y))^{-1} \left(\theta \left(F_x(u_{i-1,j}^- + u_{i+1,j}^-) + F_y(u_{i,j-1}^- + u_{i,j+1}^-) \right) \right.
$$

$$
+ u_{i,j}^{(1)} + \theta \Delta t f_{i,j}^{n+1} + (1 - \theta) \Delta t f_{i,j}^n
$$

$$
+ (1 - \theta) \left(F_x \left(u_{i-1,j}^{(1)} - 2u_{i,j}^{(1)} + u_{i+1,j}^{(1)} \right) \right.
$$

$$
\left. \left. + F_y \left(u_{i,j-1}^{(1)} - 2u_{i,j}^{(1)} + u_{i,j+1}^{(1)} \right) \right) \right) .
\tag{1.107}
$$

The final update of u applies relaxation:

$$
u_{i,j} = \omega u_{i,j}^* + (1 - \omega) u_{i,j}^- .
$$

1.6.9 Implementation of the Jacobi Method

The Jacobi method needs no coefficient matrix and right-hand side vector, but it needs an array for u in the previous iteration. We call this array u_, using the notation at the end of the previous section (at the same time level). The unknown itself is called u, while u_n is the computed solution one time level back in time. With a θ rule in time, the time loop can be coded like this:

```
for n in It[0:-1]:
    # Solve linear system by Jacobi iteration at time level n+1
    u_[:,:] = u_n  # Start value
    converged = False
    r = 0
    while not converged:
        if version == 'scalar':
            j = 0
            for i in Ix:
                u[i,j] = U_0y(t[n+1])               # Boundary
            for j in Iy[1:-1]:
                i = 0;   u[i,j] = U_0x(t[n+1])  # Boundary
                i = Nx;  u[i,j] = U_Lx(t[n+1])  # Boundary
            # Interior points
            for i in Ix[1:-1]:
                u_new = 1.0/(1.0 + 2*theta*(Fx + Fy))*(theta*(
                    Fx*(u_[i+1,j] + u_[i-1,j]) +
                    Fy*(u_[i,j+1] + u_[i,j-1])) + \
                u_n[i,j] + \
                (1-theta)*(Fx*(
                u_n[i+1,j] - 2*u_n[i,j] + u_n[i-1,j]) +
                  Fy*(
                u_n[i,j+1] - 2*u_n[i,j] + u_n[i,j-1]))\
                  + theta*dt*f(i*dx,j*dy,(n+1)*dt) + \
                (1-theta)*dt*f(i*dx,j*dy,n*dt))
                u[i,j] = omega*u_new + (1-omega)*u_[i,j]
            j = Ny
            for i in Ix:
                u[i,j] = U_Ly(t[n+1])         # Boundary

        elif version == 'vectorized':
            j = 0;  u[:,j] = U_0y(t[n+1])  # Boundary
            i = 0;  u[i,:] = U_0x(t[n+1])  # Boundary
            i = Nx; u[i,:] = U_Lx(t[n+1])  # Boundary
            j = Ny; u[:,j] = U_Ly(t[n+1])  # Boundary
        # Internal points
        f_a_np1 = f(xv, yv, t[n+1])
        f_a_n   = f(xv, yv, t[n])
        u_new = 1.0/(1.0 + 2*theta*(Fx + Fy))*(theta*(Fx*(
            u_[2:,1:-1] + u_[:-2,1:-1]) +
            Fy*(
            u_[1:-1,2:] + u_[1:-1,:-2])) +\
        u_n[1:-1,1:-1] + \
            (1-theta)*(Fx*(
            u_n[2:,1:-1] - 2*u_n[1:-1,1:-1] + u_n[:-2,1:-1]) +\
            Fy*(
            u_n[1:-1,2:] - 2*u_n[1:-1,1:-1] + u_n[1:-1,:-2]))\
            + theta*dt*f_a_np1[1:-1,1:-1] + \
            (1-theta)*dt*f_a_n[1:-1,1:-1])
```

```
        u[1:-1,1:-1] = omega*u_new + (1-omega)*u_[1:-1,1:-1]
    r += 1
    converged = np.abs(u-u_).max() < tol or r >= max_iter
    u_[:,:] = u

# Update u_n before next step
u_n, u = u, u_n
```

The vectorized version should be quite straightforward to understand once one has an understanding of how a standard 2D finite stencil is vectorized.

The first natural verification is to use the test problem in the function `quadratic` from Sect. 1.6.5. This problem is known to have no approximation error, but any iterative method will produce an approximate solution with unknown error. For a tolerance 10^{-k} in the iterative method, we can, e.g., use a slightly larger tolerance $10^{-(k-1)}$ for the difference between the exact and the computed solution.

```
def quadratic(theta, Nx, Ny):
    ...
    def assert_small_error(u, x, xv, y, yv, t, n):
        """Assert small error for iterative methods."""
        u_e = u_exact(xv, yv, t[n])
        diff = abs(u - u_e).max()
        tol = 1E-4
        msg = 'diff=%g, step %d, time=%g' % (diff, n, t[n])
        assert diff < tol, msg

    for version in 'scalar', 'vectorized':
        for theta in 1, 0.5:
            print 'testing Jacobi, %s version, theta=%g' % \
                (version, theta)
            t, cpu = solver_Jacobi(
                I=I, a=a, f=f, Lx=Lx, Ly=Ly, Nx=Nx, Ny=Ny,
                dt=dt, T=T, theta=theta,
                U_0x=0, U_0y=0, U_Lx=0, U_Ly=0,
                user_action=assert_small_error,
                version=version, iteration='Jacobi',
                omega=1.0, max_iter=100, tol=1E-5)
```

Even for a very coarse 4×4 mesh, the Jacobi method requires 26 iterations to reach a tolerance of 10^{-5}, which is quite many iterations, given that there are only 25 unknowns.

1.6.10 Test Problem: Diffusion of a Sine Hill

It can be shown that

$$u_e = A e^{-\alpha \pi^2 (L_x^{-2} + L_y^{-2}) t} \sin\left(\frac{\pi}{L_x} x\right) \sin\left(\frac{\pi}{L_y} y\right), \qquad (1.108)$$

is a solution of the 2D homogeneous diffusion equation $u_t = \alpha(u_{xx} + u_{yy})$ in a rectangle $[0, L_x] \times [0, L_y]$, for any value of the amplitude A. This solution vanishes at the boundaries, and the initial condition is the product of two sines. We may choose $A = 1$ for simplicity.

It is difficult to know if our solver based on the Jacobi method works properly since we are faced with two sources of errors: one from the discretization, E_Δ, and one from the iterative Jacobi method, E_i. The total error in the computed u can be represented as

$$E_u = E_\Delta + E_i\,.$$

One error measure is to look at the maximum value, which is obtained for the mid-point $x = L_x/2$ and $y = L_x/2$. This midpoint is represented in the discrete u if N_x and N_y are even numbers. We can then compute E_u as $E_u = |\max u_e - \max u|$, when we know an exact solution u_e of the problem.

What about E_Δ? If we use the maximum value as a measure of the error, we have in fact analytical insight into the approximation error in this particular problem. According to Sect. 1.3.9, the exact solution (1.108) of the PDE problem is also an exact solution of the discrete equations, except that the damping factor in time is different. More precisely, (1.66) and (1.67) are solutions of the discrete problem for $\theta = 1$ (Backward Euler) and $\theta = \frac{1}{2}$ (Crank-Nicolson), respectively. The factors raised to the power n is the numerical amplitude, and the errors in these factors become

$$E_\Delta = e^{-\alpha k^2 t} - \left(\frac{1 - 2(F_x \sin^2 p_x + F_x \sin^2 p_y)}{1 + 2(F_x \sin^2 p_x + F_x \sin^2 p_y)} \right)^n, \quad \theta = \frac{1}{2},$$

$$E_\Delta = e^{-\alpha k^2 t} - (1 + 4F_x \sin^2 p_x + 4F_y \sin^2 p_y)^{-n}, \quad \theta = 1\,.$$

We are now in a position to compute E_i numerically. That is, we can compute the error due to iterative solution of the linear system and see if it corresponds to the convergence tolerance used in the method. Note that the convergence is based on measuring the difference in two consecutive approximations, which is not exactly the error due to the iteration, but it is a kind of measure, and it should have about the same size as E_i.

The function `demo_classic_iterative` in `diffu2D_u0.py` implements the idea above (also for the methods in Sect. 1.6.12). The value of E_i is in particular printed at each time level. By changing the tolerance in the convergence criterion of the Jacobi method, we can see that E_i is of the same order of magnitude as the prescribed tolerance in the Jacobi method. For example: $E_\Delta \sim 10^{-2}$ with $N_x = N_y = 10$ and $\theta = \frac{1}{2}$, as long as $\max u$ has some significant size ($\max u > 0.02$). An appropriate value of the tolerance is then 10^{-3}, such that the error in the Jacobi method does not become bigger than the discretization error. In that case, E_i is around $5 \cdot 10^{-3}$. The corresponding number of Jacobi iterations (with $\omega = 1$) varies from 31 to 12 during the time simulation (for $\max u > 0.02$). Changing the tolerance to 10^{-5} causes many more iterations (61 to 42) without giving any contribution to the overall accuracy, because the total error is dominated by E_Δ.

Also, with an $N_x = N_y = 20$, the spatial accuracy increases and many more iterations are needed (143 to 45), but the dominating error is from the time discretization. However, with such a finer spatial mesh, a higher tolerance in the convergence criterion 10^{-4} is needed to keep $E_i \sim 10^{-3}$. More experiments show the disadvantage of the very simple Jacobi iteration method: the number of iterations increases with the number of unknowns, keeping the tolerance fixed, but the tolerance should also be lowered to avoid the iteration error to dominate the total error. A small adjustment of the Jacobi method, as described in Sect. 1.6.12, provides a better method.

1.6.11 The Relaxed Jacobi Method and Its Relation to the Forward Euler Method

We shall now show that solving the Poisson equation $-\alpha\nabla^2 u = f$ by the Jacobi iterative method is in fact equivalent to using a Forward Euler scheme on $u_t = \alpha\nabla^2 u + f$ and letting $t \to \infty$.

A Forward Euler discretization of the 2D diffusion equation,

$$[D_t^+ u = \alpha(D_x D_x u + D_y D_y u) + f]_{i,j}^n,$$

can be written out as

$$u_{i,j}^{n+1} = u_{i,j}^n + \frac{\Delta t}{\alpha h^2}\left(u_{i-1,j}^n + u_{i+1,j}^n + u_{i,j-1}^n + u_{i,j+1}^n - 4u_{i,j}^n + h^2 f_{i,j}\right),$$

where $h = \Delta x = \Delta y$ has been introduced for simplicity. The scheme can be reordered as

$$u_{i,j}^{n+1} = (1-\omega)u_{i,j}^n + \frac{1}{4}\omega\left(u_{i-1,j}^n + u_{i+1,j}^n + u_{i,j-1}^n + u_{i,j+1}^n - 4u_{i,j}^n + h^2 f_{i,j}\right),$$

with

$$\omega = 4\frac{\Delta t}{\alpha h^2},$$

but this latter form is nothing but the relaxed Jacobi method applied to

$$[D_x D_x u + D_y D_y u = -f]_{i,j}^n.$$

From the equivalence above we know a couple of things about the Jacobi method for solving $-\nabla^2 u = f$:

1. The method is unstable if $\omega > 1$ (since the Forward Euler method is then unstable).
2. The convergence is really slow as the iteration index increases (coming from the fact that the Forward Euler scheme requires many small time steps to reach the stationary solution).

These observations are quite disappointing: if we already have a time-dependent diffusion problem and want to take larger time steps by an implicit time discretization method, we will with the Jacobi method end up with something close to a slow Forward Euler simulation of the original problem at each time level. Nevertheless, the are two reasons for why the Jacobi method remains a fundamental building block for solving linear systems arising from PDEs: 1) a couple of iterations remove large parts of the error and this is effectively used in the very efficient class of multigrid methods; and 2) the idea of the Jacobi method can be developed into more efficient methods, especially the SOR method, which is treated next.

1.6.12 The Gauss-Seidel and SOR Methods

If we update the mesh points according to the Jacobi method (1.98) for a Backward Euler discretization with a loop over $i = 1, \ldots, N_x - 1$ and $j = 1, \ldots, N_y - 1$, we realize that when $u_{i,j}^{n+1,r+1}$ is computed, $u_{i-1,j}^{n+1,r+1}$ and $u_{i,j-1}^{n+1,r+1}$ are already computed, so these new values can be used rather than $u_{i-1,j}^{n+1,r}$ and $u_{i,j-1}^{n+1,r}$ (respectively) in the formula for $u_{i,j}^{n+1,r+1}$. This idea gives rise to the *Gauss-Seidel* iteration method, which mathematically is just a small adjustment of (1.98):

$$u_{i,j}^{n+1,r+1} =$$

$$(1 + 2F_x + 2F_y)^{-1}\left(\left(F_x\left(u_{i-1,j}^{n+1,r+1} + u_{i+1,j}^{n+1,r}\right) + F_y\left(u_{i,j-1}^{n+1,r+1} + u_{i,j+1}^{n+1,r}\right)\right)\right.$$

$$\left. + u_{i,j}^n + \Delta t f_{i,j}^{n+1}\right). \tag{3.109}$$

Observe that the way we access the mesh points in the formula (1.109) is important: points with $i - 1$ must be computed before points with i, and points with $j - 1$ must be computed before points with j. Any sequence of mesh points can be used in the Gauss-Seidel method, but the particular math formula must distinguish between already visited points in the current iteration and the points not yet visited.

The idea of relaxation (1.101) can equally well be applied to the Gauss-Seidel method. Actually, the Gauss-Seidel method with an arbitrary $0 < \omega \leq 2$ has its own name: the *Successive Over-Relaxation* method, abbreviated as SOR.

The SOR method for a θ rule discretization, with the shortened u and u^- notation, can be written

$$u_{i,j}^* = (1 + 2\theta(F_x + F_y))^{-1}\left(\theta(F_x(u_{i-1,j} + u_{i+1,j}^-) + F_y(u_{i,j-1} + u_{i,j+1}^-))\right.$$

$$+ u_{i,j}^{(1)} + \theta\Delta t f_{i,j}^{n+1} + (1 - \theta)\Delta t f_{i,j}^n$$

$$+ (1 - \theta)\left(F_x\left(u_{i-1,j}^{(1)} - 2u_{i,j}^{(1)} + u_{i+1,j}^{(1)}\right)\right.$$

$$\left.\left. + F_y\left(u_{i,j-1}^{(1)} - 2u_{i,j}^{(1)} + u_{i,j+1}^{(1)}\right)\right)\right), \tag{1.110}$$

$$u_{i,j} = \omega u_{i,j}^* + (1 - \omega)u_{i,j}^- \tag{1.111}$$

The sequence of mesh points in (1.110) is $i = 1, \ldots, N_x - 1, j = 1, \ldots, N_y - 1$ (but whether i runs faster or slower than j does not matter).

3.6.13 Scalar Implementation of the SOR Method

Since the Jacobi and Gauss-Seidel methods with relaxation are so similar, we can easily make a common code for the two:

```
for n in It[0:-1]:
    # Solve linear system by Jacobi/SOR iteration at time level n+1
    u_[:,:] = u_n  # Start value
    converged = False
    r = 0
    while not converged:
        if version == 'scalar':
            if iteration == 'Jacobi':
                u__ = u_
            elif iteration == 'SOR':
                u__ = u
            j = 0
            for i in Ix:
                u[i,j] = U_0y(t[n+1])   # Boundary
            for j in Iy[1:-1]:
                i = 0;   u[i,j] = U_0x(t[n+1])   # Boundary
                i = Nx;  u[i,j] = U_Lx(t[n+1])   # Boundary
                for i in Ix[1:-1]:
                    u_new = 1.0/(1.0 + 2*theta*(Fx + Fy))*(theta*(
                        Fx*(u_[i+1,j] + u__[i-1,j]) +
                        Fy*(u_[i,j+1] + u__[i,j-1])) + \
                    u_n[i,j] + (1-theta)*(
                      Fx*(
                    u_n[i+1,j] - 2*u_n[i,j] + u_n[i-1,j]) +
                      Fy*(
                    u_n[i,j+1] - 2*u_n[i,j] + u_n[i,j-1]))\
                      + theta*dt*f(i*dx,j*dy,(n+1)*dt) + \
                    (1-theta)*dt*f(i*dx,j*dy,n*dt))
                    u[i,j] = omega*u_new + (1-omega)*u_[i,j]
            j = Ny
            for i in Ix:
                u[i,j] = U_Ly(t[n+1])   # boundary
        r += 1
        converged = np.abs(u-u_).max() < tol or r >= max_iter
        u_[:,:] = u

    u_n, u = u, u_n  # Get ready for next iteration
```

The idea here is to introduce u__ to be used for already computed values (u) in the Gauss-Seidel/SOR version of the implementation, or just values from the previous iteration (u_) in case of the Jacobi method.

1.6.14 Vectorized Implementation of the SOR Method

Vectorizing the Gauss-Seidel iteration step turns out to be non-trivial. The problem is that vectorized operations typically imply operations on arrays where the sequence in which we visit the elements does not matter. In particular, this principle makes vectorized code trivial to parallelize. However, in the Gauss-Seidel algorithm, the sequence in which we visit the elements in the arrays does matter, and it is well known that the basic method as explained above cannot be parallelized. Therefore, also vectorization will require new thinking.

The strategy for vectorizing (and parallelizing) the Gauss-Seidel method is to use a special numbering of the mesh points called red-black numbering: every other point is red or black as in a checkerboard pattern. This numbering requires N_x and N_y to be even numbers. Here is an example of a 6×6 mesh:

```
r b r b r b r
b r b r b r b
r b r b r b r
b r b r b r b
r b r b r b r
b r b r b r b
r b r b r b r
```

The idea now is to first update all the red points. Each formula for updating a red point involves only the black neighbors. Thereafter, we update all the black points, and at each black point, only the recently computed red points are involved.

The scalar implementation of the red-black numbered Gauss-Seidel method is really compact, since we can update values directly in u (this guarantees that we use the most recently computed values). Here is the relevant code for the Backward Euler scheme in time and without a source term:

```
# Update internal points
for sweep in 'red', 'black':
    for j in range(1, Ny, 1):
        if sweep == 'red':
            start = 1 if j % 2 == 1 else 2
        elif sweep == 'black':
            start = 2 if j % 2 == 1 else 1
        for i in range(start, Nx, 2):
        u[i,j] = 1.0/(1.0 + 2*(Fx + Fy))*(
                    Fx*(u[i+1,j] + u[i-1,j]) +
                    Fy*(u[i,j+1] + u[i,j-1]) + u_n[i,j])
```

The vectorized version must be based on slices. Looking at a typical red-black pattern, e.g.,

```
r b r b r b r
b r b r b r b
r b r b r b r
b r b r b r b
r b r b r b r
b r b r b r b
r b r b r b r
```

we want to update the internal points (marking boundary points with x):

```
x x x x x x x
x r b r b r x
x b r b r b x
x r b r b r x
x b r b r b x
x r b r b r x
x x x x x x x
```

It is impossible to make one slice that picks out all the internal red points. Instead, we need two slices. The first involves points marked with R:

```
x x x x x x x
x R b R b R x
x b r b r b x
x R b R b R x
x b r b r b x
x R b R b R x
x x x x x x x
```

This slice is specified as 1::2 for i and 1::2 for j, or with slice objects:

```
i = slice(1, None, 2);   j = slice(1, None, 2)
```

The second slice involves the red points with R:

```
x x x x x x x
x r b r b r x
x b R b R b x
x r b r b r x
x b R b R b x
x r b r b r x
x x x x x x x
```

The slices are

```
i = slice(2, None, 2);   j = slice(2, None, 2)
```

For the black points, the first slice involves the B points:

```
x x x x x x x
x r B r B r x
x b r b r b x
x r B r B r x
x b r b r b x
x r B r B r x
x x x x x x x
```

with slice objects

```
i = slice(2, None, 2);   j = slice(1, None, 2)
```

The second set of black points is shown here:

```
x x x x x x x
x r b r b r x
x B r B r B x
x r b r b r x
x B r B r B x
x r b r b r x
x x x x x x x
```

with slice objects

```
i = slice(1, None, 2);   j = slice(2, None, 2)
```

That is, we need four sets of slices. The simplest way of implementing the algorithm is to make a function with variables for the slices representing i, $i - 1$, $i + 1$, j, $j - 1$, and $j + 1$, here called ic ("i center"), im1 ("i minus 1", ip1 ("i plus 1"), jc, jm1, and jp1, respectively.

```
def update(u_, u_n, ic, im1, ip1, jc, jm1, jp1):
    return \
        1.0/(1.0 + 2*theta*(Fx + Fy))*(theta*(
            Fx*(u_[ip1,jc] + u_[im1,jc]) +
            Fy*(u_[ic,jp1] + u_[ic,jm1])) +\
        u_n[ic,jc] + (1-theta)*(
            Fx*(u_n[ip1,jc] - 2*u_n[ic,jc] + u_n[im1,jc]) +\
            Fy*(u_n[ic,jp1] - 2*u_n[ic,jc] + u_n[ic,jm1]))+\
        theta*dt*f_a_np1[ic,jc] + \
        (1-theta)*dt*f_a_n[ic,jc])
```

The formula returned from update is to be compared with (1.110).

The relaxed Jacobi iteration can be implemented by

```
ic  = jc  = slice(1,-1)
im1 = jm1 = slice(0,-2)
ip1 = jp1 = slice(2,None)
u_new[ic,jc] = update(
    u_, u_n, ic, im1, ip1, jc, jm1, jp1)
u[ic,jc] = omega*u_new[ic,jc] + (1-omega)*u_[ic,jc]
```

The Gauss-Seidel (or SOR) updates need four different steps. The ic and jc slices are specified above. For each of these, we must specify the corresponding im1, ip1, jm1, and jp1 slices. The code below contains the details.

```
# Red points
ic  = slice(1,-1,2)
im1 = slice(0,-2,2)
ip1 = slice(2,None,2)
jc  = slice(1,-1,2)
jm1 = slice(0,-2,2)
jp1 = slice(2,None,2)
u_new[ic,jc] = update(
    u_new, u_n, ic, im1, ip1, jc, jm1, jp1)

ic  = slice(2,-1,2)
im1 = slice(1,-2,2)
ip1 = slice(3,None,2)
jc  = slice(2,-1,2)
jm1 = slice(1,-2,2)
jp1 = slice(3,None,2)
u_new[ic,jc] = update(
    u_new, u_n, ic, im1, ip1, jc, jm1, jp1)

# Black points
ic  = slice(2,-1,2)
im1 = slice(1,-2,2)
ip1 = slice(3,None,2)
jc  = slice(1,-1,2)
jm1 = slice(0,-2,2)
jp1 = slice(2,None,2)
u_new[ic,jc] = update(
    u_new, u_n, ic, im1, ip1, jc, jm1, jp1)

ic  = slice(1,-1,2)
im1 = slice(0,-2,2)
ip1 = slice(2,None,2)
jc  = slice(2,-1,2)
jm1 = slice(1,-2,2)
jp1 = slice(3,None,2)
u_new[ic,jc] = update(
    u_new, u_n, ic, im1, ip1, jc, jm1, jp1)

# Relax
c = slice(1,-1)
u[c,c] = omega*u_new[c,c] + (1-omega)*u_[c,c]
```

The function `solver_classic_iterative` in `diffu2D_u0.py` contains a unified implementation of the relaxed Jacobi and SOR methods in scalar and vectorized versions using the techniques explained above.

1.6.15   Direct Versus Iterative Methods

Direct methods There are two classes of methods for solving linear systems: direct methods and iterative methods. Direct methods are based on variants of the Gaussian elimination procedure and will produce an exact solution (in exact arithmetics) in an a priori known number of steps. Iterative methods, on the other hand, produce an approximate solution, and the amount of work for reaching a given accuracy is usually not known.

The most common direct method today is to use the *LU factorization* procedure to factor the coefficient matrix A as the product of a lower-triangular matrix L (with unit diagonal terms) and an upper-triangular matrix U: $A = LU$. As soon as we have L and U, a system of equations $LUc = b$ is easy to solve because of the triangular nature of L and U. We first solve $Ly = b$ for y (forward substitution), and thereafter we find c from solving $Uc = y$ (backward substitution). When A is a dense $N \times N$ matrix, the LU factorization costs $\frac{1}{3}N^3$ arithmetic operations, while the forward and backward substitution steps each require of the order N^2 arithmetic operations. That is, factorization dominates the costs, while the substitution steps are cheap.

Symmetric, positive definite coefficient matrices often arise when discretizing PDEs. In this case, the LU factorization becomes $A = LL^T$, and the associated algorithm is known as *Cholesky factorization*. Most linear algebra software offers highly optimized implementations of LU and Cholesky factorization as well as forward and backward substitution (`scipy.linalg` is the relevant Python package).

Finite difference discretizations lead to sparse coefficient matrices. An extreme case arose in Sect. 1.2.1 where A was tridiagonal. For a tridiagonal matrix, the amount of arithmetic operations in the LU and Cholesky factorization algorithms is just of the order N, not N^3. Tridiagonal matrices are special cases of *banded matrices*, where the matrices contain just a set of diagonal bands. Finite difference methods on regularly numbered rectangular and box-shaped meshes give rise to such banded matrices, with 5 bands in 2D and 7 in 3D for diffusion problems. Gaussian elimination only needs to work within the bands, leading to much more efficient algorithms.

If $A_{i,j} = 0$ for $j > i + p$ and $j < i - p$, p is the *half-bandwidth* of the matrix. We have in our 2D problem $p = N_x + 2$, while in 3D, $p = (N_x + 1)(N_y + 1) + 2$. The cost of Gaussian elimination is then $\mathcal{O}(Np^2)$, so with $p \ll N$, we see that banded matrices are much more efficient to compute with. By reordering the unknowns in clever ways, one can reduce the work of Gaussian elimination further. Fortunately, the Python programmer has access to such algorithms through the `scipy.sparse.linalg` package.

Although a direct method is an exact algorithm, rounding errors may in practice accumulate and pollute the solution. The effect grows with the size of the linear system, so both for accuracy and efficiency, iterative methods are better suited than direct methods for solving really large linear systems.

Iterative methods The Jacobi and SOR iterative methods belong to a class of iterative methods where the idea is to solve $Au = b$ by splitting A into two parts, $A = M - N$, such that solving systems $Mu = c$ is easy and efficient. With the splitting, we get a system

$$Mu = Nu + b,$$

which suggests an iterative method

$$Mu^{r+1} = Nu^r + b, \quad r = 0, 1, 2, \ldots,$$

where u^{r+1} is a new approximation to u in the $r + 1$-th iteration. To initiate the iteration, we need a start vector u^0.

The Jacobi and SOR methods are based on splitting A into a lower tridiagonal part L, the diagonal D, and an upper tridiagonal part U, such that $A = L + D + U$. The Jacobi method corresponds to $M = D$ and $N = -L - U$. The Gauss-Seidel method employs $M = L + D$ and $N = -U$, while the SOR method corresponds to

$$M = \frac{1}{\omega} D + L, \quad N = \frac{1-\omega}{\omega} D - U \,.$$

The relaxed Jacobi method has similar expressions:

$$M = \frac{1}{\omega} D, \quad N = \frac{1-\omega}{\omega} D - L - U \,.$$

With the matrix forms of the Jacobi and SOR methods as written above, we could in an implementation alternatively fill the matrix A with entries and call general implementations of the Jacobi or SOR methods that work on a system $Au = b$. However, this is almost never done since forming the matrix A requires quite some code and storing A in the computer's memory is unnecessary. It is much easier to just apply the Jacobi and SOR ideas to the finite difference stencils directly in an implementation, as we have shown in detail.

Nevertheless, the matrix formulation of the Jacobi and SOR methods have been important for analyzing their convergence behavior. One can show that the error $u^r - u$ fulfills $u^r - u = G^r(u^0 - u)$, where $G = M^{-1}N$ and G^k is a matrix exponential. For the method to converge, $\lim_{r \to \infty} ||G^r|| = 0$ is a necessary and sufficient condition. This implies that the *spectral radius* of G must be less than one. Since G is directly related to the finite difference scheme for the underlying PDE problem, one can in principle compute the spectral radius. For a given PDE problem, however, this is not a practical strategy, since it is very difficult to develop useful formulas. Analysis of model problems, usually related to the Poisson equation, reveals some trends of interest: the convergence rate of the Jacobi method goes like h^2, while that of SOR with an optimal ω goes like h, where h is the spatial spacing: $h = \Delta x = \Delta y$. That is, the efficiency of the Jacobi method quickly deteriorates with the increasing mesh resolution, and SOR is much to be preferred (even if the optimal ω remains an open question). We refer to Chapter 4 of [16] for more information on the convergence theory. One important result is that if A is symmetric and positive definite, then SOR will converge for any $0 < \omega < 2$.

The optimal ω parameter can be theoretically established for a Poisson problem as

$$\omega_o = \frac{2}{1 + \sqrt{1 - \varrho^2}}, \quad \varrho = \frac{\cos(\pi/N_x) + (\Delta x/\Delta y)^2 \cos(\pi/N_y)}{1 + (\Delta x/\Delta y)^2} \,. \tag{1.112}$$

This formula can be used as a guide also in other problems.

The Jacobi and the SOR methods have their great advantage of being trivial to implement, so they are obviously popular of this reason. However, the slow convergence of these methods limits the popularity to fairly small linear systems (i.e., coarse meshes). As soon as the matrix size grows, one is better off with more sophisticated iterative methods like the preconditioned Conjugate gradient method, which we now turn to.

Finally, we mention that there is a variant of the SOR method, called the *Symmetric Successive Over-relaxation* method, known as SSOR, where one runs a standard SOR sweep through the mesh points and then a new sweep while visiting the points in reverse order.

1.6.16 The Conjugate Gradient Method

There is no simple intuitive derivation of the Conjugate gradient method, so we refer to the many excellent expositions in the literature for the idea of the method and how the algorithm is derived. In particular, we recommend the books [1, 2, 5, 16]. A brief overview is provided in the Wikipedia article[4]. Here, we just state the pros and cons of the method from a user's perspective and how we utilize it in code.

The original Conjugate gradient method is limited to linear systems $Au = b$, where A is a symmetric and positive definite matrix. There are, however, extensions of the method to non-symmetric matrices.

A major advantage of all conjugate gradient methods is that the matrix A is only used in matrix-vector products, so we do not need form and store A if we can provide code for computing a matrix-vector product Au. Another important feature is that the algorithm is very easy to vectorize and parallelize. The primary downside of the method is that it converges slowly unless one has an effective *preconditioner* for the system. That is, instead of solving $Au = b$, we try to solve $M^{-1}Au = M^{-1}b$ in the hope that the method works better for this *preconditioned* system. The matrix M is the *preconditioner* or preconditioning matrix. Now we need to perform matrix-vector products $y = M^{-1}Au$, which is done in two steps: first the matrix-vector product $v = Au$ is carried out and then the system $My = v$ must be solved. Therefore, M must be cheap to compute and systems $My = v$ must be cheap to solve.

A perfect preconditioner is $M = A$, but in each iteration in the Conjugate gradient method one then has so solve a system with A as coefficient matrix! A key idea is to let M be some kind of *cheap approximation* to A. The simplest preconditioner is to set $M = D$, where D is the diagonal of A. This choice means running one Jacobi iteration as preconditioner. Exercise 1.8 shows that the Jacobi and SOR methods can also be viewed as preconditioners.

Constructing good preconditioners is a scientific field on its own. Here we shall treat the topic just very briefly. For a user having access to the `scipy.sparse.linalg` library, there are Conjugate gradient methods and preconditioners readily available:

- For positive definite, symmetric systems: `cg` (the Conjugate gradient method)
- For symmetric systems: `minres` (Minimum residual method)
- For non-symmetric systems:
  - `gmres` (GMRES: Generalized minimum residual method)
  - `bicg` (BiConjugate gradient method)
  - `bicgstab` (Stabilized BiConjugate gradient method)

[4] https://en.wikipedia.org/wiki/Conjugate_gradient_method

- – cgs (Conjugate gradient squared method)
- – qmr (Quasi-minimal residual iteration)
- • Preconditioner: spilu (Sparse, incomplete LU factorization)

The ILU preconditioner is an attractive all-round type of preconditioner that is suitable for most problems on serial computers. A more efficient preconditioner is the multigrid method, and algebraic multigrid is also an all-round choice as preconditioner. The Python package PyAMG[5] offers efficient implementations of the algebraic multigrid method, to be used both as a preconditioner and as a stand-alone iterative method.

The matrix arising from implicit time discretization methods applied to the diffusion equation is symmetric and positive definite. Thus, we can use the Conjugate gradient method (cg), typically in combination with an ILU preconditioner. The code is very similar to the one we created when solving the linear system by sparse Gaussian elimination, the main difference is that we now allow for calling up the Conjugate gradient function as an alternative solver.

```
def solver_sparse(
    I, a, f, Lx, Ly, Nx, Ny, dt, T, theta=0.5,
    U_0x=0, U_0y=0, U_Lx=0, U_Ly=0, user_action=None,
    method='direct', CG_prec='ILU', CG_tol=1E-5):
    """
    Full solver for the model problem using the theta-rule
    difference approximation in time. Sparse matrix with
    dedicated Gaussian elimination algorithm (method='direct')
    or ILU preconditioned Conjugate Gradients (method='CG' with
    tolerance CG_tol and preconditioner CG_prec ('ILU' or None)).
    """
    # Set up data structures as shown before

    # Precompute sparse matrix
    ...

    A = scipy.sparse.diags(
        diagonals=[main, lower, upper, lower2, upper2],
        offsets=[0, -lower_offset, lower_offset,
                 -lower2_offset, lower2_offset],
        shape=(N, N), format='csc')

    if method == 'CG':
        if CG_prec == 'ILU':
            # Find ILU preconditioner (constant in time)
            A_ilu = scipy.sparse.linalg.spilu(A)  # SuperLU defaults
            M = scipy.sparse.linalg.LinearOperator(
                shape=(N, N), matvec=A_ilu.solve)
        else:
            M = None
        CG_iter = []  # No of CG iterations at time level n

    # Time loop
    for n in It[0:-1]:
        # Compute b, vectorized version
```

[5] https://github.com/pyamg/pyamg

```
# Solve matrix system A*c = b
if method == 'direct':
    c = scipy.sparse.linalg.spsolve(A, b)
elif method == 'CG':
    x0 = u_n.T.reshape(N)  # Start vector is u_n
    CG_iter.append(0)

    def CG_callback(c_k):
        """Trick to count the no of iterations in CG."""
        CG_iter[-1] += 1

    c, info = scipy.sparse.linalg.cg(
        A, b, x0=x0, tol=CG_tol, maxiter=N, M=M,
        callback=CG_callback)

# Fill u with vector c
# Update u_n before next step
u_n, u = u, u_n
```

The number of iterations in the Conjugate gradient method is of interest, but is unfortunately not available from the cg function. Therefore, we perform a trick: in each iteration a user function CG_callback is called where we accumulate the number of iterations in a list CG_iter.

1.6.17 What Is the Recommended Method for Solving Linear Systems?

There is no clear answer to this question. If you have enough memory and computing time available, direct methods such as spsolve are to be preferred since they are easy to use and finds almost an exact solution. However, in larger 2D and in 3D problems, direct methods usually run too slowly or require too much memory, so one is forced to use iterative methods. The fastest and most reliable methods are in the Conjugate Gradient family, but these require suitable preconditioners. ILU is an all-round preconditioner, but it is not suited for parallel computing. The Jacobi and SOR iterative methods are easy to implement, and popular for that reason, but run slowly. Jacobi iteration is not an option in real problems, but SOR may be.

1.7 Random Walk

Models leading to diffusion equations, see Sect. 1.8, are usually based on reasoning with *averaged* physical quantities such as concentration, temperature, and velocity. The underlying physical processes involve complicated microscopic movement of atoms and molecules, but an average of a large number of molecules is performed in a small volume before the modeling starts, and the averaged quantity inside this volume is assigned as a point value at the centroid of the volume. This means that concentration, temperature, and velocity at a space-time point represent averages around the point in a small time interval and small spatial volume.

Random walk is a principally different kind of modeling procedure compared to the reasoning behind partial differential equations. The idea in random walk is to have a large number of "particles" that undergo random movements. Averaging can then be used afterwards to compute macroscopic quantities like concentration. The "particles" and their random movement represent a very simplified microscopic behavior of molecules, much simpler and computationally much more efficient than direct molecular simulation[6], yet the random walk model has been very powerful to describe a wide range of phenomena, including heat conduction, quantum mechanics, polymer chains, population genetics, neuroscience, hazard games, and pricing of financial instruments.

It can be shown that random walk, when averaged, produces models that are mathematically equivalent to diffusion equations. This is the primary reason why we treat random walk in this chapter: two very different algorithms (finite difference stencils and random walk) solve the same type of problems. The simplicity of the random walk algorithm makes it particularly attractive for solving diffusion equations on massively parallel computers. The exposition here is as simple as possible, and good thorough derivation of the models is provided by Hjorth-Jensen [7].

1.7.1 Random Walk in 1D

Imagine that we have some particles that perform random moves, either to the right or to the left. We may flip a coin to decide the movement of each particle, say head implies movement to the right and tail means movement to the left. Each move is one unit length. Physicists use the term *random walk* for this type of movement. The movement is also known as drunkard's walk[7]. You may try this yourself: flip the coin and make one step to the left or right, and repeat the process.

We introduce the symbol N for the number of steps in a random walk. Figure 1.16 shows four different random walks with $N = 200$.

1.7.2 Statistical Considerations

Let S_k be the stochastic variable representing a step to the left or to the right in step number k. We have that $S_k = -1$ with probability p and $S_k = 1$ with probability $q = 1 - p$. The variable S_k is known as a Bernoulli variable[8]. The expectation of S_k is

$$E[S_k] = p \cdot (-1) + q \cdot 1 = 1 - 2p,$$

and the variance is

$$\mathrm{Var}(S_k) = E[S_k^2] - E[S_k]^2 = 1 - (1 - 2p)^2 = 4p(1 - p).$$

---

[6] https://en.wikipedia.org/wiki/Molecular_dynamics
[7] https://en.wikipedia.org/wiki/The_Drunkard%27s_Walk
[8] https://en.wikipedia.org/wiki/Bernoulli_distribution

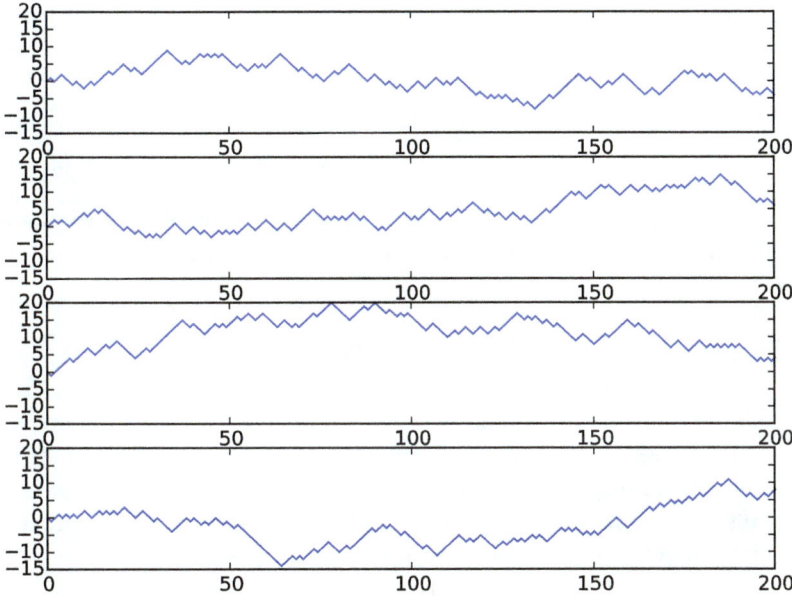

Fig. 1.16 Ensemble of 4 random walks, each with 200 steps

The position after k steps is another stochastic variable

$$\bar{X}_k = \sum_{i=0}^{k-1} S_i \,.$$

The expected position is

$$E[\bar{X}_k] = E\left(\sum_{i=0}^{k-1} S_i\right) = \sum_{i=0}^{k-1} E[S_i] = k(1-2p)\,.$$

All the S_k variables are independent. The variance therefore becomes

$$\text{Var}(\bar{X}_k) = \text{Var}\left(\sum_{i=0}^{k-1} S_i\right) = \sum_{i=0}^{k-1} \text{Var}(S_i) = k4p(1-p)\,.$$

We see that $\text{Var}(\bar{X}_k)$ is proportional with the number of steps k. For the very important case $p = q = \frac{1}{2}$, $E[\bar{X}_k] = 0$ and $\text{Var}(\bar{X}_k) = k$.

How can we estimate $E[\bar{X}_k] = 0$ and $\text{Var}(\bar{X}_k) = N$? We must have many random walks of the type in Fig. 1.16. For a given k, say $k = 100$, we find all the values of $\bar{X}_k$, name them $\bar{x}_{0,k}$, $\bar{x}_{1,k}$, $\bar{x}_{2,k}$, and so on. The empirical estimate of $E[\bar{X}_k]$ is the average,

$$E[\bar{X}_k] \approx \frac{1}{W} \sum_{j=0}^{W-1} \bar{x}_{j,k},$$

while an empirical estimate of $\text{Var}(\bar{X}_k)$ is

$$\text{Var}(\bar{X}_k) \approx \frac{1}{W} \sum_{j=0}^{W-1} (\bar{x}_{j,k})^2 - \left(\frac{1}{W} \sum_{j=0}^{W-1} \bar{x}_{j,k} \right)^2 .$$

That is, we take the statistics for a given K across the ensemble of random walks ("vertically" in Fig. 1.16). The key quantities to record are $\sum_i \bar{x}_{i,k}$ and $\sum_i \bar{x}_{i,k}^2$.

1.7.3 Playing Around with Some Code

Scalar code Python has a `random` module for drawing random numbers, and this module has a function `uniform(a, b)` for drawing a uniformly distributed random number in the interval $[a, b]$. If an event happens with probability p, we can simulate this on the computer by drawing a random number r in $[0, 1)$, because then $r \leq p$ with probability p and $r > p$ with probability $1 - p$:

```
import random
r = random.uniform(0, 1)
if r <= p:
    # Event happens
else:
    # Event does not happen
```

A random walk with N steps, starting at x_0, where we move to the left with probability p and to the right with probability $1 - p$ can now be implemented by

```
import random, numpy as np

def random_walk1D(x0, N, p):
    """1D random walk with 1 particle."""
    # Store position in step k in position[k]
    position = np.zeros(N)
    position[0] = x0
    current_pos = x0
    for k in range(N-1):
        r = random.uniform(0, 1)
        if r <= p:
            current_pos -= 1
        else:
            current_pos += 1
        position[k+1] = current_pos
    return position
```

Vectorized code Since N is supposed to be large and we want to repeat the process for many particles, we should speed up the code as much as possible. Vectorization is the obvious technique here: we draw all the random numbers at once with aid of numpy, and then we formulate vector operations to get rid of the loop over the steps (k). The `numpy.random` module has vectorized versions of the functions in

Python's built-in `random` module. For example, `numpy.random.uniform(a, b, N)` returns N random numbers uniformly distributed between a (included) and b (not included).

We can then make an array of all the steps in a random walk: if the random number is less than or equal to p, the step is -1, otherwise the step is 1:

```
r = np.random.uniform(0, 1, size=N)
steps = np.where(r <= p, -1, 1)
```

The value of `position[k]` is the sum of all steps up to step k. Such sums are often needed in vectorized algorithms and therefore available by the `numpy.cumsum` function:

```
>>> import numpy as np
>>> np.cumsum(np.array([1,3,4,6]))
array([ 1,    4,    8, 14])
```

The resulting array in this demo has elements 1, $1 + 3 = 4$, $1 + 3 + 4 = 8$, and $1 + 3 + 4 + 6 = 14$.

We can now vectorize the `random_walk1D` function:

```
def random_walk1D_vec(x0, N, p):
    """Vectorized version of random_walk1D."""
    # Store position in step k in position[k]
    position = np.zeros(N+1)
    position[0] = x0
    r = np.random.uniform(0, 1, size=N)
    steps = np.where(r <= p, -1, 1)
    position[1:] = x0 + np.cumsum(steps)
    return position
```

This code runs about 10 times faster than the scalar version. With a parallel numpy library, the code can also automatically take advantage of hardware for parallel computing because each of the four array operations can be trivially parallelized.

Fixing the random sequence During software development with random numbers it is advantageous to always generate the same sequence of random numbers, as this may help debugging processes. To fix the sequence, we set a *seed* of the random number generator to some chosen integer, e.g.,

```
np.random.seed(10)
```

Calls to `random_walk1D_vec` give positions of the particle as depicted in Fig. 3.17. The particle starts at the origin and moves with $p = \frac{1}{2}$. Since the seed is the same, the plot to the left is just a magnification of the first 1000 steps in the plot to the right.

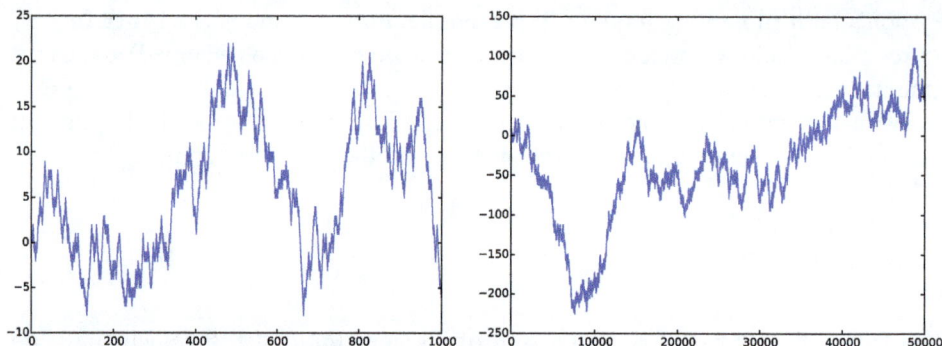

Fig. 1.17 1000 (*left*) and 50,000 (*right*) steps of a random walk

Verification When we have a scalar and a vectorized code, it is always a good idea
to develop a unit test for checking that they produce the same result. A problem
in the present context is that the two versions apply two different random number
generators. For a test to be meaningful, we need to fix the seed and use the same
generator. This means that the scalar version must either use np.random or have
this as an option. An option is the most flexible choice:

```
import random

def random_walk1D(x0, N, p, random=random):
    ...
    r = random.uniform(0, 1)
```

Using random=np.random, the r variable gets computed by np.random.uniform,
and the sequence of random numbers will be the same as in the vectorized version
that employs the same generator (given that the seed is also the same). A proper test
function may be to check that the positions in the walk are the same in the scalar
and vectorized implementations:

```
def test_random_walk1D():
    # For fixed seed, check that scalar and vectorized versions
    # produce the same result
    x0 = 2;  N = 4;  p = 0.6
    np.random.seed(10)
    scalar_computed = random_walk1D(x0, N, p, random=np.random)
    np.random.seed(10)
    vectorized_computed = random_walk1D_vec(x0, N, p)
    assert (scalar_computed == vectorized_computed).all()
```

Note that we employ == for arrays with real numbers, which is normally an inad-
equate test due to rounding errors, but in the present case, all arithmetics consists
of adding or subtracting one, so these operations are expected to have no rounding
errors. Comparing two numpy arrays with == results in a boolean array, so we need
to call the all() method to ensure that all elements are True, i.e., that all elements
in the two arrays match each other pairwise.

1.7.4   Equivalence with Diffusion

The original random walk algorithm can be said to work with dimensionless co-ordinates $\bar{x}_i = -N + i$, $i = 0, 1, \ldots, 2N + 1$ ($i \in [-N, N]$), and $\bar{t}_n = n$, $n = 0, 1, \ldots, N$. A mesh with spacings Δx and Δt with dimensions can be introduced by

$$x_i = X_0 + \bar{x}_i \Delta x, \quad t_n = \bar{t}_n \Delta t .$$

If we implement the algorithm with dimensionless coordinates, we can just use this rescaling to obtain the movement in a coordinate system without unit spacings.

Let P_i^{n+1} be the probability of finding the particle at mesh point $\bar{x}_i$ at time $\bar{t}_{n+1}$. We can reach mesh point $(i, n + 1)$ in two ways: either coming in from the left from $(i - 1, n)$ or from the right $(i + 1, n)$. Each has probability $\frac{1}{2}$ (if we assume $p = q = \frac{1}{2}$). The fundamental equation for P_i^{n+1} is

$$P_i^{n+1} = \frac{1}{2} P_{i-1}^n + \frac{1}{2} P_{i+1}^n . \tag{1.113}$$

(This equation is easiest to understand if one looks at the random walk as a Markov process and applies the transition probabilities, but this is beyond scope of the present text.)

Subtracting P_i^n from (1.113) results in

$$P_i^{n+1} - P_i^n = \frac{1}{2} \left(P_{i-1}^n - 2P_i^n + \frac{1}{2} P_{i+1}^n \right) .$$

Readers who have seen the Forward Euler discretization of a 1D diffusion equation recognize this scheme as very close to such a discretization. We have

$$\frac{\partial}{\partial t} P(x_i, t_n) = \frac{P_i^{n+1} - P_i^n}{\Delta t} + \mathcal{O}(\Delta t),$$

or in dimensionless coordinates

$$\frac{\partial}{\partial \bar{t}} P(\bar{x}_i, \bar{t}_n) \approx P_i^{n+1} - P_i^n .$$

Similarly, we have

$$\frac{\partial^2}{\partial x^2} P(x_i, t_n) = \frac{P_{i-1}^n - 2P_i^n + \frac{1}{2} P_{i+1}^n}{\Delta x^2} + \mathcal{O}(\Delta x^2),$$

$$\frac{\partial^2}{\partial x^2} P(\bar{x}_i, \bar{t}_n) \approx P_{i-1}^n - 2P_i^n + \frac{1}{2} P_{i+1}^n .$$

Equation (1.113) is therefore equivalent with the dimensionless diffusion equation

$$\frac{\partial P}{\partial \bar{t}} = \frac{1}{2} \frac{\partial^2 P}{\partial \bar{x}^2}, \tag{1.114}$$

or the diffusion equation

$$\frac{\partial P}{\partial t} = D \frac{\partial^2 P}{\partial x^2}, \tag{1.115}$$

with diffusion coefficient

$$D = \frac{\Delta x^2}{2\Delta t}.$$

This derivation shows the tight link between random walk and diffusion. If we keep track of where the particle is, and repeat the process many times, or run the algorithms for lots of particles, the histogram of the positions will approximate the solution of the diffusion equation for the local probability P_i^n.

Suppose all the random walks start at the origin. Then the initial condition for the probability distribution is the Dirac delta function $\delta(x)$. The solution of (1.114) can be shown to be

$$\bar{P}(\bar{x}, \bar{t}) = \frac{1}{\sqrt{4\pi\alpha t}} e^{-\frac{x^2}{4\alpha t}}, \qquad (1.116)$$

where $\alpha = \frac{1}{2}$.

1.7.5 Implementation of Multiple Walks

Our next task is to implement an ensemble of walks (for statistics, see Sect. 1.7.2) and also provide data from the walks such that we can compute the probabilities of the positions as introduced in the previous section. An appropriate representation of probabilities P_i^n are histograms (with i along the x axis) for a few selected values of n.

To estimate the expectation and variance of the random walks, Sect. 1.7.2 points to recording $\sum_j x_{j,k}$ and $\sum_j x_{j,k}^2$, where $x_{j,k}$ is the position at time/step level k in random walk number j. The histogram of positions needs the individual values $x_{i,k}$ for all i values and some selected k values.

We introduce `position[k]` to hold $\sum_j x_{j,k}$, `position2[k]` to hold $\sum_j (x_{j,k})^2$, and `pos_hist[i,k]` to hold $x_{i,k}$. A selection of k values can be specified by saying how many, `num_times`, and let them be equally spaced through time:

```
pos_hist_times = [(N//num_times)*i for i in range(num_times)]
```

This is one of the few situations where we want integer division ($//$) or real division rounded to an integer.

Scalar version Our scalar implementation of running `num_walks` random walks may go like this:

```
def random_walks1D(x0, N, p, num_walks=1, num_times=1,
                   random=random):
    """Simulate num_walks random walks from x0 with N steps."""
    position = np.zeros(N+1)     # Accumulated positions
    position[0] = x0*num_walks
    position2 = np.zeros(N+1)    # Accumulated positions**2
    position2[0] = x0**2*num_walks
    # Histogram at num_times selected time points
    pos_hist = np.zeros((num_walks, num_times))
    pos_hist_times = [(N//num_times)*i for i in range(num_times)]
    #print 'save hist:', post_hist_times
```

```
    for n in range(num_walks):
        num_times_counter = 0
        current_pos = x0
        for k in range(N):
            if k in pos_hist_times:
                #print 'save, k:', k, num_times_counter, n
                pos_hist[n,num_times_counter] = current_pos
                num_times_counter += 1
            # current_pos corresponds to step k+1
            r = random.uniform(0, 1)
            if r <= p:
                current_pos -= 1
            else:
                current_pos += 1
            position [k+1] += current_pos
            position2[k+1] += current_pos**2
    return position, position2, pos_hist, np.array(pos_hist_times)
```

Vectorized version We have already vectorized a single random walk. The additional challenge here is to vectorize the computation of the data for the histogram, pos_hist, but given the selected steps in pos_hist_times, we can find the corresponding positions by indexing with the list pos_hist_times: position[post_hist_times], which are to be inserted in pos_hist[n, :].

```
def random_walks1D_vec1(x0, N, p, num_walks=1, num_times=1):
    """Vectorized version of random_walks1D."""
    position  = np.zeros(N+1)    # Accumulated positions
    position2 = np.zeros(N+1)    # Accumulated positions**2
    walk = np.zeros(N+1)         # Positions of current walk
    walk[0] = x0
    # Histogram at num_times selected time points
    pos_hist = np.zeros((num_walks, num_times))
    pos_hist_times = [(N//num_times)*i for i in range(num_times)]

    for n in range(num_walks):
        r = np.random.uniform(0, 1, size=N)
        steps = np.where(r <= p, -1, 1)
        walk[1:] = x0 + np.cumsum(steps)  # Positions of this walk
        position  += walk
        position2 += walk**2
        pos_hist[n,:] = walk[pos_hist_times]
    return position, position2, pos_hist, np.array(pos_hist_times)
```

Improved vectorized version Looking at the vectorized version above, we still have one potentially long Python loop over n. Normally, num_walks will be much larger than N. The vectorization of the loop over N certainly speeds up the program, but if we think of vectorization as also a way to parallelize the code, all the independent walks (the n loop) can be executed in parallel. Therefore, we should include this loop as well in the vectorized expressions, at the expense of using more memory.

We introduce the array walks to hold the $N + 1$ steps of all the walks: each row represents the steps in one walk.

```
walks = np.zeros((num_walks, N+1))  # Positions of each walk
walks[:,0] = x0
```

Since all the steps are independent, we can just make one long vector of enough random numbers (N*num_walks), translate these numbers to ±1, then we reshape the array such that the steps of each walk are stored in the rows.

```
r = np.random.uniform(0, 1, size=N*num_walks)
steps = np.where(r <= p, -1, 1).reshape(num_walks, N)
```

The next step is to sum up the steps in each walk. We need the np.cumsum function for this, with the argument axis=1 for indicating a sum across the columns:

```
>>> a = np.arange(6).reshape(2,3)
>>> a
array([[0, 1, 2],
       [3, 4, 5]])
>>> np.cumsum(a, axis=1)
array([[ 0,  1,  3],
       [ 3,  7, 12]])
```

Now walks can be computed by

```
walks[:,1:] = x0 + np.cumsum(steps, axis=1)
```

The position vector is the sum of all the walks. That is, we want to sum all the rows, obtained by

```
position = np.sum(walks, axis=0)
```

A corresponding expression computes the squares of the positions. Finally, we need to compute pos_hist, but that is a matter of grabbing some of the walks (according to pos_hist_times):

```
pos_hist[:,:] = walks[:,pos_hist_times]
```

The complete vectorized algorithm without any loop can now be summarized:

```
def random_walks1D_vec2(x0, N, p, num_walks=1, num_times=1):
    """Vectorized version of random_walks1D; no loops."""
    position  = np.zeros(N+1)    # Accumulated positions
    position2 = np.zeros(N+1)    # Accumulated positions**2
    walks = np.zeros((num_walks, N+1))  # Positions of each walk
    walks[:,0] = x0
    # Histogram at num_times selected time points
    pos_hist = np.zeros((num_walks, num_times))
    pos_hist_times = [(N//num_times)*i for i in range(num_times)]
```

```
r = np.random.uniform(0, 1, size=N*num_walks)
steps = np.where(r <= p, -1, 1).reshape(num_walks, N)
walks[:,1:] = x0 + np.cumsum(steps, axis=1)
position  = np.sum(walks,    axis=0)
position2 = np.sum(walks**2, axis=0)
pos_hist[:,:] = walks[:,pos_hist_times]
return position, position2, pos_hist, np.array(pos_hist_times)
```

What is the gain of the vectorized implementations? One important gain is that each vectorized operation can be automatically parallelized if one applies a parallel numpy library like Numba[9]. On a single CPU, however, the speed up of the vectorized operations is also significant. With $N = 1000$ and 50,000 repeated walks, the two vectorized versions run about 25 and 18 times faster than the scalar version, with `random_walks1D_vec1` being fastest.

Remark on vectorized code and parallelization Our first attempt on vectorization removed the loop over the N steps in a single walk. However, the number of walks is usually much larger than N, because of the need for accurate statistics. Therefore, we should rather remove the loop over all walks. It turns out, from our efficiency experiments, that the function `random_walks1D_vec2` (with no loops) is slower than `random_walks1D_vec1`. This is a bit surprising and may be explained by less efficiency in the statements involving very large arrays, containing all steps for all walks at once.

From a parallelization and improved vectorization point of view, it would be more natural to switch the sequence of the loops in the serial code such that the shortest loop is the outer loop:

```
def random_walks1D2(x0, N, p, num_walks=1, num_times=1, ...):
    ...
    current_pos = x0 + np.zeros(num_walks)
    num_times_counter = -1

    for k in range(N):
        if k in pos_hist_times:
        num_times_counter += 1
        store_hist = True
    else:
        store_hist = False

        for n in range(num_walks):
            # current_pos corresponds to step k+1
            r = random.uniform(0, 1)
        if r <= p:
                current_pos[n] -= 1
            else:
                current_pos[n] += 1
            position [k+1] += current_pos[n]
            position2[k+1] += current_pos[n]**2
            if store_hist:
                pos_hist[n,num_times_counter] = current_pos[n]
    return position, position2, pos_hist, np.array(pos_hist_times)
```

---

[9] http://numba.pydata.org

The vectorized version of this code, where we just vectorize the loop over n, becomes

```python
def random_walks1D2_vec1(x0, N, p, num_walks=1, num_times=1):
    """Vectorized version of random_walks1D2."""
    position  = np.zeros(N+1)      # Accumulated positions
    position2 = np.zeros(N+1)      # Accumulated positions**2
    # Histogram at num_times selected time points
    pos_hist = np.zeros((num_walks, num_times))
    pos_hist_times = [(N//num_times)*i for i in range(num_times)]

    current_pos = np.zeros(num_walks)
    current_pos[0] = x0
    num_times_counter = -1

    for k in range(N):
        if k in pos_hist_times:
            num_times_counter += 1
            store_hist = True  # Store histogram data for this k
        else:
            store_hist = False

        # Move all walks one step
        r = np.random.uniform(0, 1, size=num_walks)
        steps = np.where(r <= p, -1, 1)
        current_pos += steps
        position[k+1]  = np.sum(current_pos)
        position2[k+1] = np.sum(current_pos**2)
        if store_hist:
            pos_hist[:,num_times_counter] = current_pos
    return position, position2, pos_hist, np.array(pos_hist_times)
```

This function runs significantly faster than the random_walks1D_vec1 function above, typically 1.7 times faster. The code is also more appropriate in a parallel computing context since each vectorized statement can work with data of size num_walks over the compute units, repeated N times (compared with data of size N, repeated num_walks times, in random_walks1D_vec1).

The scalar code with switched loops, random_walks1D2 runs a bit slower than the original code in random_walks1D, so with the longest loop as the inner loop, the vectorized function random_walks1D2_vec1 is almost 60 times faster than the scalar counterpart, while the code random_walks1D_vec2 without loops is only around 18 times faster. Taking into account the very large arrays required by the latter function, we end up with random_walks1D2_vec1 as the preferred implementation.

Test function During program development, it is highly recommended to carry out computations by hand for, e.g., N=4 and num_walks=3. Normally, this is done by executing the program with these parameters and checking with pen and paper that the computations make sense. The next step is to use this test for correctness in a formal test function.

First, we need to check that the simulation of multiple random walks reproduces the results of random_walk1D, random_walk1D_vec1, and random_walk1D_vec2 for the first walk, if the seed is the same. Second, we run three random walks (N=4) with the scalar and the two vectorized versions and check that the returned arrays are identical.

For this type of test to be successful, we must be sure that exactly the same set of random numbers are used in the three versions, a fact that requires the same random number generator and the same seed, of course, but also the same sequence of computations. This is not obviously the case with the three `random_walk1D*` functions we have presented. The critical issue in `random_walk1D_vec1` is that the first random numbers are used for the first walk, the second set of random numbers is used for the second walk and so on, to be compatible with how the random numbers are used in the function `random_walk1D`. For the function `random_walk1D_vec2` the situation is a bit more complicated since we generate all the random numbers at once. However, the critical step now is the reshaping of the array returned from `np.where`: we must reshape as (`num_walks, N`) to ensure that the first N random numbers are used for the first walk, the next N numbers are used for the second walk, and so on.

We arrive at the test function below.

```
def test_random_walks1D():
    # For fixed seed, check that scalar and vectorized versions
    # produce the same result
    x0 = 0;  N = 4;  p = 0.5

    # First, check that random_walks1D for 1 walk reproduces
    # the walk in random_walk1D
    num_walks = 1
    np.random.seed(10)
    computed = random_walks1D(
        x0, N, p, num_walks, random=np.random)
    np.random.seed(10)
    expected = random_walk1D(
        x0, N, p, random=np.random)
    assert (computed[0] == expected).all()

    # Same for vectorized versions
    np.random.seed(10)
    computed = random_walks1D_vec1(x0, N, p, num_walks)
    np.random.seed(10)
    expected = random_walk1D_vec(x0, N, p)
    assert (computed[0] == expected).all()
    np.random.seed(10)
    computed = random_walks1D_vec2(x0, N, p, num_walks)
    np.random.seed(10)
    expected = random_walk1D_vec(x0, N, p)
    assert (computed[0] == expected).all()

    # Second, check multiple walks: scalar == vectorized
    num_walks = 3
    num_times = N
    np.random.seed(10)
    serial_computed = random_walks1D(
        x0, N, p, num_walks, num_times, random=np.random)
```

```
np.random.seed(10)
vectorized1_computed = random_walks1D_vec1(
    x0, N, p, num_walks, num_times)
np.random.seed(10)
vectorized2_computed = random_walks1D_vec2(
    x0, N, p, num_walks, num_times)
# positions: [0, 1, 0, 1, 2]
# Can test without tolerance since everything is +/- 1
return_values = ['pos', 'pos2', 'pos_hist', 'pos_hist_times']
for s, v, r in zip(serial_computed,
                   vectorized1_computed,
                   return_values):
    msg = '%s: %s (serial) vs %s (vectorized)' % (r, s, v)
    assert (s == v).all(), msg
for s, v, r in zip(serial_computed,
                   vectorized2_computed,
                   return_values):
    msg = '%s: %s (serial) vs %s (vectorized)' % (r, s, v)
    assert (s == v).all(), msg
```

Such test functions are indispensable for further development of the code as we can at any time test whether the basic computations remain correct or not. This is particularly important in stochastic simulations since without test functions and fixed seeds, we always experience variations from run to run, and it can be very difficult to spot bugs through averaged statistical quantities.

1.7.6 Demonstration of Multiple Walks

Assuming now that the code works, we can just scale up the number of steps in each walk and the number of walks. The latter influences the accuracy of the statistical estimates. Figure 1.18 shows the impact of the number of walks on the expectation, which should approach zero. Figure 1.19 displays the corresponding estimate of the variance of the position, which should grow linearly with the number of steps. It does, seemingly very accurately, but notice that the scale on the y axis is so much larger than for the expectation, so irregularities due to the stochastic nature of the process become so much less visible in the variance plots. The probability of finding a particle at a certain position at time (or step) 800 is shown in Fig. 1.20. The dashed red line is the theoretical distribution (1.116) arising from solving the diffusion equation (1.114) instead. As always, we realize that one needs significantly more statistical samples to estimate a histogram accurately than the expectation or variance.

1.7.7 Ascii Visualization of 1D Random Walk

If we want to study (very) long time series of random walks, it can be convenient to plot the position in a terminal window with the time axis pointing downwards. The module avplotter in SciTools has a class Plotter for plotting functions in the terminal window with the aid of ascii symbols only. Below is the code required to visualize a simple random walk, starting at the origin, and considered over when

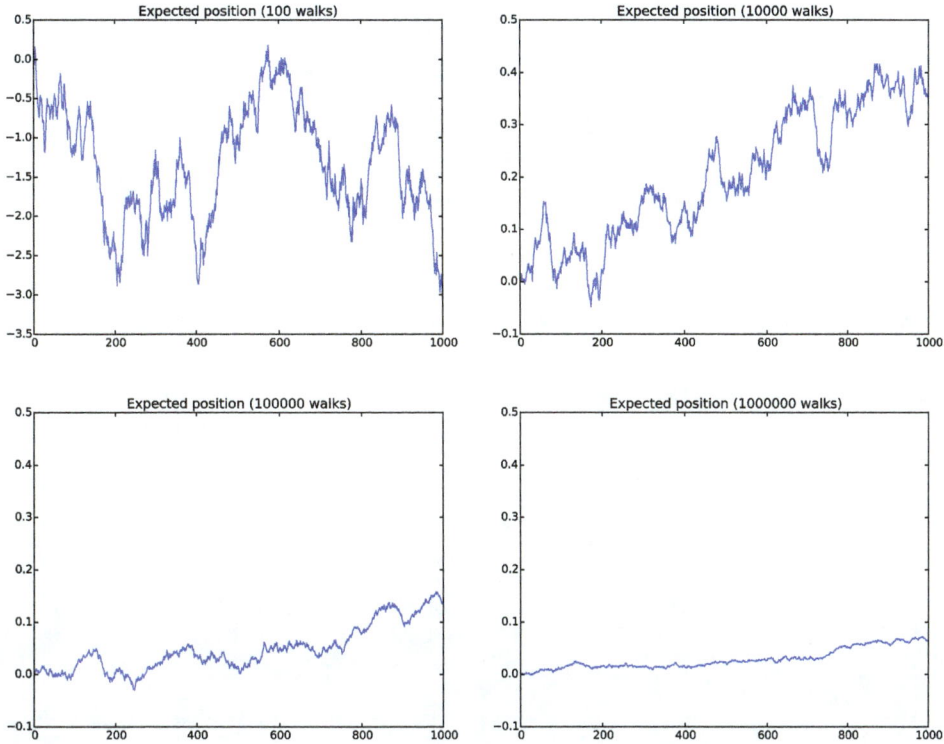

Fig. 1.18 Estimated expected value for 1000 steps, using 100 walks (*upper left*), 10,000 (*upper right*), 100,000 (*lower left*), and 1,000,000 (*lower right*)

the point $x = -1$ is reached. We use a spacing $\Delta x = 0.05$ (so $x = -1$ corresponds to $i = -20$).

```
def run_random_walk():
    from scitools.avplotter import Plotter
    import time, numpy as np
    p = Plotter(-1, 1, width=75)   # Horizontal axis: 75 chars wide
    dx = 0.05
    np.random.seed(10)

    x = 0
    while True:
        random_step = 1 if np.random.random() > 0.5 else -1
        x = x + dx*random_step
        if x < -1:
            break                  # Destination reached!
        print p.plot(0, x)

        # Allow Ctrl+c to abort the simulation
        try:
            time.sleep(0.1)  # Wait for interrupt
        except KeyboardInterrupt:
            print 'Interrupted by Ctrl+c'
            break
```

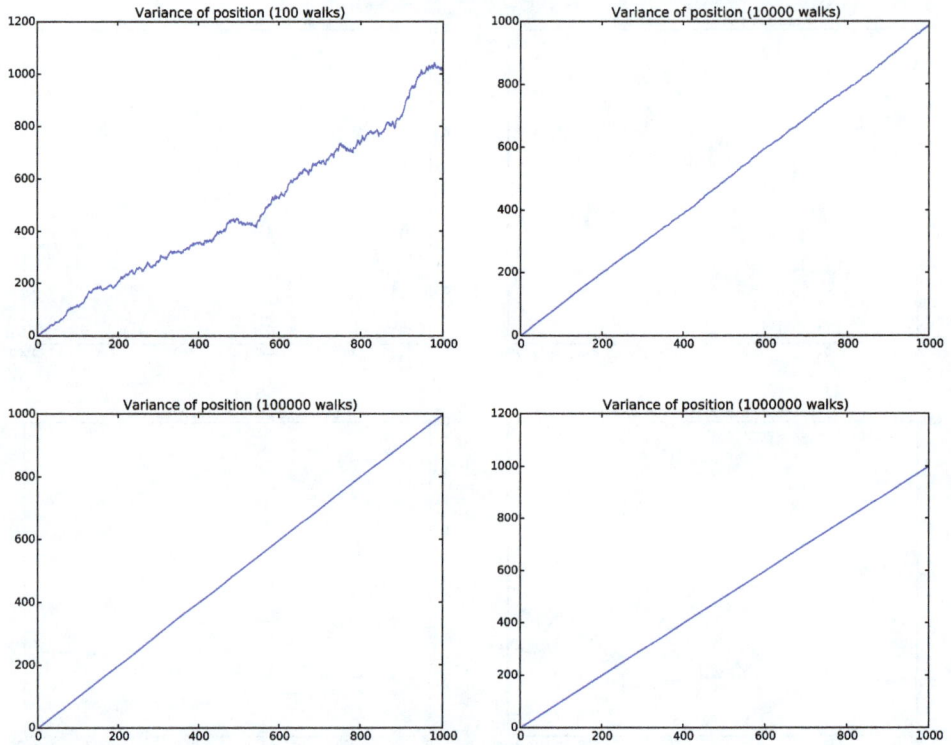

Fig. 1.19 Estimated variance over 1000 steps, using 100 walks (*upper left*), 10,000 (*upper right*), 100,000 (*lower left*), and 1,000,000 (*lower right*)

Observe that we implement an infinite loop, but allow a smooth interrupt of the program by `Ctrl+c` through Python's `KeyboardInterrupt` exception. This is a useful recipe that can be used in many occasions!

The output looks typically like

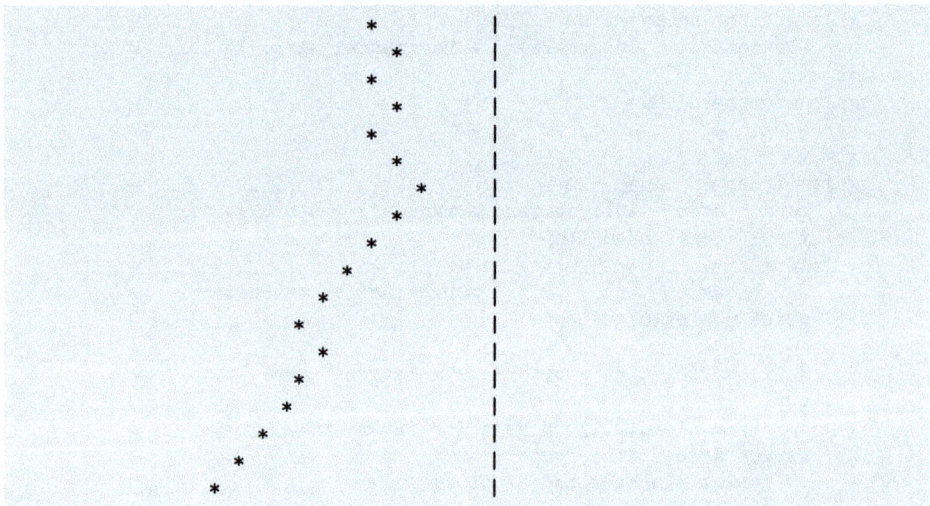

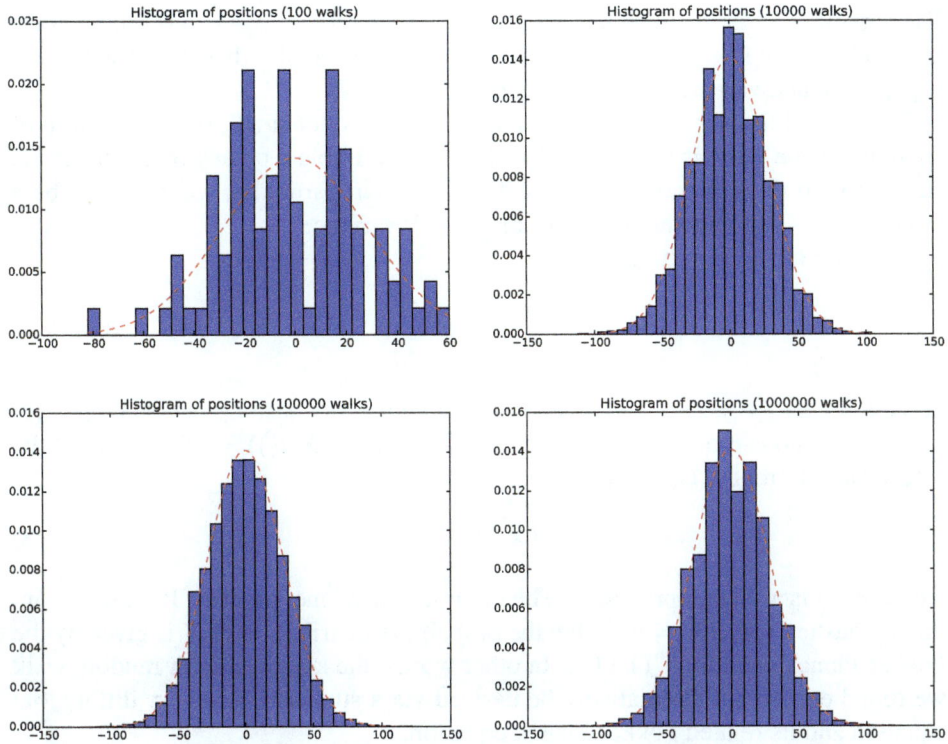

Fig. 1.20 Estimated probability distribution at step 800, using 100 walks (*upper left*), 10,000 (*upper right*), 100,000 (*lower left*), and 1,000,000 (*lower right*)

Positions beyond the limits of the x axis appear with a value. A long file[10] contains the complete ascii plot corresponding to the function `run_random_walk` above.

1.7.8 Random Walk as a Stochastic Equation

The (dimensionless) position in a random walk, $\bar{X}_k$, can be expressed as a stochastic difference equation:

$$\bar{X}_k = \bar{X}_{k-1} + s, \quad x_0 = 0, \tag{1.117}$$

where s is a Bernoulli variable[11], taking on the two values $s = -1$ and $s = 1$ with equal probability:

$$P(s = 1) = \frac{1}{2}, \quad P(s = -1) = \frac{1}{2}.$$

The s variable in a step is independent of the s variable in other steps.

The difference equation expresses essentially the sum of independent Bernoulli variables. Because of the central limit theorem, X_k, will then be normally distributed with expectation $k\mathrm{E}[s]$ and $k\mathrm{Var}(s)$. The expectation and variance of a Bernoulli variable with values $r = 0$ and $r = 1$ are p and $p(1 - p)$, respectively.

---

[10] http://bit.ly/1UbULeH
[11] https://en.wikipedia.org/wiki/Bernoulli_distribution

The variable $s = 2r - 1$ then has expectation $2E[r] - 1 = 2p - 1 = 0$ and variance $2^2 \text{Var}(r) = 4p(1 - p) = 1$. The position X_k is normally distributed with zero expectation and variance k, as we found in Sect. 1.7.2.

The central limit theorem tells that as long as k is not small, the distribution of X_k remains the same if we replace the Bernoulli variable s by any other stochastic variable with the same expectation and variance. In particular, we may let s be a standardized Gaussian variable (zero mean, unit variance).

Dividing (1.117) by Δt gives

$$\frac{\bar{X}_k - \bar{X}_{k-1}}{\Delta t} = \frac{1}{\Delta t} s .$$

In the limit $\Delta t \to 0$, $s/\Delta t$ approaches a white noise stochastic process. With $\bar{X}(t)$ as the continuous process in the limit $\Delta t \to 0$ ($X_k \to X(t_k)$), we formally get the stochastic differential equation

$$d\bar{X} = dW, \qquad\qquad (1.118)$$

where $W(t)$ is a Wiener process[12]. Then X is also a Wiener process. It follows from the stochastic ODE $dX = dW$ that the probability distribution of X is given by the Fokker-Planck equation[13] (1.114). In other words, the key results for random walk we found earlier can alternatively be derived via a stochastic ordinary differential equation and its related Fokker-Planck equation.

1.7.9 Random Walk in 2D

The most obvious generalization of 1D random walk to two spatial dimensions is to allow movements to the north, east, south, and west, with equal probability $\frac{1}{4}$.

```
def random_walk2D(x0, N, p, random=random):
    """2D random walk with 1 particle and N moves: N, E, W, S."""
    # Store position in step k in position[k]
    d = len(x0)
    position = np.zeros((N+1, d))
    position[0,:] = x0
    current_pos = np.array(x0, dtype=float)
    for k in range(N):
        r = random.uniform(0, 1)
        if r <= 0.25:
            current_pos += np.array([0, 1])    # Move north
        elif 0.25 < r <= 0.5:
            current_pos += np.array([1, 0])    # Move east
        elif 0.5 < r <= 0.75:
            current_pos += np.array([0, -1])   # Move south
        else:
            current_pos += np.array([-1, 0])   # Move west
        position[k+1,:] = current_pos
    return position
```

---

[12] https://en.wikipedia.org/wiki/Wiener_process
[13] https://en.wikipedia.org/wiki/Fokker-Planck_equation

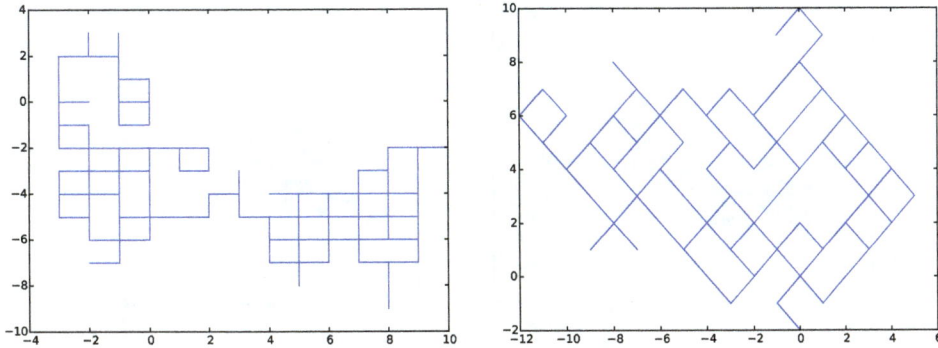

Fig. 1.21 Random walks in 2D with 200 steps: rectangular mesh (*left*) and diagonal mesh (*right*)

The left plot in Fig. 1.21 provides an example on 200 steps with this kind of walk. We may refer to this walk as a walk on a *rectangular mesh* as we move from any spatial mesh point (i, j) to one of its four neighbors in the rectangular directions: $(i + 1, j)$, $(i - 1, j)$, $(i, j + 1)$, or $(i, j - 1)$.

1.7.10 Random Walk in Any Number of Space Dimensions

From a programming point of view, especially when implementing a random walk in any number of dimensions, it is more natural to consider a walk in the diagonal directions NW, NE, SW, and SE. On a two-dimensional spatial mesh it means that we go from (i, j) to either $(i+1, j+1)$, $(i-1, j+1)$, $(i+1, j-1)$, or $(i-1, j-1)$. We can with such a *diagonal mesh* (see right plot in Fig. 1.21) draw a Bernoulli variable for the step in each spatial direction and trivially write code that works in any number of spatial directions:

```python
def random_walkdD(x0, N, p, random=random):
    """Any-D (diagonal) random walk with 1 particle and N moves."""
    # Store position in step k in position[k]
    d = len(x0)
    position = np.zeros((N+1, d))
    position[0,:] = x0
    current_pos = np.array(x0, dtype=float)
    for k in range(N):
        for i in range(d):
            r = random.uniform(0, 1)
            if r <= p:
                current_pos[i] -= 1
            else:
                current_pos[i] += 1
        position[k+1,:] = current_pos
    return position
```

A vectorized version is desired. We follow the ideas from Sect. 3.7.3, but each step is now a vector in d spatial dimensions. We therefore need to draw Nd random numbers in r, compute steps in the various directions through np.where(r <=p,

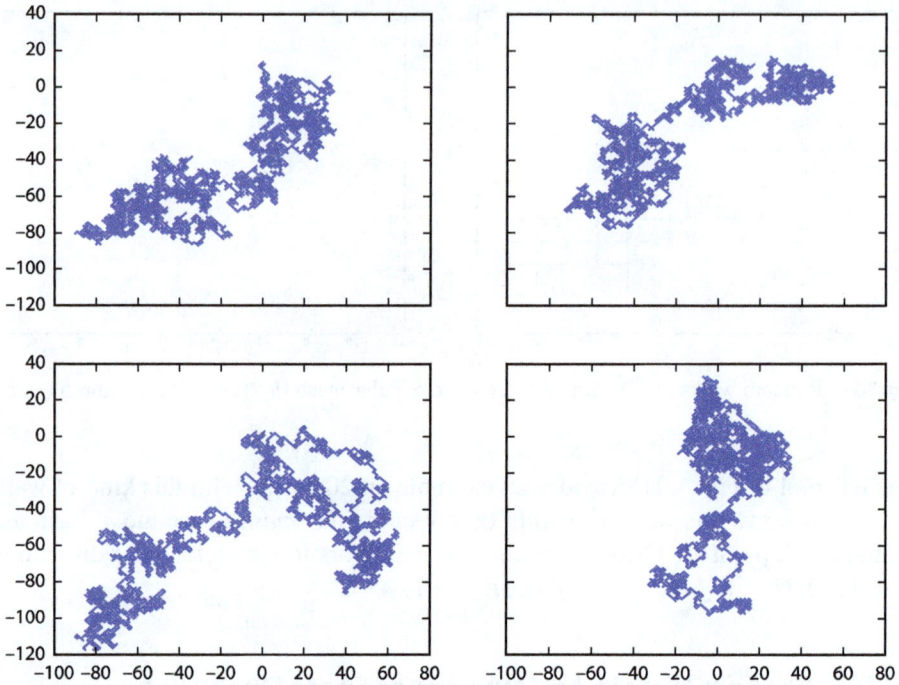

Fig. 1.22 Four random walks with 5000 steps in 2D

-1, 1) (each step being -1 or 1), and then we can reshape this array to an $N \times d$ array of step *vectors*. Doing an np.cumsum summation along axis 0 will add the vectors, as this demo shows:

```
>>> a = np.arange(6).reshape(3,2)
>>> a
array([[0, 1],
       [2, 3],
       [4, 5]])
>>> np.cumsum(a, axis=0)
array([[ 0,  1],
       [ 2,  4],
       [ 6,  9]])
```

With such summation of step vectors, we get all the positions to be filled in the position array:

```
def random_walkdD_vec(x0, N, p):
    """Vectorized version of random_walkdD."""
    d = len(x0)
    # Store position in step k in position[k]
    position = np.zeros((N+1,d))
    position[0] = np.array(x0, dtype=float)
    r = np.random.uniform(0, 1, size=N*d)
    steps = np.where(r <= p, -1, 1).reshape(N,d)
    position[1:,:] = x0 + np.cumsum(steps, axis=0)
    return position
```

1.7.11   Multiple Random Walks in Any Number of Space Dimensions

As we did in 1D, we extend one single walk to a number of walks (num_walks in the code).

Scalar code   As always, we start with implementing the scalar case:

```
def random_walksdD(x0, N, p, num_walks=1, num_times=1,
                   random=random):
    """Simulate num_walks random walks from x0 with N steps."""
    d = len(x0)
    position  = np.zeros((N+1, d))   # Accumulated positions
    position2 = np.zeros((N+1, d))   # Accumulated positions**2
    # Histogram at num_times selected time points
    pos_hist = np.zeros((num_walks, num_times, d))
    pos_hist_times = [(N//num_times)*i for i in range(num_times)]

    for n in range(num_walks):
        num_times_counter = 0
        current_pos = np.array(x0, dtype=float)
        for k in range(N):
            if k in pos_hist_times:
                pos_hist[n,num_times_counter,:] = current_pos
                num_times_counter += 1
            # current_pos corresponds to step k+1
            for i in range(d):
                r = random.uniform(0, 1)
                if r <= p:
                    current_pos[i] -= 1
                else:
                    current_pos[i] += 1
            position [k+1,:] += current_pos
            position2[k+1,:] += current_pos**2
    return position, position2, pos_hist, np.array(pos_hist_times)
```

Vectorized code   Significant speed-ups can be obtained by vectorization. We get rid of the loops in the previous function and arrive at the following vectorized code.

```
def random_walksdD_vec(x0, N, p, num_walks=1, num_times=1):
    """Vectorized version of random_walks1D; no loops."""
    d = len(x0)
    position  = np.zeros((N+1, d))  # Accumulated positions
    position2 = np.zeros((N+1, d))  # Accumulated positions**2
    walks = np.zeros((num_walks, N+1, d))  # Positions of each walk
    walks[:,0,:] = x0
    # Histogram at num_times selected time points
    pos_hist = np.zeros((num_walks, num_times, d))
    pos_hist_times = [(N//num_times)*i for i in range(num_times)]

    r = np.random.uniform(0, 1, size=N*num_walks*d)
    steps = np.where(r <= p, -1, 1).reshape(num_walks, N, d)
    walks[:,1:,:] = x0 + np.cumsum(steps, axis=1)
    position  = np.sum(walks,    axis=0)
    position2 = np.sum(walks**2, axis=0)
    pos_hist[:,:,:] = walks[:,pos_hist_times,:]
    return position, position2, pos_hist, np.array(pos_hist_times)
```

1.8   Applications

1.8.1   Diffusion of a Substance

The first process to be considered is a substance that gets transported through a fluid at rest by pure diffusion. We consider an arbitrary volume V of this fluid, containing the substance with concentration function $c(\boldsymbol{x}, t)$. Physically, we can think of a very small volume with centroid $\boldsymbol{x}$ at time t and assign the ratio of the volume of the substance and the total volume to $c(\boldsymbol{x}, t)$. This means that the mass of the substance in a small volume ΔV is approximately $\varrho c \Delta V$, where ϱ is the density of the substance. Consequently, the total mass of the substance inside the volume V is the sum of all $\varrho c \Delta V$, which becomes the volume integral $\int_V \varrho c \, dV$.

Let us reason how the mass of the substance changes and thereby derive a PDE governing the concentration c. Suppose the substance flows out of V with a flux $\boldsymbol{q}$. If ΔS is a small part of the boundary ∂V of V, the volume of the substance flowing out through dS in a small time interval Δt is $\varrho \boldsymbol{q} \cdot \boldsymbol{n} \Delta t \Delta S$, where $\boldsymbol{n}$ is an outward unit normal to the boundary ∂V, see Fig. 1.23. We realize that only the normal component of $\boldsymbol{q}$ is able to transport mass in and out of V. The total outflow of the mass of the substance in a small time interval Δt becomes the surface integral

$$\int_{\partial V} \varrho \boldsymbol{q} \cdot \boldsymbol{n} \Delta t \, dS \, .$$

Assuming conservation of mass, this outflow of mass must be balanced by a loss of mass inside the volume. The increase of mass inside the volume, during a small time interval Δt, is

$$\int_V \varrho (c(\boldsymbol{x}, t + \Delta t) - c(\boldsymbol{x}, t)) dV,$$

assuming ϱ is constant, which is reasonable. The outflow of mass balances the loss of mass in V, which is the increase with a minus sign. Setting the two contributions equal to each other ensures balance of mass inside V. Dividing by Δt gives

$$\int_V \varrho \frac{c(\boldsymbol{x}, t + \Delta t) - c(\boldsymbol{x}, t)}{\Delta t} dV = - \int_{\partial V} \varrho \boldsymbol{q} \cdot \boldsymbol{n} \, dS \, .$$

Note the minus sign on the right-hand side: the left-hand side expresses loss of mass, while the integral on the right-hand side is the gain of mass.

Fig. 1.23 An arbitrary volume of a fluid

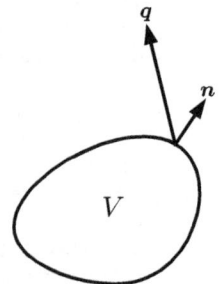

Now, letting $\Delta t \to 0$, we have

$$\frac{c(\boldsymbol{x}, t + \Delta t) - c(\boldsymbol{x}, t)}{\Delta t} \to \frac{\partial c}{\partial t},$$

so

$$\int_V \varrho \frac{\partial c}{\partial t} dV + \int_{\partial V} \varrho \boldsymbol{q} \cdot \boldsymbol{n} \, dS = 0. \qquad (1.119)$$

To arrive at a PDE, we express the surface integral as a volume integral using Gauss' divergence theorem:

$$\int_V \left(\varrho \frac{\partial c}{\partial t} + \nabla \cdot (\varrho \boldsymbol{q}) \right) dV = 0.$$

Since ϱ is constant, we can divide by this quantity. If the integral is to vanish for an arbitrary volume V, the integrand must vanish too, and we get the mass conservation PDE for the substance:

$$\frac{\partial c}{\partial t} + \nabla \cdot \boldsymbol{q} = 0. \qquad (1.120)$$

A fundamental problem is that this is a scalar PDE for four unknowns: c and the three components of $\boldsymbol{q}$. We therefore need additional equations. Here, Fick's law comes at rescue: it models how the flux $\boldsymbol{q}$ of the substance is related to the concentration c. Diffusion is recognized by mass flowing from regions with high concentration to regions of low concentration. This principle suggests that $\boldsymbol{q}$ is proportional to the negative gradient of c:

$$\boldsymbol{q} = -\alpha \nabla c, \qquad (1.121)$$

where α is an empirically determined constant. The relation (1.121) is known as Fick's law. Inserting (1.121) in (1.120) gives a scalar PDE for the concentration c:

$$\frac{\partial c}{\partial t} = \alpha \nabla^2 c. \qquad (1.122)$$

1.8.2   Heat Conduction

Heat conduction is a well-known diffusion process. The governing PDE is in this case based on the first law of thermodynamics: the increase in energy of a system is equal to the work done on the system, plus the supplied heat. Here, we shall consider media at rest and neglect work done on the system. The principle then reduces to a balance between increase in internal energy and supplied heat flow by conduction.

Let $e(x, t)$ be the *internal energy* per unit mass. The increase of the internal energy in a small volume ΔV in a small time interval Δt is then

$$\varrho(e(\boldsymbol{x}, t + \Delta t) - e(\boldsymbol{x}, t))\Delta V,$$

where ϱ is the density of the material subject to heat conduction. In an arbitrary volume V, as depicted in Fig. 1.23, the corresponding increase in internal energy becomes the volume integral

$$\int_V \varrho(e(x,t+\Delta t) - e(x,t))dV.$$

This increase in internal energy is balanced by heat supplied by conduction. Let q be the heat flow per time unit. Through the surface ∂V of V the following amount of heat flows out of V during a time interval Δt:

$$\int_{\partial V} q \cdot n\Delta t\, dS.$$

The simplified version of the first law of thermodynamics then states that

$$\int_V \varrho(e(x,t+\Delta t) - e(x,t))dV = -\int_{\partial V} q \cdot n\Delta t\, dS.$$

The minus sign on the right-hand side ensures that the integral there models net *inflow* of heat (since n is an outward unit normal, $q \cdot n$ models *outflow*). Dividing by Δt and notifying that

$$\lim_{\Delta t \to 0} \frac{e(x,t+\Delta t) - e(x,t)}{\Delta t} = \frac{\partial e}{\partial t},$$

we get (in the limit $\Delta t \to 0$)

$$\int_V \varrho\frac{\partial e}{\partial t}dV + \int_{\partial V} q \cdot n\Delta t\, dS = 0.$$

This is the integral equation for heat conduction, but we aim at a PDE. The next step is therefore to transform the surface integral to a volume integral via Gauss' divergence theorem. The result is

$$\int_V \left(\varrho\frac{\partial e}{\partial t} + \nabla \cdot q\right) dV = 0.$$

If this equality is to hold for all volumes V, the integrand must vanish, and we have the PDE

$$\varrho\frac{\partial e}{\partial t} = -\nabla \cdot q. \tag{1.123}$$

Sometimes the supplied heat can come from the medium itself. This is the case, for instance, when radioactive rock generates heat. Let us add this effect. If $f(x,t)$ is the supplied heat per unit volume per unit time, the heat supplied in a small volume is $f \Delta t \Delta V$, and inside an arbitrary volume V the supplied generated heat becomes

$$\int_V f \Delta t \, dV .$$

Adding this to the integral statement of the (simplified) first law of thermodynamics, and continuing the derivation, leads to the PDE

$$\varrho \frac{\partial e}{\partial t} = -\nabla \cdot \boldsymbol{q} + f . \qquad (1.124)$$

There are four unknown scalar fields: e and $\boldsymbol{q}$. Moreover, the temperature T, which is our primary quantity to compute, does not enter the model yet. We need an additional equation, called the *equation of state*, relating e, $V = 1/\varrho =$, and T: $e = e(V, T)$. By the chain rule we have

$$\frac{\partial e}{\partial t} = \frac{\partial e}{\partial T}\bigg|_V \frac{\partial T}{\partial t} + \frac{\partial e}{\partial V}\bigg|_T \frac{\partial V}{\partial t} .$$

The first coefficient $\partial e / \partial T$ is called *specific heat capacity at constant volume*, denoted by c_v:

$$c_v = \frac{\partial e}{\partial T}\bigg|_V .$$

The specific heat capacity will in general vary with T, but taking it as a constant is a good approximation in many applications.

The term $\partial e / \partial V$ models effects due to compressibility and volume expansion. These effects are often small and can be neglected. We shall do so here. Using $\partial e / \partial t = c_v \partial T / \partial t$ in the PDE gives

$$\varrho c_v \frac{\partial T}{\partial t} = -\nabla \cdot \boldsymbol{q} + f .$$

We still have four unknown scalar fields (T and $\boldsymbol{q}$). To close the system, we need a relation between the heat flux $\boldsymbol{q}$ and the temperature T called *Fourier's law*:

$$\boldsymbol{q} = -k \nabla T,$$

which simply states that heat flows from hot to cold areas, along the path of greatest variation. In a solid medium, k depends on the material of the medium, and in multi-material media one must regard k as spatially dependent. In a fluid, it is common to assume that k is constant. The value of k reflects how easy heat is conducted through the medium, and k is named the *coefficient of heat conduction*.

We now have one scalar PDE for the unknown temperature field $T(\boldsymbol{x}, t)$:

$$\varrho c_v \frac{\partial T}{\partial t} = \nabla \cdot (k \nabla T) + f . \qquad (1.125)$$

1.8.3   Porous Media Flow

The requirement of mass balance for flow of a single, incompressible fluid through a deformable (elastic) porous medium leads to the equation

$$S\frac{\partial p}{\partial t} + \nabla \cdot \left(\boldsymbol{q} - \alpha \frac{\partial \boldsymbol{u}}{\partial t} \right) = 0,$$

where p is the fluid pressure, $\boldsymbol{q}$ is the fluid velocity, $\boldsymbol{u}$ is the displacement (deformation) of the medium, S is the storage coefficient of the medium (related to the compressibility of the fluid and the material in the medium), and α is another coefficient. In many circumstances, the last term with $\boldsymbol{u}$ can be neglected, an assumption that decouples the equation above from a model for the deformation of the medium. The famous *Darcy's law* relates $\boldsymbol{q}$ to p:

$$\boldsymbol{q} = -\frac{K}{\mu}(\nabla p - \varrho \boldsymbol{g}),$$

where K is the permeability of the medium, μ is the dynamic viscosity of the fluid, ϱ is the density of the fluid, and $\boldsymbol{g}$ is the acceleration of gravity, here taken as $\boldsymbol{g} = -g\boldsymbol{k}$. Combining the two equations results in the diffusion model

$$S\frac{\partial p}{\partial t} = \mu^{-1}\nabla(K\nabla p) + \frac{\varrho g}{\mu}\frac{\partial K}{\partial z}. \qquad (1.126)$$

Boundary conditions consist of specifying p or $\boldsymbol{q} \cdot \boldsymbol{n}$ (i.e., normal velocity) at each point of the boundary.

1.8.4   Potential Fluid Flow

Let $\boldsymbol{v}$ be the velocity of a fluid. The condition $\nabla \times \boldsymbol{v} = 0$ is relevant for many flows, especially in geophysics when viscous effects are negligible. From vector calculus it is known that $\nabla \times \boldsymbol{v} = 0$ implies that v can be derived from a scalar potential field ϕ: $\boldsymbol{v} = \nabla\phi$. If the fluid is incompressible, $\nabla \cdot \boldsymbol{v} = 0$, it follows that $\nabla \cdot \nabla\phi = 0$, or

$$\nabla^2 \phi = 0. \qquad (1.127)$$

This Laplace equation is sufficient for determining ϕ and thereby describe the fluid motion. This type of flow is known as potential flow[14]. One very important application where potential flow is a good model is water waves. As boundary condition we must prescribe $\boldsymbol{v} \cdot \boldsymbol{n} = \partial\phi/\partial n$. This gives rise to what is known as a pure Neumann problem and will cause numerical difficulties because ϕ and ϕ plus any constant are two solutions of the problem. The simplest remedy is to fix the value of ϕ at a point.

---

[14] https://en.wikipedia.org/wiki/Potential_flow

1.8.5 Streamlines for 2D Fluid Flow

The streamlines in a two-dimensional stationary fluid flow are lines tangential to the flow. The stream function[15] ψ is often introduced in two-dimensional flow such that its contour lines, $\psi = $ const, gives the streamlines. The relation between ψ and the velocity field $\boldsymbol{v} = (u, v)$ is

$$u = \frac{\partial \psi}{\partial y}, \quad v = -\frac{\partial \psi}{\partial x}.$$

It follows that $\nabla \boldsymbol{v} = \psi_{yx} - \psi_{xy} = 0$, so the stream function can only be used for incompressible flows. Since

$$\nabla \times \boldsymbol{v} = \left(\frac{\partial v}{\partial y} - \frac{\partial u}{\partial x} \right) \boldsymbol{k} \equiv \omega \boldsymbol{k},$$

we can derive the relation

$$\nabla^2 \psi = -\omega, \tag{1.128}$$

which is a governing equation for the stream function $\psi(x, y)$ if the vorticity ω is known.

1.8.6 The Potential of an Electric Field

Under the assumption of time independence, Maxwell's equations for the electric field $\boldsymbol{E}$ become

$$\nabla \cdot \boldsymbol{E} = \frac{\rho}{\epsilon_0},$$

$$\nabla \times \boldsymbol{E} = 0,$$

where ρ is the electric charge density and ϵ_0 is the electric permittivity of free space (i.e., vacuum). Since $\nabla \times \boldsymbol{E} = 0$, $\boldsymbol{E}$ can be derived from a potential φ, $\boldsymbol{E} = -\nabla\varphi$. The electric field potential is therefore governed by the Poisson equation

$$\nabla^2 \varphi = -\frac{\rho}{\epsilon_0}. \tag{1.129}$$

If the medium is heterogeneous, ρ will depend on the spatial location $\boldsymbol{r}$. Also, ϵ_0 must be exchanged with an electric permittivity function $\epsilon(\boldsymbol{r})$.

Each point of the boundary must be accompanied by, either a Dirichlet condition $\varphi(\boldsymbol{r}) = \varphi_D(\boldsymbol{r})$, or a Neumann condition $\frac{\partial \varphi(\boldsymbol{r})}{\partial n} = \varphi_N(\boldsymbol{r})$.

1.8.7 Development of Flow Between Two Flat Plates

Diffusion equations may also arise as simplified versions of other mathematical models, especially in fluid flow. Consider a fluid flowing between two flat, parallel

---

[15] https://en.wikipedia.org/wiki/Stream_function

plates. The velocity is uni-directional, say along the z axis, and depends only on the distance x from the plates; $\boldsymbol{u} = u(x,t)\boldsymbol{k}$. The flow is governed by the Navier-Stokes equations,

$$\varrho \frac{\partial \boldsymbol{u}}{\partial t} + \varrho \boldsymbol{u} \cdot \nabla \boldsymbol{u} = -\nabla p + \mu \nabla^2 \boldsymbol{u} + \varrho \boldsymbol{f},$$

$$\nabla \cdot \boldsymbol{u} = 0,$$

where p is the pressure field, unknown along with the velocity $\boldsymbol{u}$, ϱ is the fluid density, μ the dynamic viscosity, and $\boldsymbol{f}$ is some external body force. The geometric restrictions of flow between two flat plates puts restrictions on the velocity, $\boldsymbol{u} = u(x,t)\boldsymbol{i}$, and the z component of the Navier-Stokes equations collapses to a diffusion equation:

$$\varrho \frac{\partial u}{\partial t} = -\frac{\partial p}{\partial z} + \mu \frac{\partial^2 u}{\partial z^2} + \varrho f_z,$$

if f_z is the component of $\boldsymbol{f}$ in the z direction.

The boundary conditions are derived from the fact that the fluid sticks to the plates, which means $\boldsymbol{u} = 0$ at the plates. Say the location of the plates are $z = 0$ and $z = L$. We then have

$$u(0,t) = u(L,t) = 0.$$

One can easily show that $\partial p/\partial z$ must be a constant or just a function of time t. We set $\partial p/\partial z = -\beta(t)$. The body force could be a component of gravity, if desired, set as $f_z = \gamma g$. Switching from z to x as independent variable gives a very standard one-dimensional diffusion equation:

$$\varrho \frac{\partial u}{\partial t} = \mu \frac{\partial^2 u}{\partial x^2} + \beta(t) + \varrho \gamma g, \quad x \in [0,L], \ t \in (0,T].$$

The boundary conditions are

$$u(0,t) = u(L,t) = 0,$$

while some initial condition

$$u(x,0) = I(x)$$

must also be prescribed.

The flow is driven by either the pressure gradient β or gravity, or a combination of both. One may also consider one moving plate that drives the fluid. If the plate at $x = L$ moves with velocity $U_L(t)$, we have the adjusted boundary condition

$$u(L,t) = U_L(t).$$

1.8.8 Flow in a Straight Tube

Now we consider viscous fluid flow in a straight tube with radius R and rigid walls. The governing equations are the Navier-Stokes equations, but as in Sect. 1.8.7, it is natural to assume that the velocity is directed along the tube, and that it is axi-symmetric. These assumptions reduced the velocity field to $\boldsymbol{u} = u(r, x, t)\boldsymbol{i}$, if the x axis is directed along the tube. From the equation of continuity, $\nabla \cdot \boldsymbol{u} = 0$, we see that u must be independent of x. Inserting $\boldsymbol{u} = u(r, t)\boldsymbol{i}$ in the Navier-Stokes equations, expressed in axi-symmetric cylindrical coordinates, results in

$$\varrho \frac{\partial u}{\partial t} = \mu \frac{1}{r} \frac{\partial}{\partial r} \left(r \frac{\partial u}{\partial r} \right) + \beta(t) + \varrho \gamma g, \quad r \in [0, R], \ t \in (0, T]. \tag{1.130}$$

Here, $\beta(t) = -\partial p / \partial x$ is the pressure gradient along the tube. The associated boundary condition is $u(R, t) = 0$.

1.8.9 Tribology: Thin Film Fluid Flow

Thin fluid films are extremely important inside machinery to reduce friction be-tween gliding surfaces. The mathematical model for the fluid motion takes the form of a diffusion problem and is quickly derived here. We consider two solid surfaces whose distance is described by a gap function $h(x, y)$. The space between these surfaces is filled with a fluid with dynamic viscosity μ. The fluid may move partially because of pressure gradients and partially because the surfaces move. Let $U\boldsymbol{i} + V\boldsymbol{j}$ be the relative velocity of the two surfaces and p the pressure in the fluid. The mathematical model builds on two principles: 1) conservation of mass, 2) assumption of locally quasi-static flow between flat plates.

The conservation of mass equation reads $\nabla \cdot \boldsymbol{u}$, where $\boldsymbol{u}$ is the local fluid velocity. For thin films the detailed variation between the surfaces is not of interest, so $\nabla \cdot \boldsymbol{u} = 0$ is integrated (average) in the direction perpendicular to the surfaces. This gives rise to the alternative mass conservation equation

$$\nabla \cdot \boldsymbol{q} = 0, \quad \boldsymbol{q} = \int_0^{h(x,y)} \boldsymbol{u} \, dz,$$

where z is the coordinate perpendicular to the surfaces, and $\boldsymbol{q}$ is then the volume flux in the fluid gap.

Locally, we may assume that we have steady flow between two flat surfaces, with a pressure gradient and where the lower surface is at rest and the upper moves with velocity $U\boldsymbol{i} + V\boldsymbol{j}$. The corresponding mathematical problem is actually the limit problem in Sect. 1.8.7 as $t \to \infty$. The limit problem can be solved analytically, and the local volume flux becomes

$$\boldsymbol{q}(x, y, z) = \int_0^h \boldsymbol{u}(x, y, z) dz = -\frac{h^3}{12\mu} \nabla p + \frac{1}{2} U h \boldsymbol{i} + \frac{1}{2} V h \boldsymbol{j} \, .$$

The idea is to use this expression locally also when the surfaces are not flat, but slowly varying, and if U, V, or p varies in time, provided the time variation is sufficiently slow. This is a common quasi-static approximation, much used in math-ematical modeling.

Inserting the expression for q via p, U, and V in the equation $\nabla q = 0$ gives a diffusion PDE for p:

$$\nabla \cdot \left(\frac{h^3}{12\mu} \nabla p \right) = \frac{1}{2} \frac{\partial}{\partial x} (hU) + \frac{1}{2} \frac{\partial}{\partial x} (hV). \qquad (1.131)$$

The boundary conditions must involve p or q at the boundary.

1.8.10 Propagation of Electrical Signals in the Brain

One can make a model of how electrical signals are propagated along the neuronal fibers that receive synaptic inputs in the brain. The signal propagation is one-dimensional and can, in the simplest cases, be governed by the Cable equation[16]:

$$c_m \frac{\partial V}{\partial t} = \frac{1}{r_l} \frac{\partial^2 V}{\partial x^2} - \frac{1}{r_m} V \qquad (1.132)$$

where $V(x,t)$ is the voltage to be determined, c_m is capacitance of the neuronal fiber, while r_l and r_m are measures of the resistance. The boundary conditions are often taken as $V = 0$ at a short circuit or open end, $\partial V/\partial x = 0$ at a sealed end, or $\partial V/\partial x \propto V$ where there is an injection of current.

1.9 Exercises

Exercise 3.6: Stabilizing the Crank-Nicolson method by Rannacher time stepping

It is well known that the Crank-Nicolson method may give rise to non-physical oscillations in the solution of diffusion equations if the initial data exhibit jumps (see Sect. 1.3.6). Rannacher [15] suggested a stabilizing technique consisting of using the Backward Euler scheme for the first two time steps with step length $\frac{1}{2}\Delta t$. One can generalize this idea to taking $2m$ time steps of size $\frac{1}{2}\Delta t$ with the Backward Euler method and then continuing with the Crank-Nicolson method, which is of second-order in time. The idea is that the high frequencies of the initial solution are quickly damped out, and the Backward Euler scheme treats these high frequencies correctly. Thereafter, the high frequency content of the solution is gone and the Crank-Nicolson method will do well.

Test this idea for $m = 1, 2, 3$ on a diffusion problem with a discontinuous initial condition. Measure the convergence rate using the solution (1.45) with the boundary conditions (1.46)–(1.47) for t values such that the conditions are in the vicinity of ± 1. For example, $t < 5a1.6 \cdot 10^{-2}$ makes the solution diffusion from a step to almost a straight line. The program `diffu_erf_sol.py` shows how to compute the analytical solution.

Project 1.7: Energy estimates for diffusion problems

This project concerns so-called *energy estimates* for diffusion problems that can be used for qualitative analytical insight and for verification of implementations.

---

[16] http://en.wikipedia.org/wiki/Cable_equation

a) We start with a 1D homogeneous diffusion equation with zero Dirichlet conditions:

$$u_t = \alpha u_x x, \qquad x \in \Omega = (0, L),\ t \in (0, T], \qquad (1.133)$$
$$u(0, t) = u(L, t) = 0, \qquad t \in (0, T], \qquad (1.134)$$
$$u(x, 0) = I(x), \qquad x \in [0, L]. \qquad (1.135)$$

The energy estimate for this problem reads

$$\|u\|_{L^2} \le \|I\|_{L^2}, \qquad (1.136)$$

where the $\|\cdot\|_{L^2}$ norm is defined by

$$\|g\|_{L^2} = \sqrt{\int_0^L g^2 dx}. \qquad (1.137)$$

The quantify $\|u\|_{L^2}$ or $\frac{1}{2}\|u\|_{L^2}$ is known as the *energy* of the solution, although it is not the physical energy of the system. A mathematical tradition has introduced the notion *energy* in this context.

The estimate (1.136) says that the "size of u" never exceeds that of the initial condition, or more precisely, it says that the area under the u curve decreases with time.

To show (1.136), multiply the PDE by u and integrate from 0 to L. Use that $u u_t$ can be expressed as the time derivative of u^2 and that $u_x x u$ can integrated by parts to form an integrand u_x^2. Show that the time derivative of $\|u\|_{L^2}^2$ must be less than or equal to zero. Integrate this expression and derive (1.136).

b) Now we address a slightly different problem,

$$u_t = \alpha u_x x + f(x, t), \qquad x \in \Omega = (0, L),\ t \in (0, T], \qquad (1.138)$$
$$u(0, t) = u(L, t) = 0, \qquad t \in (0, T], \qquad (1.139)$$
$$u(x, 0) = 0, \qquad x \in [0, L]. \qquad (1.140)$$

The associated energy estimate is

$$\|u\|_{L^2} \le \|f\|_{L^2}. \qquad (1.141)$$

(This result is more difficult to derive.)

Now consider the compound problem with an initial condition $I(x)$ and a right-hand side $f(x, t)$:

$$u_t = \alpha u_x x + f(x, t), \qquad x \in \Omega = (0, L),\ t \in (0, T], \qquad (1.142)$$
$$u(0, t) = u(L, t) = 0, \qquad t \in (0, T], \qquad (1.143)$$
$$u(x, 0) = I(x), \qquad x \in [0, L]. \qquad (1.144)$$

Show that if w_1 fulfills (1.133)–(1.135) and w_2 fulfills (1.138)–(1.140), then $u = w_1 + w_2$ is the solution of (1.142)–(1.144). Using the triangle inequality for norms,

$$\|a + b\| \le \|a\| + \|b\|,$$

show that the energy estimate for (1.142)–(1.144) becomes

$$||u||_{L^2} \le ||I||_{L^2} + ||f||_{L^2} . \tag{1.145}$$

c) One application of (1.145) is to prove uniqueness of the solution. Suppose u_1 and u_2 both fulfill (1.142)–(1.144). Show that $u = u_1 - u_2$ then fulfills (1.142)–(1.144) with $f = 0$ and $I = 0$. Use (1.145) to deduce that the energy must be zero for all times and therefore that $u_1 = u_2$, which proves that the solution is unique.

d) Generalize (1.145) to a 2D/3D diffusion equation $u_t = \nabla \cdot (\alpha \nabla u)$ for $x \in \Omega$.

Hint Use integration by parts in multi dimensions:

$$\int_\Omega u \nabla \cdot (\alpha \nabla u)\, dx = -\int_\Omega \alpha \nabla u \cdot \nabla u\, dx + \int_{\partial\Omega} u\alpha \frac{\partial u}{\partial n},$$

where $\frac{\partial u}{\partial n} = \boldsymbol{n} \cdot \nabla u$, $\boldsymbol{n}$ being the outward unit normal to the boundary $\partial \Omega$ of the domain Ω.

e) Now we also consider the multi-dimensional PDE $u_t = \nabla \cdot (\alpha \nabla u)$. Integrate both sides over Ω and use Gauss' divergence theorem, $\int_\Omega \nabla \cdot \boldsymbol{q}\, dx = \int_{\partial\Omega} \boldsymbol{q} \cdot \boldsymbol{n}\, ds$ for a vector field $\boldsymbol{q}$. Show that if we have homogeneous Neumann conditions on the boundary, $\partial u/\partial n = 0$, area under the u surface remains constant in time and

$$\int_\Omega u\, dx = \int_\Omega I\, dx . \tag{3.146}$$

f) Establish a code in 1D, 2D, or 3D that can solve a diffusion equation with a source term f, initial condition I, and zero Dirichlet or Neumann conditions on the whole boundary.
We can use (1.145) and (1.146) as a partial verification of the code. Choose some functions f and I and check that (1.145) is obeyed at any time when zero Dirichlet conditions are used. Iterate over the same I functions and check that (1.146) is fulfilled when using zero Neumann conditions.

g) Make a list of some possible bugs in the code, such as indexing errors in arrays, failure to set the correct boundary conditions, evaluation of a term at a wrong time level, and similar. For each of the bugs, see if the verification tests from the previous subexercise pass or fail. This investigation shows how strong the energy estimates and the estimate (1.146) are for pointing out errors in the implementation.

Filename: `diffu_energy`.

Exercise 1.8: Splitting methods and preconditioning

In Sect. 1.6.15, we outlined a class of iterative methods for $Au = b$ based on splitting A into $A = M - N$ and introducing the iteration

$$Mu^k = Nu^k + b .$$

The very simplest splitting is $M = I$, where I is the identity matrix. Show that this choice corresponds to the iteration

$$u^k = u^{k-1} + r^{k-1}, \quad r^{k-1} = b - Au^{k-1}, \tag{1.147}$$

where r^{k-1} is the residual in the linear system in iteration $k - 1$. The formula (1.147) is known as Richardson's iteration. Show that if we apply the simple iteration method (1.147) to the *preconditioned* system $M^{-1}Au = M^{-1}b$, we arrive at the Jacobi method by choosing $M = D$ (the diagonal of A) as preconditioner and the SOR method by choosing $M = \omega^{-1}D + L$ (L being the lower triangular part of A). This equivalence shows that we can apply one iteration of the Jacobi or SOR method as preconditioner.

Problem 1.9: Oscillating surface temperature of the earth

Consider a day-and-night or seasonal variation in temperature at the surface of the earth. How deep down in the ground will the surface oscillations reach? For simplicity, we model only the vertical variation along a coordinate x, where $x = 0$ at the surface, and x increases as we go down in the ground. The temperature is governed by the heat equation

$$\varrho c_v \frac{\partial T}{\partial t} = \nabla \cdot (k\nabla T),$$

in some spatial domain $x \in [0, L]$, where L is chosen large enough such that we can assume that T is approximately constant, independent of the surface oscillations, for $x > L$. The parameters ϱ, c_v, and k are the density, the specific heat capacity at constant volume, and the heat conduction coefficient, respectively.

a) Derive the mathematical model for computing $T(x,t)$. Assume the surface oscillations to be sinusoidal around some mean temperature T_m. Let $T = T_m$ initially. At $x = L$, assume $T \approx T_m$.

b) Scale the model in a) assuming k is constant. Use a time scale $t_c = \omega^{-1}$ and a length scale $x_c = \sqrt{2\alpha/\omega}$, where $\alpha = k/(\varrho c_v)$. The primary unknown can be scaled as $\frac{T-T_m}{2A}$.
Show that the scaled PDE is

$$\frac{\partial u}{\partial \bar{t}} = \frac{1}{2}\frac{\partial^2 u}{\partial x^2},$$

with initial condition $u(\bar{x}, 0) = 0$, left boundary condition $u(0, \bar{t}) = \sin(\bar{t})$, and right boundary condition $u(\bar{L}, \bar{t}) = 0$. The bar indicates a dimensionless quantity.
Show that $u(\bar{x}, \bar{t}) = e^{-\bar{x}}\sin(\bar{x} - \bar{t})$ is a solution that fulfills the PDE and the boundary condition at $\bar{x} = 0$ (this is the solution we will experience as $\bar{t} \to \infty$ and $L \to \infty$). Conclude that an appropriate domain for x is $[0, 4]$ if a damping $e^{-4} \approx 0.18$ is appropriate for implementing $\bar{u} \approx$ const; increasing to $[0, 6]$ damps $\bar{u}$ to 0.0025.

c) Compute the scaled temperature and make animations comparing two solutions with $\bar{L} = 4$ and $\bar{L} = 8$, respectively (keep Δx the same).

Problem 1.10: Oscillating and pulsating flow in tubes

We consider flow in a straight tube with radius R and straight walls. The flow is driven by a pressure gradient $\beta(t)$. The effect of gravity can be neglected. The mathematical problem reads

$$\varrho \frac{\partial u}{\partial t} = \mu \frac{1}{r} \frac{\partial}{\partial r} \left(r \frac{\partial u}{\partial r} \right) + \beta(t), \qquad r \in [0, R], \ t \in (0, T], \qquad (1.148)$$

$$u(r, 0) = I(r), \qquad r \in [0, R], \qquad (1.149)$$

$$u(R, t) = 0, \qquad t \in (0, T], \qquad (1.150)$$

$$\frac{\partial u}{\partial r}(0, t) = 0, \qquad t \in (0, T]. \qquad (1.151)$$

We consider two models for $\beta(t)$. One plain, sinusoidal oscillation:

$$\beta = A \sin(\omega t), \qquad (1.152)$$

and one with periodic pulses,

$$\beta = A \sin^{16}(\omega t). \qquad (1.153)$$

Note that both models can be written as $\beta = A \sin^m(\omega t)$, with $m = 1$ and $m = 16$, respectively.

a) Scale the mathematical model, using the viscous time scale $\varrho R^2 / \mu$.
b) Implement the scaled model from a), using the unifying θ scheme in time and centered differences in space.
c) Verify the implementation in b) using a manufactured solution that is quadratic in r and linear in t. Make a corresponding test function.

Hint You need to include an extra source term in the equation to allow for such tests. Let the spatial variation be $1 - r^2$ such that the boundary condition is fulfilled.

d) Make animations for $m = 1, 16$ and $\alpha = 1, 0.1$. Choose T such that the motion has reached a steady state (non-visible changes from period to period in u).
e) For $\alpha \gg 1$, the scaling in a) is not good, because the characteristic time for changes (due to the pressure) is much smaller than the viscous diffusion time scale (α becomes large). We should in this case base the short time scale on $1/\omega$. Scale the model again, and make an animation for $m = 1, 16$ and $\alpha = 10$.

Filename: `axisymm_flow`.

Problem 1.11: Scaling a welding problem

Welding equipment makes a very localized heat source that moves in time. We shall investigate the heating due to welding and choose, for maximum simplicity, a one-dimensional heat equation with a fixed temperature at the ends, and we neglect melting. We shall scale the problem, and besides solving such a problem numerically, the aim is to investigate the appropriateness of alternative scalings.

The governing PDE problem reads

$$\varrho c \frac{\partial u}{\partial t} = k \frac{\partial^2 u}{\partial x^2} + f, \qquad x \in (0, L),\ t \in (0, T),$$
$$u(x, 0) = U_s, \qquad\qquad x \in [0, L],$$
$$u(0, t) = u(L, t) = 0, \quad t \in (0, T].$$

Here, u is the temperature, ϱ the density of the material, c a heat capacity, k the heat conduction coefficient, f is the heat source from the welding equipment, and U_s is the initial constant (room) temperature in the material.

A possible model for the heat source is a moving Gaussian function:

$$f = A \exp\left(-\frac{1}{2}\left(\frac{x - vt}{\sigma}\right)^2\right),$$

where A is the strength, σ is a parameter governing how peak-shaped (or localized in space) the heat source is, and v is the velocity (in positive x direction) of the source.

a) Let x_c, t_c, u_c, and f_c be scales, i.e., characteristic sizes, of x, t, u, and f, respectively. The natural choice of x_c and f_c is L and A, since these make the scaled x and f in the interval $[0, 1]$. If each of the three terms in the PDE are equally important, we can find t_c and u_c by demanding that the coefficients in the scaled PDE are all equal to unity. Perform this scaling. Use scaled quantities in the arguments for the exponential function in f too and show that

$$\bar{f} = e^{-\frac{1}{2}\beta^2(\bar{x} - \gamma \bar{t})^2},$$

where β and γ are dimensionless numbers. Give an interpretation of β and γ.

b) Argue that for large γ we should base the time scale on the movement of the heat source. Show that this gives rise to the scaled PDE

$$\frac{\partial \bar{u}}{\partial \bar{t}} = \gamma^{-1} \frac{\partial^2 \bar{u}}{\partial \bar{x}^2} + \bar{f},$$

and

$$\bar{f} = \exp\left(-\frac{1}{2}\beta^2(\bar{x} - \bar{t})^2\right).$$

Discuss when the scalings in a) and b) are appropriate.

c) One aim with scaling is to get a solution that lies in the interval $[-1, 1]$. This is not always the case when u_c is based on a scale involving a source term, as we do in a) and b). However, from the scaled PDE we realize that if we replace $\bar{f}$ with $\delta \bar{f}$, where δ is a dimensionless factor, this corresponds to replacing u_c by u_c/δ. So, if we observe that $\bar{u} \sim 1/\delta$ in simulations, we can just replace $\bar{f}$ by $\delta \bar{f}$ in the scaled PDE.

Use this trick and implement the two scaled models. Reuse software for the diffusion equation (e.g., the `solver` function in `diffu1D_vc.py`). Make a function   `run(gamma, beta=10, delta=40, scaling=1, animate=False)` that runs the model with the given γ, β, and δ parameters as well as an indicator scaling that is 1 for the scaling in a) and 2 for the scaling in b). The last argument can be used to turn screen animations on or off.

Experiments show that with $\gamma = 1$ and $\beta = 10$, $\delta = 20$ is appropriate. Then max $|\bar{u}|$ will be larger than 4 for $\gamma = 40$, but that is acceptable.

Equip the run function with visualization, both animation of $\bar{u}$ and $\bar{f}$, and plots with $\bar{u}$ and $\bar{f}$ for $t = 0.2$ and $t = 0.5$.

Hint Since the amplitudes of $\bar{u}$ and $\bar{f}$ differs by a factor δ, it is attractive to plot $\bar{f}/\delta$ together with $\bar{u}$.

d) Use the software in c) to investigate $\gamma = 0.2, 1, 5, 40$ for the two scalings. Discuss the results.

Filename: `welding`.

Exercise 1.12: Implement a Forward Euler scheme for axi-symmetric diffusion

Based on the discussion in Sect. 1.5.6, derive in detail the discrete equations for a Forward Euler in time, centered in space, finite difference method for axi-symmetric diffusion. The diffusion coefficient may be a function of the radial coordinate. At the outer boundary $r = R$, we may have either a Dirichlet or Robin condition. Implement this scheme. Construct appropriate test problems.

Filename: `FE_axisym`.

2

Advection Equations

Wave (Chap. 2) and diffusion (Chap. 3) equations are solved reliably by finite difference methods. As soon as we add a first-order derivative in space, representing *advective* transport (also known as *convective* transport), the numerics gets more complicated and intuitively attractive methods no longer work well. We shall show how and why such methods fail and provide remedies. The present chapter builds on basic knowledge about finite difference methods for diffusion and wave equations, including the analysis by Fourier components, truncation error analysis (Appendix B), and compact difference notation.

Remark on terminology

It is common to refer to movement of a fluid as convection, while advection is the transport of some material dissolved or suspended in the fluid. We shall mostly choose the word advection here, but both terms are in heavy use, and for mass transport of a substance the PDE has an advection term, while the similar term for the heat equation is a convection term.

Much more comprehensive discussion of dispersion analysis for advection problems can be found in the book by Duran [3]. This is a an excellent resource for further studies on the topic of advection PDEs, with emphasis on generalizations to real geophysical problems. The book by Fletcher [4] also has a good overview of methods for advection and convection problems.

2.1 One-Dimensional Time-Dependent Advection Equations

We consider the pure advection model

$$\frac{\partial u}{\partial t} + v \frac{\partial u}{\partial x} = 0, \qquad\qquad x \in (0, L),\ t \in (0, T], \qquad (2.1)$$

$$u(x, 0) = I(x), \qquad\qquad x \in (0, L), \qquad (2.2)$$

$$u(0, t) = U_0, \qquad\qquad t \in (0, T]. \qquad (2.3)$$

In (2.1), v is a given parameter, typically reflecting the transport velocity of a quantity u with a flow. There is only one boundary condition (2.3) since the spatial derivative is only first order in the PDE (2.1). The information at $x = 0$ and the initial condition get transported in the positive x direction if $v > 0$ through the domain.

It is easiest to find the solution of (2.1) if we remove the boundary condition and consider a process on the infinite domain $(-\infty, \infty)$. The solution is simply

$$u(x, t) = I(x - vt). \tag{2.4}$$

This is also the solution we expect locally in a finite domain before boundary conditions have reflected or modified the wave.

A particular feature of the solution (2.4) is that

$$u(x_i, t_{n+1}) = u(x_{i-1}, t_n), \tag{2.5}$$

if $x_i = i\Delta x$ and $t_n = n\Delta t$ are points in a uniform mesh. We see this relation from

$$
\begin{aligned}
u(i\Delta x, (n+1)\Delta t) &= I(i\Delta x - v(n+1)\Delta t) \\
&= I((i-1)\Delta x - vn\Delta t - v\Delta t + \Delta x) \\
&= I((i-1)\Delta x - vn\Delta t) \\
&= u((i-1)\Delta x, n\Delta t),
\end{aligned}
$$

provided $v = \Delta x/\Delta t$. So, whenever we see a scheme that collapses to

$$u_i^{n+1} = u_{i-1}^n, \tag{2.6}$$

for the PDE in question, we have in fact a scheme that reproduces the analytical solution, and many of the schemes to be presented possess this nice property!

Finally, we add that a discussion of appropriate boundary conditions for the advection PDE in multiple dimensions is a challenging topic beyond the scope of this text.

2.1.1   Simplest Scheme: Forward in Time, Centered in Space

Method A first attempt to solve a PDE like (2.1) will normally be to look for a time-discretization scheme that is explicit so we avoid solving systems of linear equations. In space, we anticipate that centered differences are most accurate and therefore best. These two arguments lead us to a Forward Euler scheme in time and centered differences in space:

$$[D_t^+ u + v D_{2x} u = 0]_i^n. \tag{2.7}$$

Written out, we see that this expression implies that

$$u^{n+1} = u^n - \frac{1}{2}C(u_{i+1}^n - u_{i-1}^n),$$

with C as the Courant number

$$C = \frac{v\Delta t}{\Delta x}.$$

Implementation A solver function for our scheme goes as follows.

```python
import numpy as np
import matplotlib.pyplot as plt

def solver_FECS(I, U0, v, L, dt, C, T, user_action=None):
    Nt = int(round(T/float(dt)))
    t = np.linspace(0, Nt*dt, Nt+1)    # Mesh points in time
    dx = v*dt/C
    Nx = int(round(L/dx))
    x = np.linspace(0, L, Nx+1)         # Mesh points in space
    # Make sure dx and dt are compatible with x and t
    dx = x[1] - x[0]
    dt = t[1] - t[0]
    C = v*dt/dx

    u   = np.zeros(Nx+1)
    u_n = np.zeros(Nx+1)

    # Set initial condition u(x,0) = I(x)
    for i in range(0, Nx+1):
        u_n[i] = I(x[i])

    if user_action is not None:
        user_action(u_n, x, t, 0)

    for n in range(0, Nt):
        # Compute u at inner mesh points
        for i in range(1, Nx):
            u[i] = u_n[i] - 0.5*C*(u_n[i+1] - u_n[i-1])

        # Insert boundary condition
        u[0] = U0

        if user_action is not None:
            user_action(u, x, t, n+1)

        # Switch variables before next step
        u_n, u = u, u_n
```

Test cases The typical solution u has the shape of I and is transported at velocity v to the right (if $v > 0$). Let us consider two different initial conditions, one smooth (Gaussian pulse) and one non-smooth (half-truncated cosine pulse):

$$u(x, 0) = A e^{-\frac{1}{2}\left(\frac{x - L/10}{\sigma}\right)^2}, \tag{2.8}$$

$$u(x, 0) = A \cos\left(\frac{5\pi}{L}\left(x - \frac{L}{10}\right)\right), \quad x < \frac{L}{5} \text{ else } 0. \tag{2.9}$$

The parameter A is the maximum value of the initial condition.

Before doing numerical simulations, we scale the PDE problem and introduce $\bar{x} = x/L$ and $\bar{t} = vt/L$, which gives

$$\frac{\partial \bar{u}}{\partial \bar{t}} + \frac{\partial \bar{u}}{\partial \bar{x}} = 0.$$

The unknown u is scaled by the maximum value of the initial condition: $\bar{u} = u/\max|I(x)|$ such that $|\bar{u}(\bar{x}, 0)| \in [0, 1]$. The scaled problem is solved by setting $v = 1$, $L = 1$, and $A = 1$. From now on we drop the bars.

To run our test cases and plot the solution, we make the function

```python
def run_FECS(case):
    """Special function for the FECS case."""
    if case == 'gaussian':
        def I(x):
            return np.exp(-0.5*((x-L/10)/sigma)**2)
    elif case == 'cosinehat':
        def I(x):
            return np.cos(np.pi*5/L*(x - L/10)) if x < L/5 else 0

    L = 1.0
    sigma = 0.02
    legends = []

    def plot(u, x, t, n):
        """Animate and plot every m steps in the same figure."""
        plt.figure(1)
        if n == 0:
            lines = plot(x, u)
        else:
            lines[0].set_ydata(u)
            plt.draw()
            #plt.savefig()
        plt.figure(2)
        m = 40
        if n % m != 0:
            return
        print 't=%g, n=%d, u in [%g, %g] w/%d points' % \
              (t[n], n, u.min(), u.max(), x.size)
        if np.abs(u).max() > 3:  # Instability?
            return
        plt.plot(x, u)
        legends.append('t=%g' % t[n])
        if n > 0:
            plt.hold('on')

    plt.ion()
    U0 = 0
    dt = 0.001
    C = 1
    T = 1
    solver(I=I, U0=U0, v=1.0, L=L, dt=dt, C=C, T=T,
           user_action=plot)
    plt.legend(legends, loc='lower left')
    plt.savefig('tmp.png'); plt.savefig('tmp.pdf')
    plt.axis([0, L, -0.75, 1.1])
    plt.show()
```

Bug? Running either of the test cases, the plot becomes a mess, and the printout of u values in the plot function reveals that u grows very quickly. We may reduce Δt and make it very small, yet the solution just grows. Such behavior points to a bug in the code. However, choosing a coarse mesh and performing one time step by hand calculations produces the same numbers as the code, so the implementation seems to be correct. The hypothesis is therefore that the solution is unstable.

2.1.2 Analysis of the Scheme

It is easy to show that a typical Fourier component

$$u(x,t) = B \sin(k(x - ct))$$

is a solution of our PDE for any spatial wave length $\lambda = 2\pi/k$ and any amplitude B. (Since the PDE to be investigated by this method is homogeneous and linear, B will always cancel out, so we tend to skip this amplitude, but keep it here in the beginning for completeness.)

A general solution may be viewed as a collection of long and short waves with different amplitudes. Algebraically, the work simplifies if we introduce the complex Fourier component

$$u(x,t) = A_e e^{ikx},$$

with

$$A_e = B e^{-ikv\Delta t} = B e^{-iCk\Delta x}.$$

Note that $|A_e| \leq 1$.

It turns out that many schemes also allow a Fourier wave component as solution, and we can use the numerically computed values of A_e (denoted A) to learn about the quality of the scheme. Hence, to analyze the difference scheme we have just implemented, we look at how it treats the Fourier component

$$u_q^n = A^n e^{ikq\Delta x}.$$

Inserting the numerical component in the scheme,

$$[D_t^+ A e^{ikq\Delta x} + v D_{2x} A e^{ikq\Delta x} = 0]_q^n,$$

and making use of (A.25) results in

$$\left[e^{ikq\Delta x} \left(\frac{A-1}{\Delta t} + v \frac{1}{\Delta x} i \sin(k\Delta x) \right) = 0 \right]_q^n,$$

which implies

$$A = 1 - iC \sin(k\Delta x).$$

The numerical solution features the formula A^n. To find out whether A^n means growth in time, we rewrite A in polar form: $A = A_r e^{i\phi}$, for real numbers A_r and ϕ, since we then have $A^n = A_r^n e^{i\phi n}$. The magnitude of A^n is A_r^n. In our case, $A_r = (1 + C^2 \sin^2(kx))^{1/2} > 1$, so A_r^n will increase in time, whereas the exact solution will not. Regardless of Δt, we get unstable numerical solutions.

2.1.3   Leapfrog in Time, Centered Differences in Space

Method   Another explicit scheme is to do a "leapfrog" jump over $2\Delta t$ in time and combine it with central differences in space:

$$[D_{2t}u + vD_{2x}u = 0]_i^n,$$

which results in the updating formula

$$u_i^{n+1} = u_i^{n-1} - C(u_{i+1}^n - u_{i-1}^n).$$

A special scheme is needed to compute u^1, but we leave that problem for now. Anyway, this special scheme can be found in advec1D.py.

Implementation   We now need to work with three time levels and must modify our solver a bit:

```
Nt = int(round(T/float(dt)))
t = np.linspace(0, Nt*dt, Nt+1)    # Mesh points in time
...
u   = np.zeros(Nx+1)
u_1 = np.zeros(Nx+1)
u_2 = np.zeros(Nx+1)
...
for n in range(0, Nt):
    if scheme == 'FE':
        for i in range(1, Nx):
            u[i] = u_1[i] - 0.5*C*(u_1[i+1] - u_1[i-1])
    elif scheme == 'LF':
        if n == 0:
            # Use some scheme for the first step
            for i in range(1, Nx):
                ...
        else:
            for i in range(1, Nx+1):
                u[i] = u_2[i] - C*(u_1[i] - u_1[i-1])

    # Switch variables before next step
    u_2, u_1, u = u_1, u, u_2
```

Running a test case   Let us try a coarse mesh such that the smooth Gaussian initial condition is represented by 1 at mesh node 1 and 0 at all other nodes. This triangular initial condition should then be advected to the right. Choosing scaled variables as $\Delta t = 0.1$, $T = 1$, and $C = 1$ gives the plot in Fig. 2.1, which is in fact identical to the exact solution (!).

Running more test cases   We can run two types of initial conditions for $C = 0.8$: one very smooth with a Gaussian function (Fig. 2.2) and one with a discontinuity in the first derivative (Fig. 2.3). Unless we have a very fine mesh, as in the left plots in the figures, we get small ripples behind the main wave, and this main wave has the amplitude reduced.

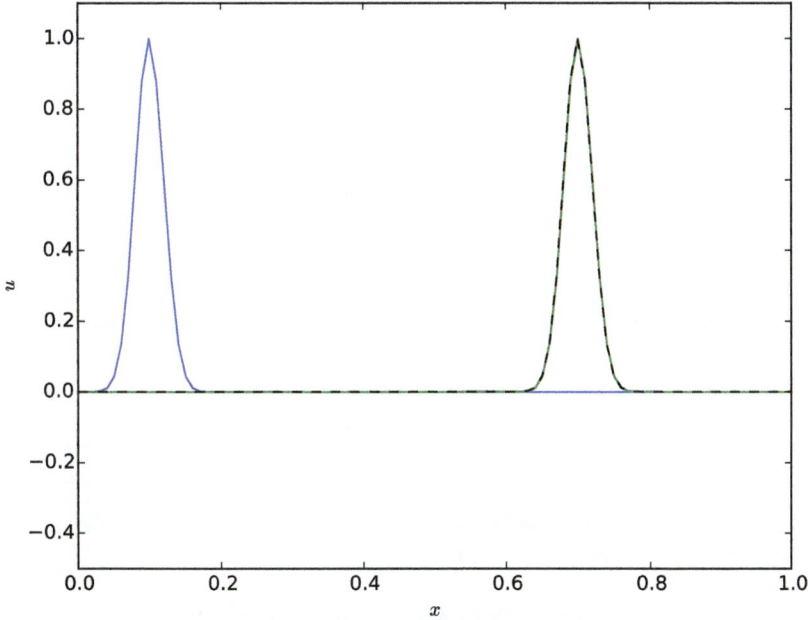

Fig. 2.1 Exact solution obtained by Leapfrog scheme with $\Delta t = 0.1$ and $C = 1$

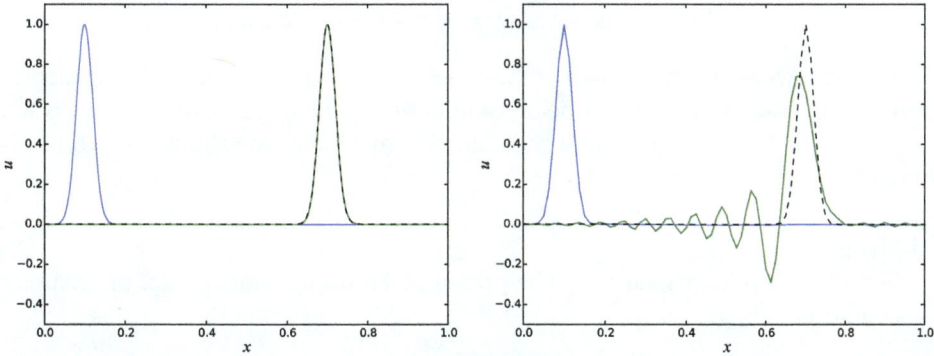

Fig. 2.2 Advection of a Gaussian function with a leapfrog scheme and $C = 0.8$, $\Delta t = 0.001$ (*left*) and $\Delta t = 0.01$ (*right*)

Advection of the Gaussian function with a leapfrog scheme, using $C = 0.8$ and $\Delta t = 0.01$ can be seen in a movie file[1]. Alternatively, with $\Delta t = 0.001$, we get this movie file[2].

Advection of the cosine hat function with a leapfrog scheme, using $C = 0.8$ and $\Delta t = 0.01$ can be seen in a movie file[3]. Alternatively, with $\Delta t = 0.001$, we get this movie file[4].

---

[1] http://tinyurl.com/gokgkov/mov-advec/gaussian/LF/C08_dt01.ogg

[2] http://tinyurl.com/gokgkov/mov-advec/gaussian/LF/C08_dt001.ogg

[3] http://tinyurl.com/gokgkov/mov-advec/cosinehat/LF/C08_dt01.ogg

[4] http://tinyurl.com/gokgkov/mov-advec/cosinehat/LF/C08_dt001.ogg

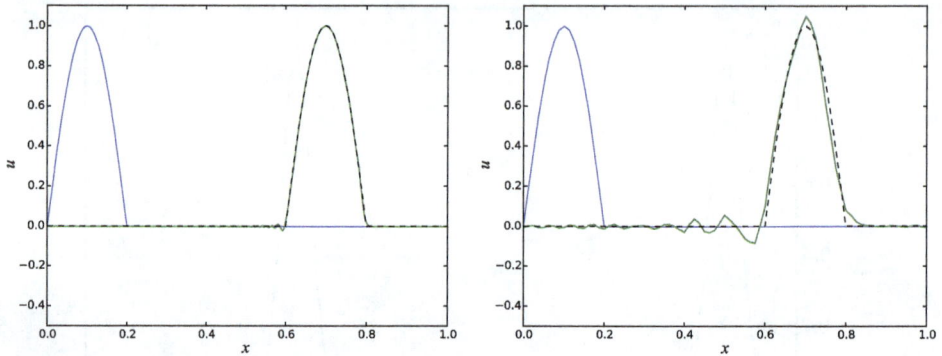

Fig. 2.3 Advection of half a cosine function with a leapfrog scheme and $C = 0.8$, $\Delta t = 0.001$ (*left*) and $\Delta t = 0.01$ (*right*)

Analysis We can perform a Fourier analysis again. Inserting the numerical Fourier component in the Leapfrog scheme, we get

$$A^2 - i2C\sin(k\Delta x)A - 1 = 0,$$

and

$$A = -iC\sin(k\Delta x) \pm \sqrt{1 - C^2\sin^2(k\Delta x)}\,.$$

Rewriting to polar form, $A = A_r e^{i\phi}$, we see that $A_r = 1$, so the numerical component is neither increasing nor decreasing in time, which is exactly what we want. However, for $C > 1$, the square root can become complex valued, so stability is obtained only as long as $C \leq 1$.

Stability

For all the working schemes to be presented in this chapter, we get the stability condition $C \leq 1$:

$$\Delta t \leq \frac{\Delta x}{v}\,.$$

This is called the CFL condition and applies almost always to successful schemes for advection problems. Of course, one can use Crank-Nicolson or Backward Euler schemes for increased and even unconditional stability (no Δt restrictions), but these have other less desired damping problems.

We introduce $p = k\Delta x$. The amplification factor now reads

$$A = -iC\sin p \pm \sqrt{1 - C^2\sin^2 p},$$

and is to be compared to the exact amplification factor

$$A_{\mathrm{e}} = e^{-ikv\Delta t} = e^{-ikC\Delta x} = e^{-iCp}\,.$$

Section 2.1.9 compares numerical amplification factors of many schemes with the exact expression.

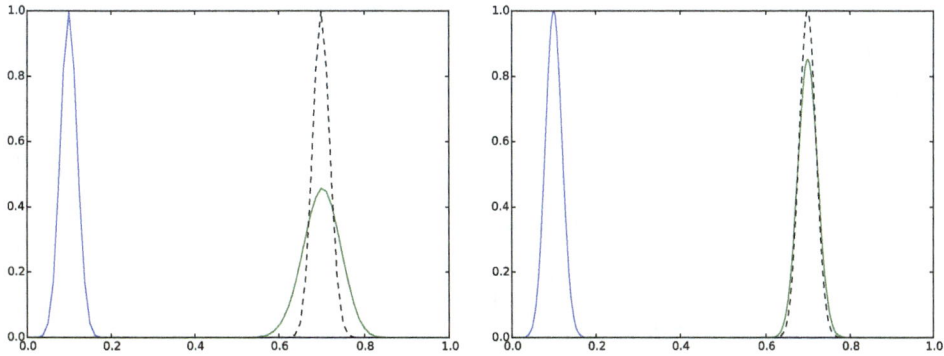

Fig. 2.4 Advection of a Gaussian function with a forward in time, upwind in space scheme and $C = 0.8$, $\Delta t = 0.01$ (*left*) and $\Delta t = 0.001$ (*right*)

2.1.4 Upwind Differences in Space

Since the PDE reflects transport of information along with a flow in positive x direction, when $v > 0$, it could be natural to go (what is called) upstream and not downstream in the spatial derivative to collect information about the change of the function. That is, we approximate

$$\frac{\partial u}{\partial x}(x_i, t_n) \approx [D_x^- u]_i^n = \frac{u_i^n - u_{i-1}^n}{\Delta x}.$$

This is called an *upwind difference* (the corresponding difference in the time direction would be called a backward difference, and we could use that name in space too, but *upwind* is the common name for a difference against the flow in advection problems). This spatial approximation does magic compared to the scheme we had with Forward Euler in time and centered difference in space. With an upwind difference,

$$[D_t^+ u + v D_x^- u = 0]_i^n, \tag{4.10}$$

written out as

$$u_i^{n+1} = u_i^n - C(u_i^n - u_{i-1}^n),$$

gives a generally popular and robust scheme that is stable if $C \leq 1$. As with the Leapfrog scheme, it becomes exact if $C = 1$, exactly as shown in Fig. 2.1. This is easy to see since $C = 1$ gives the property (2.6). However, any $C < 1$ gives a significant reduction in the amplitude of the solution, which is a purely numerical effect, see Fig. 2.4 and 2.5. Experiments show, however, that reducing Δt or Δx, while keeping C reduces the error.

Advection of the Gaussian function with a forward in time, upwind in space scheme, using $C = 0.8$ and $\Delta t = 0.01$ can be seen in a movie file[5]. Alternatively, with $\Delta t = 0.005$, we get this movie file[6].

---

[5] http://tinyurl.com/gokgkov/mov-advec/gaussian/UP/C08_dt001/movie.ogg
[6] http://tinyurl.com/gokgkov/mov-advec/gaussian/UP/C08_dt0005/movie.ogg

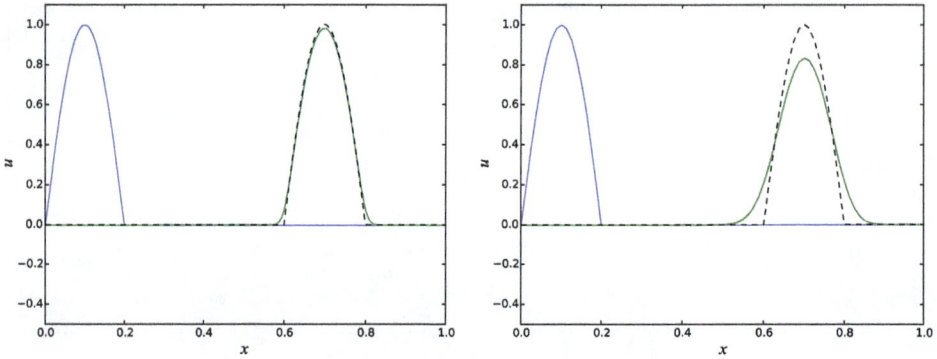

Fig. 2.5 Advection of half a cosine function with a forward in time, upwind in space scheme and $C = 0.8$, $\Delta t = 0.001$ (*left*) and $\Delta t = 0.01$ (*right*)

Advection of the cosine hat function with a forward in time, upwind in space scheme, using $C = 0.8$ and $\Delta t = 0.01$ can be seen in a movie file[7]. Alternatively, with $\Delta t = 0.001$, we get this movie file[8].

The amplification factor can be computed using the formula (A.23),

$$\frac{A-1}{\Delta t} + \frac{v}{\Delta x}(1 - e^{-ik\Delta x}) = 0,$$

which means

$$A = 1 - C(1 - \cos(p) - i \sin(p)).$$

For $C < 1$ there is, unfortunately, non-physical damping of discrete Fourier components, giving rise to reduced amplitude of u_i^n as in Fig. 2.4 and 2.5. The damping seen in these figures is quite severe. Stability requires $C \leq 1$.

Interpretation of upwind difference as artificial diffusion

One can interpret the upwind difference as extra, artificial diffusion in the equation. Solving

$$\frac{\partial u}{\partial t} + v\frac{\partial u}{\partial x} = \nu\frac{\partial^2 u}{\partial x^2},$$

by a forward difference in time and centered differences in space,

$$D_t^+ u + v D_{2x}u = \nu D_x D_x u]_i^n,$$

actually gives the upwind scheme (2.10) if $\nu = v\Delta x/2$. That is, solving the PDE $u_t + vu_x = 0$ by centered differences in space and forward difference in time is unsuccessful, but by adding some artificial diffusion νu_{xx}, the method becomes stable:

$$\frac{\partial u}{\partial t} + v\frac{\partial u}{\partial x} = \left(\alpha + \frac{v\Delta x}{2}\right)\frac{\partial^2 u}{\partial x^2}.$$

---

[7] http://tinyurl.com/gokgkov/mov-advec/cosinehat/UP/C08_dt01.ogg
[8] http://tinyurl.com/gokgkov/mov-advec/cosinehat/UP/C08_dt001.ogg

2.1.5 Periodic Boundary Conditions

So far, we have given the value on the left boundary, u_0^n, and used the scheme to propagate the solution signal through the domain. Often, we want to follow such signals for long time series, and periodic boundary conditions are then relevant since they enable a signal that leaves the right boundary to immediately enter the left boundary and propagate through the domain again.

The periodic boundary condition is

$$u(0,t) = u(L,t), \quad u_0^n = u_{N_x}^n.$$

It means that we in the first equation, involving u_0^n, insert $u_{N_x}^n$, and that we in the last equation, involving $u_{N_x}^{n+1}$ insert u_0^{n+1}. Normally, we can do this in the simple way that u_1[0] is updated as u_1[Nx] at the beginning of a new time level.

In some schemes we may need $u_{N_x+1}^n$ and u_{-1}^n. Periodicity then means that these values are equal to u_1^n and $u_{N_x-1}^n$, respectively. For the upwind scheme, it is sufficient to set u_1[0]=u_1[Nx] at a new time level before computing u[1]. This ensures that u[1] becomes right and at the next time level u[0] at the current time level is correctly updated. For the Leapfrog scheme we must update u[0] and u[Nx] using the scheme:

```
if periodic_bc:
    i = 0
    u[i] = u_2[i] - C*(u_1[i+1] - u_1[Nx-1])
for i in range(1, Nx):
    u[i] = u_2[i] - C*(u_1[i+1] - u_1[i-1])
if periodic_bc:
    u[Nx] = u[0]
```

2.1.6 Implementation

Test condition Analytically, we can show that the integral in space under the $u(x,t)$ curve is constant:

$$\int_0^L \left(\frac{\partial u}{\partial t} + v \frac{\partial u}{\partial x} \right) dx = 0$$

$$\frac{\partial}{\partial t} \int_0^L u \, dx = - \int_0^L v \frac{\partial u}{\partial x} dx$$

$$\frac{\partial u}{\partial t} \int_0^L u \, dx = [vu]_0^L = 0$$

as long as $u(0) = u(L) = 0$. We can therefore use the property

$$\int_0^L u(x,t)dx = \text{const}$$

as a partial verification during the simulation. Now, any numerical method with $C \neq 1$ will deviate from the constant, expected value, so the integral is a measure of the error in the scheme. The integral can be computed by the Trapezoidal integration rule

```
dx*(0.5*u[0] + 0.5*u[Nx] + np.sum(u[1:-1]))
```

if u is an array holding the solution.

The code An appropriate solver function for multiple schemes may go as shown below.

```
def solver(I, U0, v, L, dt, C, T, user_action=None,
           scheme='FE', periodic_bc=True):

    Nt = int(round(T/float(dt)))
    t = np.linspace(0, Nt*dt, Nt+1)    # Mesh points in time
    dx = v*dt/C
    Nx = int(round(L/dx))
    x = np.linspace(0, L, Nx+1)        # Mesh points in space
    # Make sure dx and dt are compatible with x and t
    dx = x[1] - x[0]
    dt = t[1] - t[0]
    C = v*dt/dx
    print 'dt=%g, dx=%g, Nx=%d, C=%g' % (dt, dx, Nx, C)


    u   = np.zeros(Nx+1)
    u_n = np.zeros(Nx+1)
    u_nm1 = np.zeros(Nx+1)
    integral = np.zeros(Nt+1)


    # Set initial condition u(x,0) = I(x)
    for i in range(0, Nx+1):
        u_n[i] = I(x[i])

    # Insert boundary condition
    u[0] = U0

    # Compute the integral under the curve
    integral[0] = dx*(0.5*u_n[0] + 0.5*u_n[Nx] + np.sum(u_n[1:-1]))

    if user_action is not None:
        user_action(u_n, x, t, 0)
```

```
    for n in range(0, Nt):
        if scheme == 'FE':
            if periodic_bc:
                i = 0
                u[i] = u_n[i] - 0.5*C*(u_n[i+1] - u_n[Nx])
                u[Nx] = u[0]
            for i in range(1, Nx):
                u[i] = u_n[i] - 0.5*C*(u_n[i+1] - u_n[i-1])
        elif scheme == 'LF':
            if n == 0:
                # Use upwind for first step
                if periodic_bc:
                    i = 0
                    u_n[i] = u_n[Nx]
                for i in range(1, Nx+1):
                    u[i] = u_n[i] - C*(u_n[i] - u_n[i-1])
            else:
                if periodic_bc:
                    i = 0
                    u[i] = u_nm1[i] - C*(u_n[i+1] - u_n[Nx-1])
                for i in range(1, Nx):
                    u[i] = u_nm1[i] - C*(u_n[i+1] - u_n[i-1])
                if periodic_bc:
                    u[Nx] = u[0]
        elif scheme == 'UP':
            if periodic_bc:
                u_n[0] = u_n[Nx]
            for i in range(1, Nx+1):
                u[i] = u_n[i] - C*(u_n[i] - u_n[i-1])
        else:
            raise ValueError('scheme="%s" not implemented' % scheme)


        if not periodic_bc:
            # Insert boundary condition
            u[0] = U0

        # Compute the integral under the curve
        integral[n+1] = dx*(0.5*u[0] + 0.5*u[Nx] +  np.sum(u[1:-1]))

        if user_action is not None:
            user_action(u, x, t, n+1)

        # Switch variables before next step
        u_nm1, u_n, u = u_n, u, u_nm1
    return integral
```

Solving a specific problem We need to call up the `solver` function in some kind of administering problem solving function that can solve specific problems and make appropriate visualization. The function below makes both static plots, screen animation, and hard copy videos in various formats.

```
def run(scheme='UP', case='gaussian', C=1, dt=0.01):
    """General admin routine for explicit and implicit solvers."""

    if case == 'gaussian':
        def I(x):
            return np.exp(-0.5*((x-L/10)/sigma)**2)
    elif case == 'cosinehat':
        def I(x):
            return np.cos(np.pi*5/L*(x - L/10)) if x < L/5 else 0

    L = 1.0
    sigma = 0.02
    global lines  # needs to be saved between calls to plot

    def plot(u, x, t, n):
        """Plot t=0 and t=0.6 in the same figure."""
        plt.figure(1)
        global lines
        if n == 0:
            lines = plt.plot(x, u)
            plt.axis([x[0], x[-1], -0.5, 1.5])
            plt.xlabel('x'); plt.ylabel('u')
            plt.axes().set_aspect(0.15)
            plt.savefig('tmp_%04d.png' % n)
            plt.savefig('tmp_%04d.pdf' % n)

        else:
            lines[0].set_ydata(u)
            plt.axis([x[0], x[-1], -0.5, 1.5])
            plt.title('C=%g, dt=%g, dx=%g' %
                      (C, t[1]-t[0], x[1]-x[0]))
            plt.legend(['t=%.3f' % t[n]])
            plt.xlabel('x'); plt.ylabel('u')
            plt.draw()
            plt.savefig('tmp_%04d.png' % n)
        plt.figure(2)
        eps = 1E-14
        if abs(t[n] - 0.6) > eps and abs(t[n] - 0) > eps:
            return
        print 't=%g, n=%d, u in [%g, %g] w/%d points' % \
              (t[n], n, u.min(), u.max(), x.size)
        if np.abs(u).max() > 3:   # Instability?
            return
        plt.plot(x, u)
        plt.hold('on')
        plt.draw()
        if n > 0:
            y = [I(x_-v*t[n]) for x_ in x]
            plt.plot(x, y, 'k--')
            if abs(t[n] - 0.6) < eps:
                filename = ('tmp_%s_dt%s_C%s' % \
                            (scheme, t[1]-t[0], C)).replace('.', '')
                np.savez(filename, x=x, u=u, u_e=y)
```

```
    plt.ion()
    UO = 0
    T = 0.7
    v = 1
    # Define video formats and libraries
    codecs = dict(flv='flv', mp4='libx264', webm='libvpx',
                  ogg='libtheora')
    # Remove video files
    import glob, os
    for name in glob.glob('tmp_*.png'):
        os.remove(name)
    for ext in codecs:
        name = 'movie.%s' % ext
        if os.path.isfile(name):
            os.remove(name)

integral = solver(
    I=I, UO=UO, v=v, L=L, dt=dt, C=C, T=T,
    scheme=scheme, user_action=plot)
# Finish up figure(2)
plt.figure(2)
plt.axis([0, L, -0.5, 1.1])
plt.xlabel('$x$'); plt.ylabel('$u$')
plt.savefig('tmp1.png'); plt.savefig('tmp1.pdf')
plt.show()
# Make videos from figure(1) animation files
for codec in codecs:
    cmd = 'ffmpeg -i tmp_%%04d.png -r 25 -vcodec %s movie.%s' % \
          (codecs[codec], codec)
    os.system(cmd)
print 'Integral of u:', integral.max(), integral.min()
```

The complete code is found in the file advec1D.py.

2.1.7 A Crank-Nicolson Discretization in Time and Centered Differences in Space

Another obvious candidate for time discretization is the Crank-Nicolson method combined with centered differences in space:

$$[D_t u]_i^n + v\frac{1}{2}([D_{2x}u]_i^{n+1} + [D_{2x}u]_i^n) = 0.$$

It can be nice to include the Backward Euler scheme too, via the θ-rule,

$$[D_t u]_i^n + v\theta[D_{2x}u]_i^{n+1} + v(1-\theta)[D_{2x}u]_i^n = 0.$$

When θ is different from zero, this gives rise to an *implicit* scheme,

$$u_i^{n+1} + \frac{\theta}{2}C(u_{i+1}^{n+1} - u_{i-1}^{n+1}) = u_i^n - \frac{1-\theta}{2}C(u_{i+1}^n - u_{i-1}^n)$$

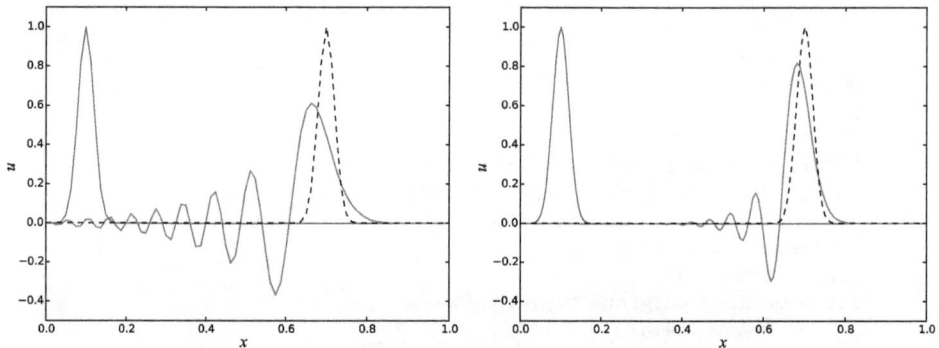

Fig. 2.6 Crank-Nicolson in time, centered in space, Gaussian profile, $C = 0.8$, $\Delta t = 0.01$ (*left*) and $\Delta t = 0.005$ (*right*)

for $i = 1, \ldots, N_x - 1$. At the boundaries we set $u = 0$ and simulate just to the point of time when the signal hits the boundary (and gets reflected).

$$u_0^{n+1} = u_{N_x}^{n+1} = 0.$$

The elements on the diagonal in the matrix become:

$$A_{i,i} = 1, \quad i = 0, \ldots, N_x.$$

On the subdiagonal and superdiagonal we have

$$A_{i-1,i} = -\frac{\theta}{2}C, \quad A_{i+1,i} = \frac{\theta}{2}C, \quad i = 1, \ldots, N_x - 1,$$

with $A_{0,1} = 0$ and $A_{N_x-1,N_x} = 0$ due to the known boundary conditions. And finally, the right-hand side becomes

$$b_0 = u_{N_x}^n$$
$$b_i = u_i^n - \frac{1-\theta}{2}C(u_{i+1}^n - u_{i-1}^n), \quad i = 1, \ldots, N_x - 1$$
$$b_{N_x} = u_0^n.$$

The dispersion relation follows from inserting $u_q^n = A^n e^{ikx}$ and using the formula (A.25) for the spatial differences:

$$A = \frac{1 - (1-\theta)iC \sin p}{1 + \theta iC \sin p}.$$

Movie 1 Crank-Nicolson in time, centered in space, $C = 0.8$, $\Delta t = 0.005$.
https://raw.githubusercontent.com/hplgit/fdm-book/master/doc/pub/book/html/mov-advec/
gaussian/CN/C08_dt0005/movie.ogg

Movie 2 Backward-Euler in time, centered in space, $C = 0.8$, $\Delta t = 0.005$.
https://raw.githubusercontent.com/hplgit/fdm-book/master/doc/pub/book/html/mov-advec/
cosinehat/BE/C_08_dt005.ogg

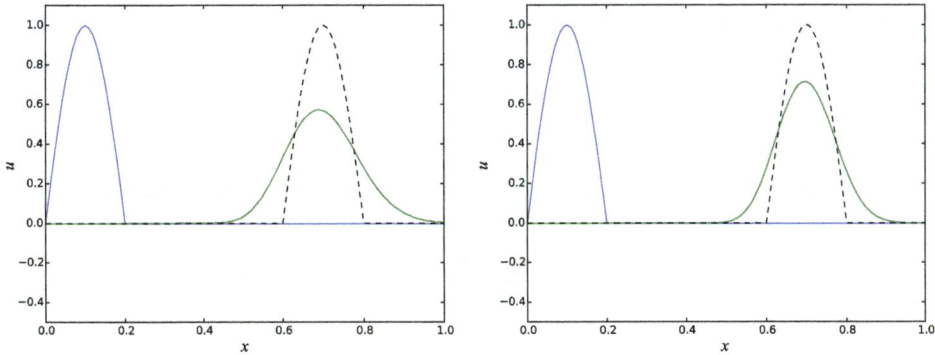

Fig. 2.7 Backward-Euler in time, centered in space, half a cosine profile, $C = 0.8$, $\Delta t = 0.01$ (*left*) and $\Delta t = 0.005$ (*right*)

Figure 2.6 depicts a numerical solution for $C = 0.8$ and the Crank-Nicolson with severe oscillations behind the main wave. These oscillations are damped as the mesh is refined. Switching to the Backward Euler scheme removes the oscillations, but the amplitude is significantly reduced. One could expect that the discontinuous derivative in the initial condition of the half a cosine wave would make even stronger demands on producing a smooth profile, but Fig. 2.7 shows that also here, Backward-Euler is capable of producing a smooth profile. All in all, there are no major differences between the Gaussian initial condition and the half a cosine condition for any of the schemes.

2.1.8 The Lax-Wendroff Method

The Lax-Wendroff method is based on three ideas:

1. Express the new unknown u_i^{n+1} in terms of known quantities at $t = t_n$ by means of a Taylor polynomial of second degree.
2. Replace time-derivatives at $t = t_n$ by spatial derivatives, using the PDE.
3. Discretize the spatial derivatives by second-order differences so we achieve a scheme of accuracy $\mathcal{O}(\Delta t^2) + \mathcal{O}(\Delta x^2)$.

Let us follow the recipe. First we have the three-term Taylor polynomial,

$$u_i^{n+1} = u_i^n + \Delta t \left(\frac{\partial u}{\partial t} \right)_i^n + \frac{1}{2} \Delta t^2 \left(\frac{\partial^2 u}{\partial t^2} \right)_i^n .$$

From the PDE we have that temporal derivatives can be substituted by spatial derivatives:

$$\frac{\partial u}{\partial t} = -v \frac{\partial u}{\partial x},$$

and furthermore,

$$\frac{\partial^2 u}{\partial t^2} = v^2 \frac{\partial^2 u}{\partial x^2} .$$

Inserted in the Taylor polynomial formula, we get

$$u_i^{n+1} = u_i^n - v\Delta t \left(\frac{\partial u}{\partial x}\right)_i^n + \frac{1}{2}\Delta t^2 v^2 \left(\frac{\partial^2 u}{\partial x^2}\right)_i^n.$$

To obtain second-order accuracy in space we now use central differences:

$$u_i^{n+1} = u_i^n - v\Delta t [D_{2x}u]_i^n + \frac{1}{2}\Delta t^2 v^2 [D_x D_x u]_i^n,$$

or written out,

$$u_i^{n+1} = u_i^n - \frac{1}{2}C(u_{i+1}^n - u_{i-1}^n) + \frac{1}{2}C^2(u_{i+1}^n - 2u_i^n + u_{i-1}^n).$$

This is the explicit Lax-Wendroff scheme.

Lax-Wendroff works because of artificial viscosity

From the formulas above, we notice that the Lax-Wendroff method is nothing but a Forward Euler, central difference in space scheme, which we have shown to be useless because of chronic instability, plus an artificial diffusion term of strength $\frac{1}{2}\Delta t v^2$. It means that we can take an unstable scheme and add some diffusion to stabilize it. This is a common trick to deal with advection problems. Sometimes, the real physical diffusion is not sufficiently large to make schemes stable, so then we also add artificial diffusion.

From an analysis similar to the ones carried out above, we get an amplification factor for the Lax-Wendroff method that equals

$$A = 1 - iC \sin p - 2C^2 \sin^2(p/2).$$

This means that $|A| = 1$ and also that we have an exact solution if $C = 1$!

2.1.9 Analysis of Dispersion Relations

We have developed expressions for $A(C, p)$ in the exact solution $u_q^n = A^n e^{ikq\Delta x}$ of the discrete equations. Note that the Fourier component that solves the original PDE problem has no damping and moves with constant velocity v. There are two basic errors in the numerical Fourier component: there may be damping and the wave velocity may depend on C and $p = k\Delta x$.

The shortest wavelength that can be represented is $\lambda = 2\Delta x$. The corresponding k is $k = 2\pi/\lambda = \pi/\Delta x$, so $p = k\Delta x \in (0, \pi]$.

Given a complex A as a function of C and p, how can we visualize it? The two key ingredients in A is the magnitude, reflecting damping or growth of the wave, and the angle, closely related to the velocity of the wave. The Fourier component

$$D^n e^{ik(x-ct)}$$

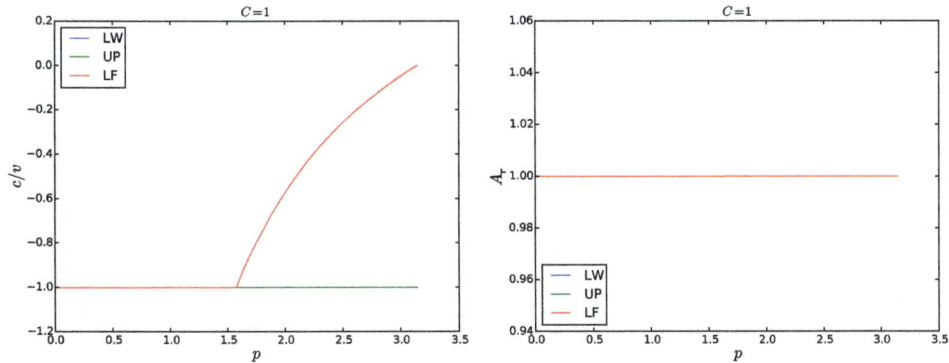

Fig. 2.8 Dispersion relations for $C = 1$

has damping D and wave velocity c. Let us express our A in polar form, $A = A_r e^{-i\phi}$, and insert this expression in our discrete component $u_q^n = A^n e^{ikq\Delta x} = A^n e^{ikx}$:

$$u_q^n = A_r^n e^{-i\phi n} e^{ikx} = A_r^n e^{i(kx - n\phi)} = A_r^n e^{i(k(x - ct))},$$

for

$$c = \frac{\phi}{k\Delta t}.$$

Now,

$$k\Delta t = \frac{Ck\Delta x}{v} = \frac{Cp}{v},$$

so

$$c = \frac{\phi v}{Cp}.$$

An appropriate dimensionless quantity to plot is the scaled wave velocity c/v:

$$\frac{c}{v} = \frac{\phi}{Cp}.$$

Figures 2.8–2.13 contain dispersion curves, velocity and damping, for various values of C. The horizontal axis shows the dimensionless frequency p of the wave, while the figures to the left illustrate the error in wave velocity c/v (should ideally be 1 for all p), and the figures to the right display the absolute value (magnitude) of the damping factor A_r. The curves are labeled according to the table below.

Label	Method
FE	Forward Euler in time, centered difference in space
LF	Leapfrog in time, centered difference in space
UP	Forward Euler in time, upwind difference in space
CN	Crank-Nicolson in time, centered difference in space
LW	Lax-Wendroff's method
BE	Backward Euler in time, centered difference in space

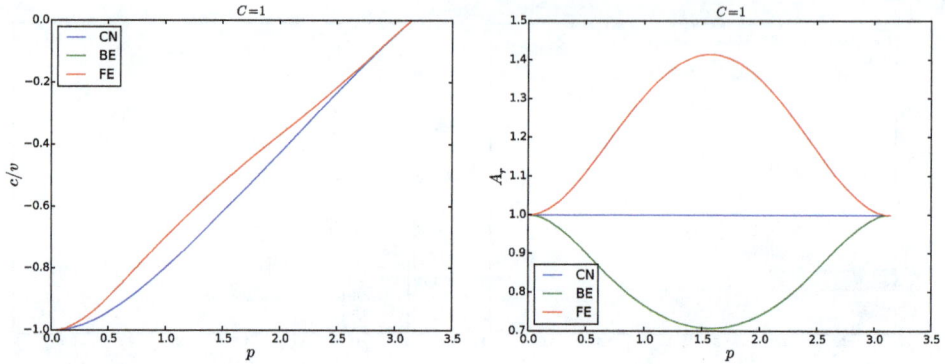

Fig. 2.9 Dispersion relations for $C = 1$

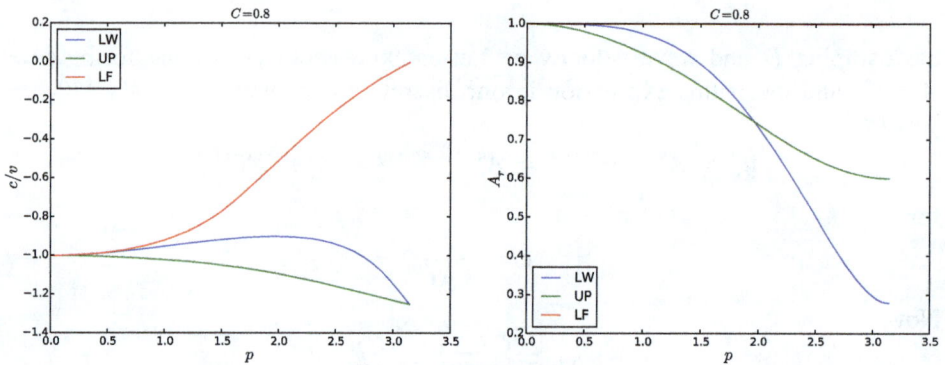

Fig. 2.10 Dispersion relations for $C = 0.8$

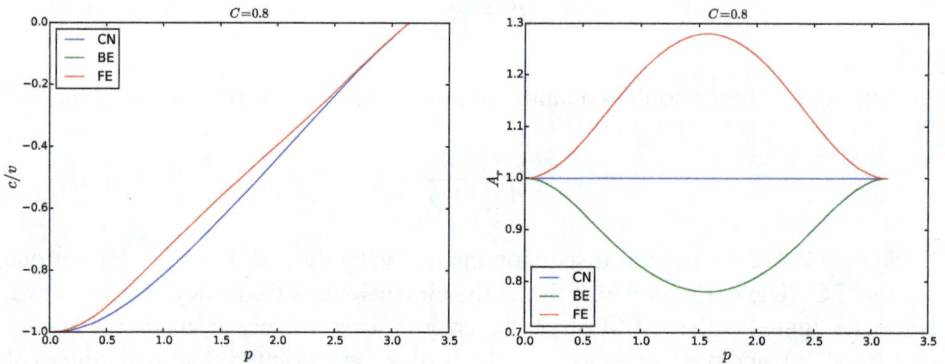

Fig. 2.11 Dispersion relations for $C = 0.8$

The total damping after some time $T = n\Delta t$ is reflected by $A_r(C, p)^n$. Since normally $A_r < 1$, the damping goes like $A_r^{1/\Delta t}$ and approaches zero as $\Delta t \to 0$. The only way to reduce damping is to increase C and/or the mesh resolution.

We can learn a lot from the dispersion relation plots. For example, looking at the plots for $C = 1$, the schemes LW, UP, and LF has no amplitude reduction, but LF has wrong phase velocity for the shortest wave in the mesh. This wave does not

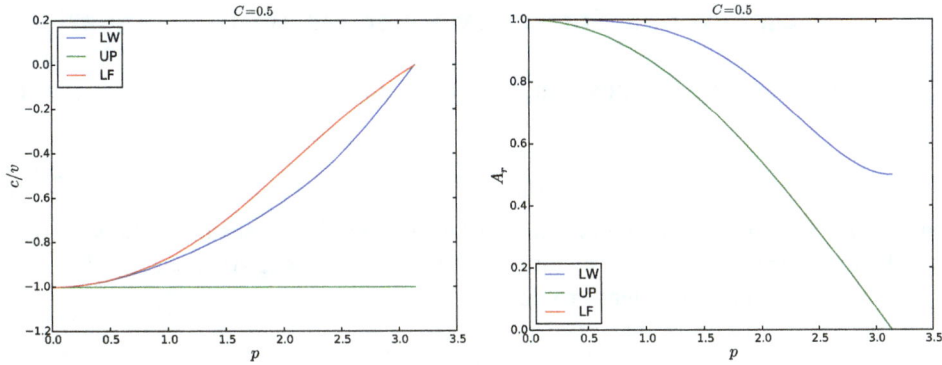

Fig. 2.12 Dispersion relations for $C = 0.5$

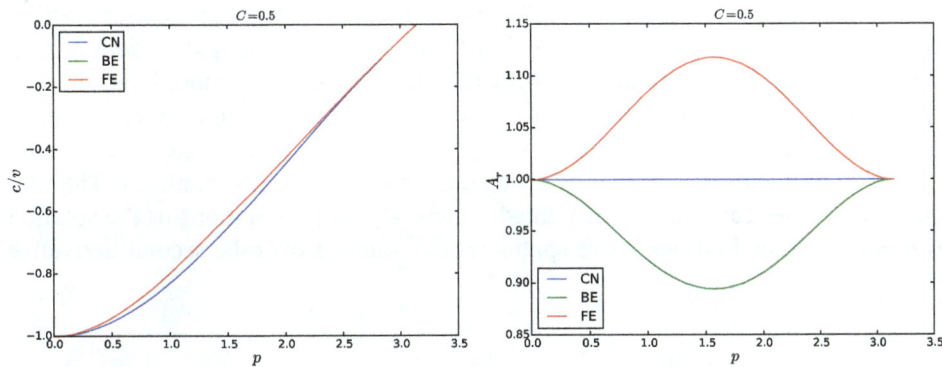

Fig. 2.13 Dispersion relations for $C = 0.5$

(normally) have enough amplitude to be seen, so for all practical purposes, there is no damping or wrong velocity of the individual waves, so the total shape of the wave is also correct. For the CN scheme, see Fig. 2.6, each individual wave has its amplitude, but they move with different velocities, so after a while, we see some of these waves lagging behind. For the BE scheme, see Fig. 2.7, all the shorter waves are so heavily dampened that we cannot see them after a while. We see only the longest waves, which have slightly wrong velocity, but visible amplitudes are sufficiently equal to produce what looks like a smooth profile.

Another feature was that the Leapfrog method produced oscillations, while the upwind scheme did not. Since the Leapfrog method does not dampen the shorter waves, which have wrong wave velocities of order 10 percent, we can see these waves as noise. The upwind scheme, however, dampens these waves. The same effect is also present in the Lax-Wendroff scheme, but the damping of the intermediate waves is hardly present, so there is visible noise in the total signal.

We realize that, compared to pure truncation error analysis, dispersion analysis sheds more light on the behavior of the computational schemes. Truncation analysis just says that Lax-Wendroff is better than upwind, because of the increased order in time, but most people would say upwind is the better one when looking at the plots.

2.2   One-Dimensional Stationary Advection-Diffusion Equation

Now we pay attention to a physical process where advection (or convection) is in balance with diffusion:

$$v\frac{du}{dx} = \alpha\frac{d^2u}{dx^2}\,.$$

(2.11)

For simplicity, we assume v and α to be constant, but the extension to the variable-coefficient case is trivial. This equation can be viewed as the stationary limit of the corresponding time-dependent problem

$$\frac{\partial u}{\partial t} + v\frac{\partial u}{\partial x} = \alpha\frac{\partial^2 u}{\partial x^2}\,.$$

(2.12)

Equations of the form (2.11) or (2.12) arise from transport phenomena, either mass or heat transport. One can also view the equations as a simple model problem for the Navier-Stokes equations. With the chosen boundary conditions, the differential equation problem models the phenomenon of a *boundary layer*, where the solution changes rapidly very close to the boundary. This is a characteristic of many fluid flow problems, which makes strong demands to numerical methods. The fundamental numerical difficulty is related to non-physical oscillations of the solution (instability) if the first-derivative spatial term dominates over the second-derivative term.

2.2.1   A Simple Model Problem

We consider (2.11) on $[0, L]$ equipped with the boundary conditions $u(0) = U_0$, $u(L) = U_L$. By scaling we can reduce the number of parameters in the problem. We scale x by $\bar{x} = x/L$, and u by

$$\bar{u} = \frac{u - U_0}{U_L - U_0}\,.$$

Inserted in the governing equation we get

$$\frac{v(U_L - U_0)}{L}\frac{d\bar{u}}{d\bar{x}} = \frac{\alpha(U_L - U_0)}{L^2}\frac{d^2\bar{u}}{d\bar{x}^2}, \quad \bar{u}(0) = 0,\ \bar{u}(1) = 1\,.$$

Dropping the bars is common. We can then simplify to

$$\frac{du}{dx} = \epsilon\frac{d^2u}{dx^2}, \quad u(0) = 0,\ u(1) = 1\,.$$

(2.13)

There are two competing effects in this equation: the advection term transports signals to the right, while the diffusion term transports signals to the left and the right. The value $u(0) = 0$ is transported through the domain if ϵ is small, and $u \approx 0$ except in the vicinity of $x = 1$, where $u(1) = 1$ and the diffusion transports some information about $u(1) = 1$ to the left. For large ϵ, diffusion dominates and the u takes on the "average" value, i.e., u gets a linear variation from 0 to 1 throughout the domain.

It turns out that we can find an exact solution to the differential equation problem and also to many of its discretizations. This is one reason why this model problem has been so successful in designing and investigating numerical methods for mixed convection/advection and diffusion. The exact solution reads

$$u_e(x) = \frac{e^{x/\epsilon} - 1}{e^{1/\epsilon} - 1}.$$

The forthcoming plots illustrate this function for various values of ϵ.

2.2.2 A Centered Finite Difference Scheme

The most obvious idea to solve (2.13) is to apply centered differences:

$$[D_{2x}u = \epsilon D_x D_x u]_i$$

for $i = 1, \ldots, N_x - 1$, with $u_0 = 0$ and $u_{N_x} = 1$. Note that this is a coupled system of algebraic equations involving $u_0, \ldots, u_{N_x}$.

Written out, the scheme becomes a tridiagonal system

$$A_{i-1,i}u_{i-1} + A_{i,i}u_i + A_{i+1,i}u_{i+1} = 0,$$

for $i = 1, \ldots, N_x - 1$

$$A_{0,0} = 1,$$

$$A_{i-1,i} = -\frac{1}{\Delta x} - \epsilon\frac{1}{\Delta x^2},$$

$$A_{i,i} = 2\epsilon\frac{1}{\Delta x^2},$$

$$A_{i,i+1} = \frac{1}{\Delta x} - \epsilon\frac{1}{\Delta x^2},$$

$$A_{N_x,N_x} = 1.$$

The right-hand side of the linear system is zero except $b_{N_x} = 1$.

Figure 2.14 shows reasonably accurate results with $N_x = 20$ and $N_x = 40$ cells in x direction and a value of $\epsilon = 0.1$. Decreasing ϵ to 0.01 leads to oscillatory solutions as depicted in Fig. 2.15. This is, unfortunately, a typical phenomenon in this type of problem: non-physical oscillations arise for small ϵ unless the resolution N_x is big enough. Exercise 2.1 develops a precise criterion: u is oscillation-free if

$$\Delta x \leq \frac{2}{\epsilon}.$$

If we take the present model as a simplified model for a *viscous boundary layer* in real, industrial fluid flow applications, $\epsilon \sim 10^{-6}$ and millions of cells are required to resolve the boundary layer. Fortunately, this is not strictly necessary as we have methods in the next section to overcome the problem!

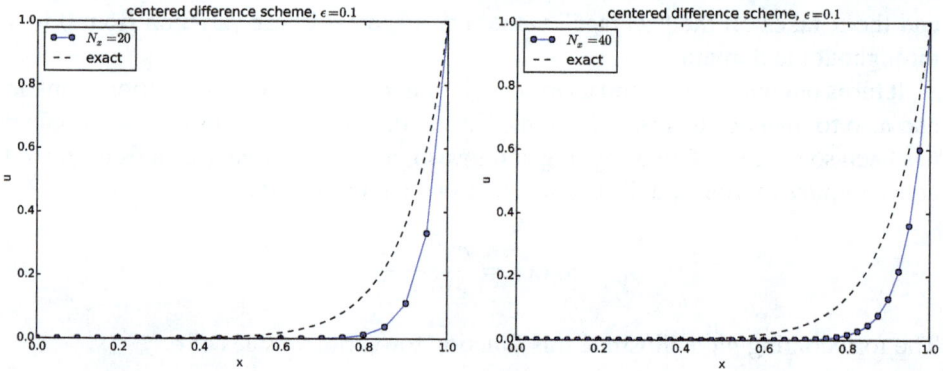

Fig. 2.14 Comparison of exact and numerical solution for $\epsilon = 0.1$ and $N_x = 20, 40$ with centered differences

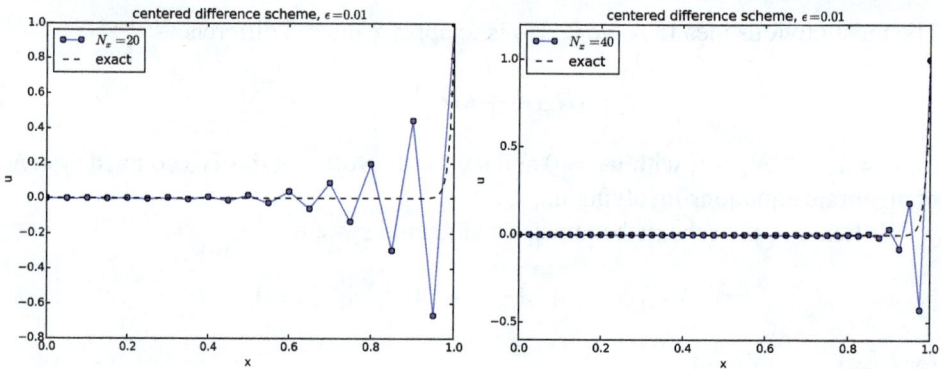

Fig. 2.15 Comparison of exact and numerical solution for $\epsilon = 0.01$ and $N_x = 20, 40$ with centered differences

Solver

A suitable solver for doing the experiments is presented below.

```
import numpy as np

def solver(eps, Nx, method='centered'):
    """
    Solver for the two point boundary value problem u'=eps*u'',
    u(0)=0, u(1)=1.
    """
    x = np.linspace(0, 1, Nx+1)        # Mesh points in space
    # Make sure dx and dt are compatible with x and t
    dx = x[1] - x[0]
    u   = np.zeros(Nx+1)

    # Representation of sparse matrix and right-hand side
    diagonal = np.zeros(Nx+1)
    lower    = np.zeros(Nx)
    upper    = np.zeros(Nx)
    b        = np.zeros(Nx+1)
```

```
# Precompute sparse matrix (scipy format)
if method == 'centered':
    diagonal[:] = 2*eps/dx**2
    lower[:] = -1/dx - eps/dx**2
    upper[:] =  1/dx - eps/dx**2
elif method == 'upwind':
    diagonal[:] = 1/dx + 2*eps/dx**2
    lower[:] =  1/dx - eps/dx**2
    upper[:] = - eps/dx**2

# Insert boundary conditions
upper[0] = 0
lower[-1] = 0
diagonal[0] = diagonal[-1] = 1
b[-1] = 1.0

# Set up sparse matrix and solve
diags = [0, -1, 1]
import scipy.sparse
import scipy.sparse.linalg
A = scipy.sparse.diags(
    diagonals=[diagonal, lower, upper],
    offsets=[0, -1, 1], shape=(Nx+1, Nx+1),
    format='csr')
u[:] = scipy.sparse.linalg.spsolve(A, b)
return u, x
```

2.2.3 Remedy: Upwind Finite Difference Scheme

The scheme can be stabilized by letting the advective transport term, which is the dominating term, collect its information in the flow direction, i.e., upstream or upwind of the point in question. So, instead of using a centered difference

$$\frac{du}{dx}_i \approx \frac{u_{i+1} - u_{i-1}}{2\Delta x},$$

we use the one-sided *upwind* difference

$$\frac{du}{dx}_i \approx \frac{u_i - u_{i-1}}{\Delta x},$$

in case $v > 0$. For $v < 0$ we set

$$\frac{du}{dx}_i \approx \frac{u_{i+1} - u_i}{\Delta x},$$

On compact operator notation form, our upwind scheme can be expressed as

$$[D_x^- u = \epsilon D_x D_x u]_i$$

provided $v > 0$ (and $\epsilon > 0$).

We write out the equations and implement them as shown in the program in Sect. 2.2.2. The results appear in Fig. 2.16 and 2.17: no more oscillations!

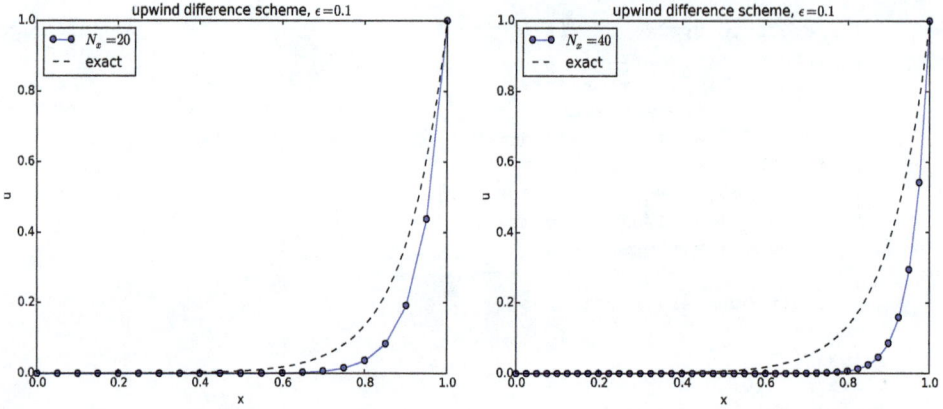

Fig. 2.16 Comparison of exact and numerical solution for $\epsilon = 0.1$ and $N_x = 20, 40$ with upwind difference

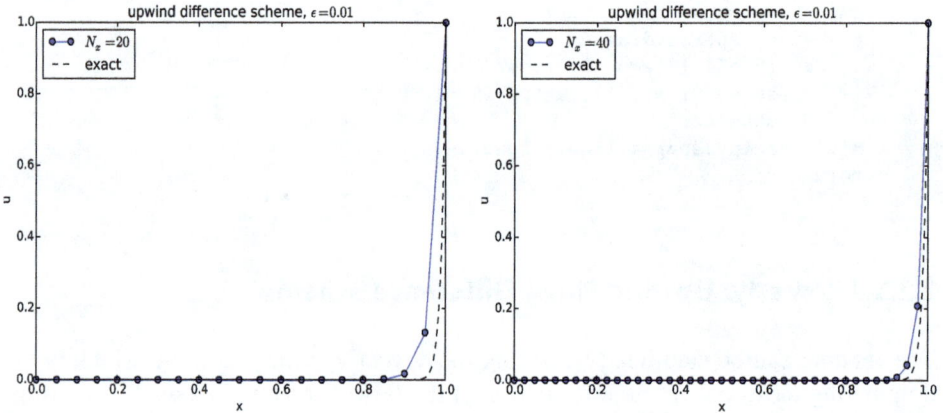

Fig. 2.17 Comparison of exact and numerical solution for $\epsilon = 0.01$ and $N_x = 20, 40$ with upwind difference

We see that the upwind scheme is always stable, but it gives a thicker boundary layer when the centered scheme is also stable. Why the upwind scheme is always stable is easy to understand as soon as we undertake the mathematical analysis in Exercise 2.1. Moreover, the thicker layer (seemingly larger diffusion) can be understood by doing Exercise 2.2.

Exact solution for this model problem

It turns out that one can introduce a linear combination of the centered and upwind differences for the first-derivative term in this model problem. One can then adjust the weight in the linear combination so that the numerical solution becomes identical to the analytical solution of the differential equation problem at any mesh point.

2.3   Time-dependent Convection-Diffusion Equations

Now it is time to combine time-dependency, convection (advection) and diffusion into one equation:

$$\frac{\partial u}{\partial t} + v\frac{\partial u}{\partial x} = \alpha\frac{\partial^2 u}{\partial x^2}. \tag{2.14}$$

Analytical insight   The diffusion is now dominated by convection, a wave, and diffusion, a loss of amplitude. One possible analytical solution is a traveling Gaussian function

$$u(x,t) = B\exp\left(-\left(\frac{x-vt}{4at}\right)\right).$$

This function moves with velocity $v > 0$ to the right ($v < 0$ to the left) due to convection, but at the same time we have a damping $e^{-16a^2t^2}$ from diffusion.

2.3.1   Forward in Time, Centered in Space Scheme

The Forward Euler for the diffusion equation is a successful scheme, but it has a very strict stability condition. The similar Forward in time, centered in space strategy always gives unstable solutions for the advection PDE. What happens when we have both diffusion and advection present at once?

$$[D_t u + vD_{2x}u = \alpha D_x D_x u + f]_i^n.$$

We expect that diffusion will stabilize the scheme, but that advection will destabilize it.

Another problem is non-physical oscillations, but not growing amplitudes, due to centered differences in the advection term. There will hence be two types of instabilities to consider. Our analysis showed that pure advection with centered differences in space needs some artificial diffusion to become stable (and then it produces upwind differences for the advection term). Adding more physical diffusion should further help the numerics to stabilize the non-physical oscillations.

The scheme is quickly implemented, but suffers from the need for small space and time steps, according to this reasoning. A better approach is to get rid of the non-physical oscillations in space by simply applying an upwind difference on the advection term.

2.3.2   Forward in Time, Upwind in Space Scheme

A good approximation for the pure advection equation is to use upwind discretization of the advection term. We also know that centered differences are good for the diffusion term, so let us combine these two discretizations:

$$[D_t u + vD_x^- u = \alpha D_x D_x u + f]_i^n, \tag{2.15}$$

for $v > 0$. Use vD^+u if $v < 0$. In this case the physical diffusion and the extra numerical diffusion $v\Delta x/2$ will stabilize the solution, but give an overall too large reduction in amplitude compared with the exact solution.

We may also interpret the upwind difference as artificial numerical diffusion and centered differences in space everywhere, so the scheme can be expressed as

$$\left[D_t u + v D_{2x}^- u = \alpha \frac{v\Delta x}{2} D_x D_x u + f\right]_i^n .\tag{2.16}$$

2.4 Applications of Advection Equations

There are two major areas where advection and convection applications arise: transport of a substance and heat transport *in a fluid*. To derive the models, we may look at the similar derivations of diffusion models in Sect. 3.8, but change the assumption from a solid to fluid medium. This gives rise to the extra advection or convection term $v \cdot \nabla u$. We briefly show how this is done.

Normally, transport in a fluid is dominated by the fluid flow and not diffusion, so we can neglect diffusion compared to advection or convection. The end result is anyway an equation of the form

$$\frac{\partial u}{\partial t} + v \cdot \nabla u = 0 .$$

2.4.1 Transport of a Substance

The diffusion of a substance in Sect. 1.8.1 takes place in a solid medium, but in a fluid we can have two transport mechanisms: one by diffusion and one by advection. The latter arises from the fact that the substance particles are moved with the fluid velocity v such that the effective flux now consists of two and not only one component as in (1.121):

$$q = -\alpha\nabla c + vc .$$

Inserted in the equation $\nabla \cdot q = 0$ we get the extra advection term $\nabla \cdot (vc)$. Very often we deal with incompressible flows, $\nabla \cdot v = 0$ such that the advective term becomes $v \cdot \nabla c$. The mass transport equation for a substance then reads

$$\frac{\partial c}{\partial t} + v \cdot \nabla c = \alpha\nabla^2 c .\tag{2.17}$$

2.4.2 Transport of Heat in Fluids

The derivation of the heat equation in Sect. 3.8.2 is limited to heat transport in solid bodies. If we turn the attention to heat transport in fluids, we get a material derivative of the internal energy in (3.123),

$$\frac{De}{dt} = -\nabla \cdot q,$$

and more terms if work by stresses is also included, where

$$\frac{De}{dt} = \frac{\partial e}{\partial t} + v \cdot \nabla e,$$

v being the velocity of the fluid. The convective term $v \cdot \nabla e$ must therefore be added to the governing equation, resulting typically in

$$\varrho c \left(\frac{\partial T}{\partial t} + v \cdot \nabla T \right) = \nabla \cdot (k \nabla T) + f, \tag{2.18}$$

where f is some external heating inside the medium.

2.5 Exercises

Exercise 4.1: Analyze 1D stationary convection-diffusion problem
Explain the observations in the numerical experiments from Sect. 4.2.2 and 4.2.3 by finding exact numerical solutions.

Hint The difference equations allow solutions on the form A^i, where A is an unknown constant and i is a mesh point counter. There are two solutions for A, so the general solution is a linear combination of the two, where the constants in the linear combination are determined from the boundary conditions.
Filename: `twopt_BVP_analysis1`.

Exercise 2.2: Interpret upwind difference as artificial diffusion
Consider an upwind, one-sided difference approximation to a term du/dx in a differential equation. Show that this formula can be expressed as a centered difference plus an artificial diffusion term of strength proportional to Δx. This means that introducing an upwind difference also means introducing extra diffusion of order $\mathcal{O}(\Delta x)$.
Filename: `twopt_BVP_analysis2`.

3

Nonlinear Differential Equations

3.1 Introduction of Basic Concepts

3.1.1 Linear Versus Nonlinear Equations

Algebraic equations A linear, scalar, algebraic equation in x has the form

$$ax + b = 0,$$

for arbitrary real constants a and b. The unknown is a number x. All other algebraic equations, e.g., $x^2 + ax + b = 0$, are nonlinear. The typical feature in a nonlinear algebraic equation is that the unknown appears in products with itself, like x^2 or $e^x = 1 + x + \frac{1}{2}x^2 + \frac{1}{3!}x^3 + \ldots$

We know how to solve a linear algebraic equation, $x = -b/a$, but there are no general methods for finding the exact solutions of nonlinear algebraic equations, except for very special cases (quadratic equations constitute a primary example). A nonlinear algebraic equation may have no solution, one solution, or many solutions. The tools for solving nonlinear algebraic equations are *iterative methods*, where we construct a series of linear equations, which we know how to solve, and hope that the solutions of the linear equations converge to a solution of the nonlinear equation we want to solve. Typical methods for nonlinear algebraic equation equations are Newton's method, the Bisection method, and the Secant method.

Differential equations The unknown in a differential equation is a function and not a number. In a linear differential equation, all terms involving the unknown function are linear in the unknown function or its derivatives. Linear here means that the unknown function, or a derivative of it, is multiplied by a number or a known function. All other differential equations are non-linear.

The easiest way to see if an equation is nonlinear, is to spot nonlinear terms where the unknown function or its derivatives are multiplied by each other. For example, in

$$u'(t) = -a(t)u(t) + b(t),$$

the terms involving the unknown function u are linear: u' contains the derivative of the unknown function multiplied by unity, and au contains the unknown function multiplied by a known function. However,

$$u'(t) = u(t)(1 - u(t)),$$

is nonlinear because of the term $-u^2$ where the unknown function is multiplied by itself. Also

$$\frac{\partial u}{\partial t} + u \frac{\partial u}{\partial x} = 0,$$

is nonlinear because of the term $u u_x$ where the unknown function appears in a product with its derivative. (Note here that we use different notations for derivatives: u' or du/dt for a function $u(t)$ of one variable, $\frac{\partial u}{\partial t}$ or u_t for a function of more than one variable.)

Another example of a nonlinear equation is

$$u'' + \sin(u) = 0,$$

because $\sin(u)$ contains products of u, which becomes clear if we expand the function in a Taylor series:

$$\sin(u) = u - \frac{1}{3} u^3 + \ldots$$

Mathematical proof of linearity

To really prove mathematically that some differential equation in an unknown u is linear, show for each term $T(u)$ that with $u = au_1 + bu_2$ for constants a and b,

$$T(au_1 + bu_2) = aT(u_1) + bT(u_2).$$

For example, the term $T(u) = (\sin^2 t)u'(t)$ is linear because

$$
\begin{aligned}
T(au_1 + bu_2) &= (\sin^2 t)(au_1(t) + bu_2(t)) \\
&= a(\sin^2 t)u_1(t) + b(\sin^2 t)u_2(t) \\
&= aT(u_1) + bT(u_2).
\end{aligned}
$$

However, $T(u) = \sin u$ is nonlinear because

$$T(au_1 + bu_2) = \sin(au_1 + bu_2) \neq a \sin u_1 + b \sin u_2.$$

3.1.2   A Simple Model Problem

A series of forthcoming examples will explain how to tackle nonlinear differential equations with various techniques. We start with the (scaled) logistic equation as model problem:

$$u'(t) = u(t)(1 - u(t)). \tag{3.1}$$

This is a nonlinear ordinary differential equation (ODE) which will be solved by different strategies in the following. Depending on the chosen time discretization of (5.1), the mathematical problem to be solved at every time level will either be a linear algebraic equation or a nonlinear algebraic equation. In the former case, the time discretization method transforms the nonlinear ODE into linear subproblems at each time level, and the solution is straightforward to find since linear algebraic equations are easy to solve. However, when the time discretization leads to nonlinear algebraic equations, we cannot (except in very rare cases) solve these without turning to approximate, iterative solution methods.

The next subsections introduce various methods for solving nonlinear differential equations, using (5.1) as model. We shall go through the following set of cases:

- explicit time discretization methods (with no need to solve nonlinear algebraic equations)
- implicit Backward Euler time discretization, leading to nonlinear algebraic equations solved by
  - an exact analytical technique
  - Picard iteration based on manual linearization
  - a single Picard step
  - Newton's method
- implicit Crank-Nicolson time discretization and linearization via a geometric mean formula

Thereafter, we compare the performance of the various approaches. Despite the simplicity of (3.1), the conclusions reveal typical features of the various methods in much more complicated nonlinear PDE problems.

3.1.3   Linearization by Explicit Time Discretization

Time discretization methods are divided into explicit and implicit methods. Explicit methods lead to a closed-form formula for finding new values of the unknowns, while implicit methods give a linear or nonlinear system of equations that couples (all) the unknowns at a new time level. Here we shall demonstrate that explicit methods constitute an efficient way to deal with nonlinear differential equations.

The Forward Euler method is an explicit method. When applied to (3.1), sampled at $t = t_n$, it results in

$$\frac{u^{n+1} - u^n}{\Delta t} = u^n (1 - u^n),$$

which is a *linear* algebraic equation for the unknown value u^{n+1} that we can easily solve:

$$u^{n+1} = u^n + \Delta t \, u^n (1 - u^n).$$

In this case, the nonlinearity in the original equation poses no difficulty in the discrete algebraic equation. Any other explicit scheme in time will also give only linear algebraic equations to solve. For example, a typical 2nd-order Runge-Kutta method for (3.1) leads to the following formulas:

$$u^* = u^n + \Delta t u^n (1 - u^n),$$

$$u^{n+1} = u^n + \Delta t \frac{1}{2} \left(u^n (1 - u^n) + u^* (1 - u^*) \right).$$

The first step is linear in the unknown u^*. Then u^* is known in the next step, which is linear in the unknown u^{n+1}.

3.1.4   Exact Solution of Nonlinear Algebraic Equations

Switching to a Backward Euler scheme for (3.1),

$$\frac{u^n - u^{n-1}}{\Delta t} = u^n(1 - u^n),\tag{3.2}$$

results in a nonlinear algebraic equation for the unknown value u^n. The equation is of quadratic type:

$$\Delta t(u^n)^2 + (1 - \Delta t)u^n - u^{n-1} = 0,$$

and may be solved exactly by the well-known formula for such equations. Before we do so, however, we will introduce a shorter, and often cleaner, notation for nonlinear algebraic equations at a given time level. The notation is inspired by the natural notation (i.e., variable names) used in a program, especially in more advanced partial differential equation problems. The unknown in the algebraic equation is denoted by u, while $u^{(1)}$ is the value of the unknown at the previous time level (in general, $u^{(\ell)}$ is the value of the unknown ℓ levels back in time). The notation will be frequently used in later sections. What is meant by u should be evident from the context: u may either be 1) the exact solution of the ODE/PDE problem, 2) the numerical approximation to the exact solution, or 3) the unknown solution at a certain time level.

The quadratic equation for the unknown u^n in (3.2) can, with the new notation, be written

$$F(u) = \Delta t u^2 + (1 - \Delta t)u - u^{(1)} = 0.\tag{3.3}$$

The solution is readily found to be

$$u = \frac{1}{2\Delta t}\left(-1 + \Delta t \pm \sqrt{(1 - \Delta t)^2 - 4\Delta t u^{(1)}}\right).\tag{3.4}$$

Now we encounter a fundamental challenge with nonlinear algebraic equations: the equation may have more than one solution. How do we pick the right solution? This is in general a hard problem. In the present simple case, however, we can analyze the roots mathematically and provide an answer. The idea is to expand the roots in a series in Δt and truncate after the linear term since the Backward Euler scheme will introduce an error proportional to Δt anyway. Using sympy, we find the following Taylor series expansions of the roots:

```
>>> import sympy as sym
>>> dt, u_1, u = sym.symbols('dt u_1 u')
>>> r1, r2 = sym.solve(dt*u**2 + (1-dt)*u - u_1, u)  # find roots
>>> r1
(dt - sqrt(dt**2 + 4*dt*u_1 - 2*dt + 1) - 1)/(2*dt)
>>> r2
(dt + sqrt(dt**2 + 4*dt*u_1 - 2*dt + 1) - 1)/(2*dt)
>>> print r1.series(dt, 0, 2)    # 2 terms in dt, around dt=0
-1/dt + 1 - u_1 + dt*(u_1**2 - u_1) + O(dt**2)
>>> print r2.series(dt, 0, 2)
u_1 + dt*(-u_1**2 + u_1) + O(dt**2)
```

We see that the r1 root, corresponding to a minus sign in front of the square root in (5.4), behaves as $1/\Delta t$ and will therefore blow up as $\Delta t \to 0$! Since we know that u takes on finite values, actually it is less than or equal to 1, only the r2 root is of relevance in this case: as $\Delta t \to 0$, $u \to u^{(1)}$, which is the expected result.

For those who are not well experienced with approximating mathematical formulas by series expansion, an alternative method of investigation is simply to compute the limits of the two roots as $\Delta t \to 0$ and see if a limit appears unreasonable:

```
>>> print r1.limit(dt, 0)
-oo
>>> print r2.limit(dt, 0)
u_1
```

3.1.5 Linearization

When the time integration of an ODE results in a nonlinear algebraic equation, we must normally find its solution by defining a sequence of linear equations and hope that the solutions of these linear equations converge to the desired solution of the nonlinear algebraic equation. Usually, this means solving the linear equation repeatedly in an iterative fashion. Alternatively, the nonlinear equation can sometimes be approximated by one linear equation, and consequently there is no need for iteration.

Constructing a linear equation from a nonlinear one requires *linearization* of each nonlinear term. This can be done manually as in Picard iteration, or fully algorithmically as in Newton's method. Examples will best illustrate how to linearize nonlinear problems.

3.1.6 Picard Iteration

Let us write (3.3) in a more compact form

$$F(u) = au^2 + bu + c = 0,$$

with $a = \Delta t$, $b = 1 - \Delta t$, and $c = -u^{(1)}$. Let u^- be an available approximation of the unknown u. Then we can linearize the term u^2 simply by writing $u^- u$. The resulting equation, $\hat{F}(u) = 0$, is now linear and hence easy to solve:

$$F(u) \approx \hat{F}(u) = au^- u + bu + c = 0.$$

Since the equation $\hat{F} = 0$ is only approximate, the solution u does not equal the exact solution u_e of the exact equation $F(u_e) = 0$, but we can hope that u is closer to u_e than u^- is, and hence it makes sense to repeat the procedure, i.e., set $u^- = u$ and solve $\hat{F}(u) = 0$ again. There is no guarantee that u is closer to u_e than u^-, but this approach has proven to be effective in a wide range of applications.

The idea of turning a nonlinear equation into a linear one by using an approximation u^- of u in nonlinear terms is a widely used approach that goes under many names: *fixed-point iteration*, the method of *successive substitutions*, *non-linear Richardson iteration*, and *Picard iteration*. We will stick to the latter name.

Picard iteration for solving the nonlinear equation arising from the Backward Euler discretization of the logistic equation can be written as

$$u = -\frac{c}{au^- + b}, \quad u^- \leftarrow u.$$

The $\leftarrow$ symbols means assignment (we set u^- equal to the value of u). The iteration is started with the value of the unknown at the previous time level: $u^- = u^{(1)}$.

Some prefer an explicit iteration counter as superscript in the mathematical notation. Let u^k be the computed approximation to the solution in iteration k. In iteration $k + 1$ we want to solve

$$au^k u^{k+1} + bu^{k+1} + c = 0 \quad \Rightarrow \quad u^{k+1} = -\frac{c}{au^k + b}, \quad k = 0, 1, \ldots$$

Since we need to perform the iteration at every time level, the time level counter is often also included:

$$au^{n,k} u^{n,k+1} + bu^{n,k+1} - u^{n-1} = 0 \quad \Rightarrow \quad u^{n,k+1} = \frac{u^n}{au^{n,k} + b}, \quad k = 0, 1, \ldots,$$

with the start value $u^{n,0} = u^{n-1}$ and the final converged value $u^n = u^{n,k}$ for sufficiently large k.

However, we will normally apply a mathematical notation in our final formulas that is as close as possible to what we aim to write in a computer code and then it becomes natural to use u and u^- instead of u^{k+1} and u^k or $u^{n,k+1}$ and $u^{n,k}$.

Stopping criteria The iteration method can typically be terminated when the change in the solution is smaller than a tolerance ϵ_u:

$$|u - u^-| \leq \epsilon_u,$$

or when the residual in the equation is sufficiently small ($< \epsilon_r$),

$$|F(u)| = |au^2 + bu + c| < \epsilon_r.$$

A single Picard iteration Instead of iterating until a stopping criterion is fulfilled, one may iterate a specific number of times. Just one Picard iteration is popular as this corresponds to the intuitive idea of approximating a nonlinear term like $(u^n)^2$ by $u^{n-1}u^n$. This follows from the linearization u^-u^n and the initial choice of $u^- = u^{n-1}$ at time level t_n. In other words, a single Picard iteration corresponds to using the solution at the previous time level to linearize nonlinear terms. The resulting discretization becomes (using proper values for a, b, and c)

$$\frac{u^n - u^{n-1}}{\Delta t} = u^n(1 - u^{n-1}), \tag{3.5}$$

which is a linear algebraic equation in the unknown u^n, making it easy to solve for u^n without any need for an alternative notation.

We shall later refer to the strategy of taking one Picard step, or equivalently, linearizing terms with use of the solution at the previous time step, as the *Picard1* method. It is a widely used approach in science and technology, but with some limitations if Δt is not sufficiently small (as will be illustrated later).

Notice

Equation (3.5) does not correspond to a "pure" finite difference method where the equation is sampled at a point and derivatives replaced by differences (because the u^{n-1} term on the right-hand side must then be u^n). The best interpretation of the scheme (3.5) is a Backward Euler difference combined with a single (perhaps insufficient) Picard iteration at each time level, with the value at the previous time level as start for the Picard iteration.

3.1.7 Linearization by a Geometric Mean

We consider now a Crank-Nicolson discretization of (3.1). This means that the time derivative is approximated by a centered difference,

$$[D_t u = u(1-u)]^{n+\frac{1}{2}},$$

written out as

$$\frac{u^{n+1} - u^n}{\Delta t} = u^{n+\frac{1}{2}} - (u^{n+\frac{1}{2}})^2 . \tag{3.6}$$

The term $u^{n+\frac{1}{2}}$ is normally approximated by an arithmetic mean,

$$u^{n+\frac{1}{2}} \approx \frac{1}{2}(u^n + u^{n+1}),$$

such that the scheme involves the unknown function only at the time levels where we actually intend to compute it. The same arithmetic mean applied to the nonlinear term gives

$$(u^{n+\frac{1}{2}})^2 \approx \frac{1}{4}(u^n + u^{n+1})^2,$$

which is nonlinear in the unknown u^{n+1}. However, using a *geometric mean* for $(u^{n+\frac{1}{2}})^2$ is a way of linearizing the nonlinear term in (3.6):

$$(u^{n+\frac{1}{2}})^2 \approx u^n u^{n+1} .$$

Using an arithmetic mean on the linear $u^{n+\frac{1}{2}}$ term in (3.6) and a geometric mean for the second term, results in a linearized equation for the unknown u^{n+1}:

$$\frac{u^{n+1} - u^n}{\Delta t} = \frac{1}{2}(u^n + u^{n+1}) + u^n u^{n+1},$$

which can readily be solved:

$$u^{n+1} = \frac{1 + \frac{1}{2}\Delta t}{1 + \Delta t u^n - \frac{1}{2}\Delta t} u^n .$$

This scheme can be coded directly, and since there is no nonlinear algebraic equation to iterate over, we skip the simplified notation with u for u^{n+1} and $u^{(1)}$ for u^n. The technique with using a geometric average is an example of transforming a nonlinear algebraic equation to a linear one, without any need for iterations.

The geometric mean approximation is often very effective for linearizing quadratic nonlinearities. Both the arithmetic and geometric mean approximations have truncation errors of order Δt^2 and are therefore compatible with the truncation error $\mathcal{O}(\Delta t^2)$ of the centered difference approximation for u' in the Crank-Nicolson method.

Applying the operator notation for the means and finite differences, the linearized Crank-Nicolson scheme for the logistic equation can be compactly expressed as

$$\left[D_t u = \overline{u}^t + \overline{u^2}^{t,g} \right]^{n+\frac{1}{2}} .$$

Remark

If we use an arithmetic instead of a geometric mean for the nonlinear term in (3.6), we end up with a nonlinear term $(u^{n+1})^2$. This term can be linearized as $u^- u^{n+1}$ in a Picard iteration approach and in particular as $u^n u^{n+1}$ in a Picard1 iteration approach. The latter gives a scheme almost identical to the one arising from a geometric mean (the difference in u^{n+1} being $\frac{1}{4}\Delta t u^n (u^{n+1} - u^n) \approx \frac{1}{4}\Delta t^2 u' u$, i.e., a difference of size Δt^2).

3.1.8 Newton's Method

The Backward Euler scheme (3.2) for the logistic equation leads to a nonlinear algebraic equation (3.3). Now we write any nonlinear algebraic equation in the general and compact form

$$F(u) = 0 .$$

Newton's method linearizes this equation by approximating $F(u)$ by its Taylor series expansion around a computed value u^- and keeping only the linear part:

$$F(u) = F(u^-) + F'(u^-)(u - u^-) + \frac{1}{2}F''(u^-)(u - u^-)^2 + \cdots$$
$$\approx F(u^-) + F'(u^-)(u - u^-) = \hat{F}(u) .$$

The linear equation $\hat{F}(u) = 0$ has the solution

$$u = u^- - \frac{F(u^-)}{F'(u^-)} .$$

Expressed with an iteration index in the unknown, Newton's method takes on the more familiar mathematical form

$$u^{k+1} = u^k - \frac{F(u^k)}{F'(u^k)}, \quad k = 0, 1, \ldots$$

It can be shown that the error in iteration $k + 1$ of Newton's method is proportional to the square of the error in iteration k, a result referred to as *quadratic convergence*. This means that for small errors the method converges very fast, and in particular much faster than Picard iteration and other iteration methods. (The proof of this result is found in most textbooks on numerical analysis.) However, the quadratic convergence appears only if u^k is sufficiently close to the solution. Further away from the solution the method can easily converge very slowly or diverge. The reader is encouraged to do Exercise 3.3 to get a better understanding for the behavior of the method.

Application of Newton's method to the logistic equation discretized by the Backward Euler method is straightforward as we have

$$F(u) = au^2 + bu + c, \quad a = \Delta t, \ b = 1 - \Delta t, \ c = -u^{(1)},$$

and then

$$F'(u) = 2au + b.$$

The iteration method becomes

$$u = u^- + \frac{a(u^-)^2 + bu^- + c}{2au^- + b}, \quad u^- \leftarrow u. \tag{3.7}$$

At each time level, we start the iteration by setting $u^- = u^{(1)}$. Stopping criteria as listed for the Picard iteration can be used also for Newton's method.

An alternative mathematical form, where we write out a, b, and c, and use a time level counter n and an iteration counter k, takes the form

$$u^{n,k+1} = u^{n,k} + \frac{\Delta t(u^{n,k})^2 + (1 - \Delta t)u^{n,k} - u^{n-1}}{2\Delta t u^{n,k} + 1 - \Delta t}, \quad u^{n,0} = u^{n-1}, \tag{3.8}$$

for $k = 0, 1, \dots$. A program implementation is much closer to (3.7) than to (3.8), but the latter is better aligned with the established mathematical notation used in the literature.

3.1.9 Relaxation

One iteration in Newton's method or Picard iteration consists of solving a linear problem $\hat{F}(u) = 0$. Sometimes convergence problems arise because the new solution u of $\hat{F}(u) = 0$ is "too far away" from the previously computed solution u^-. A remedy is to introduce a relaxation, meaning that we first solve $\hat{F}(u^*) = 0$ for a suggested value u^* and then we take u as a weighted mean of what we had, u^-, and what our linearized equation $\hat{F} = 0$ suggests, u^*:

$$u = \omega u^* + (1 - \omega)u^-.$$

The parameter ω is known as a *relaxation parameter*, and a choice $\omega < 1$ may prevent divergent iterations.

Relaxation in Newton's method can be directly incorporated in the basic iteration formula:

$$u = u^- - \omega \frac{F(u^-)}{F'(u^-)}. \tag{3.9}$$

3.1.10   Implementation and Experiments

The program `logistic.py` contains implementations of all the methods described above. Below is an extract of the file showing how the Picard and Newton methods are implemented for a Backward Euler discretization of the logistic equation.

```
def BE_logistic(u0, dt, Nt, choice='Picard',
                eps_r=1E-3, omega=1, max_iter=1000):
    if choice == 'Picard1':
        choice = 'Picard'
        max_iter = 1

    u = np.zeros(Nt+1)
    iterations = []
    u[0] = u0
    for n in range(1, Nt+1):
        a = dt
        b = 1 - dt
        c = -u[n-1]

        if choice == 'Picard':

            def F(u):
                return a*u**2 + b*u + c

            u_ = u[n-1]
            k = 0
            while abs(F(u_)) > eps_r and k < max_iter:
                u_ = omega*(-c/(a*u_ + b)) + (1-omega)*u_
                k += 1
            u[n] = u_
            iterations.append(k)

        elif choice == 'Newton':

            def F(u):
                return a*u**2 + b*u + c

            def dF(u):
                return 2*a*u + b

            u_ = u[n-1]
            k = 0
            while abs(F(u_)) > eps_r and k < max_iter:
                u_ = u_ - F(u_)/dF(u_)
                k += 1
            u[n] = u_
            iterations.append(k)
    return u, iterations
```

The Crank-Nicolson method utilizing a linearization based on the geometric mean gives a simpler algorithm:

```
def CN_logistic(u0, dt, Nt):
    u = np.zeros(Nt+1)
    u[0] = u0
    for n in range(0, Nt):
        u[n+1] = (1 + 0.5*dt)/(1 + dt*u[n] - 0.5*dt)*u[n]
    return u
```

We may run experiments with the model problem (3.1) and the different strategies for dealing with nonlinearities as described above. For a quite coarse time resolution, $\Delta t = 0.9$, use of a tolerance $\epsilon_r = 0.1$ in the stopping criterion introduces an iteration error, especially in the Picard iterations, that is visibly much larger than the time discretization error due to a large Δt. This is illustrated by comparing the upper two plots in Fig. 3.1. The one to the right has a stricter tolerance $\epsilon = 10^{-3}$, which causes all the curves corresponding to Picard and Newton iteration to be on top of each other (and no changes can be visually observed by reducing ϵ_r further). The reason why Newton's method does much better than Picard iteration in the upper left plot is that Newton's method with one step comes far below the ϵ_r tolerance, while the Picard iteration needs on average 7 iterations to bring the residual down to $\epsilon_r = 10^{-1}$, which gives insufficient accuracy in the solution of the nonlinear equation. It is obvious that the Picard1 method gives significant errors in

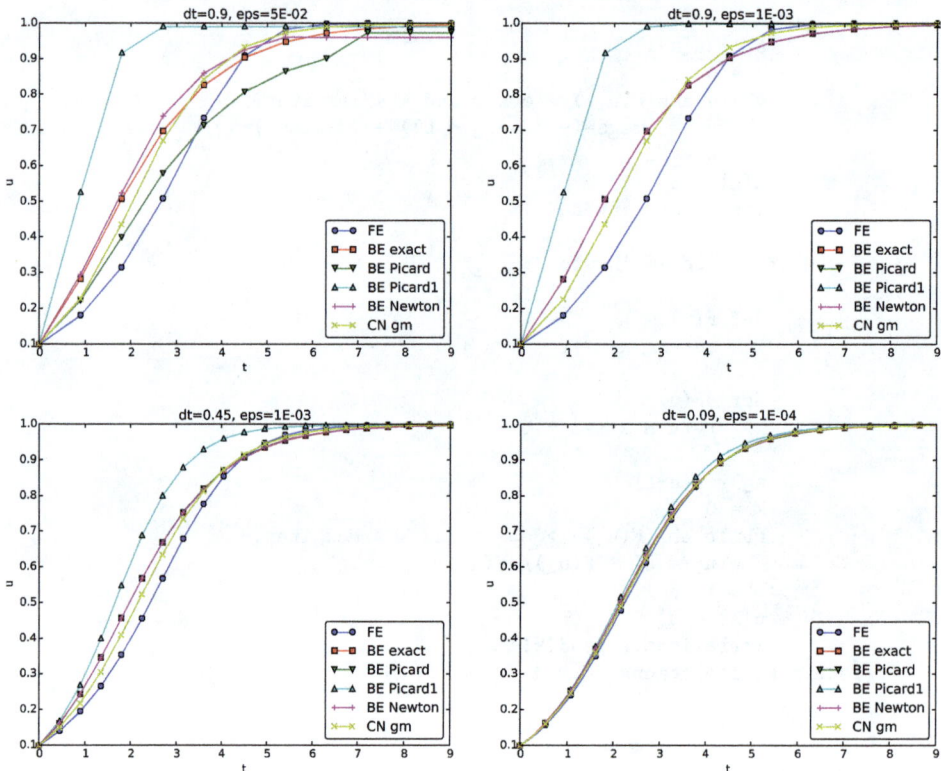

Fig. 3.1 Impact of solution strategy and time step length on the solution

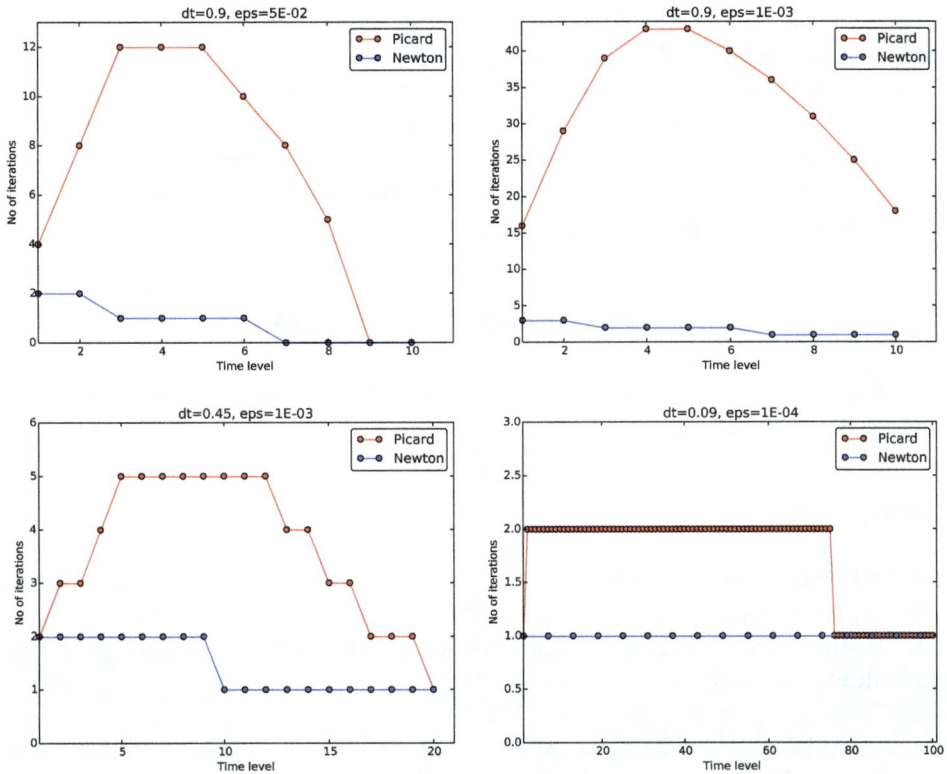

Fig. 3.2 Comparison of the number of iterations at various time levels for Picard and Newton iteration

addition to the time discretization unless the time step is as small as in the lower right plot.

The *BE exact* curve corresponds to using the exact solution of the quadratic equation at each time level, so this curve is only affected by the Backward Euler time discretization. The *CN gm* curve corresponds to the theoretically more accurate Crank-Nicolson discretization, combined with a geometric mean for linearization. This curve appears more accurate, especially if we take the plot in the lower right with a small Δt and an appropriately small ϵ_r value as the exact curve.

When it comes to the need for iterations, Fig. 3.2 displays the number of iterations required at each time level for Newton's method and Picard iteration. The smaller Δt is, the better starting value we have for the iteration, and the faster the convergence is. With $\Delta t = 0.9$ Picard iteration requires on average 32 iterations per time step, but this number is dramatically reduced as Δt is reduced.

However, introducing relaxation and a parameter $\omega = 0.8$ immediately reduces the average of 32 to 7, indicating that for the large $\Delta t = 0.9$, Picard iteration takes too long steps. An approximately optimal value for ω in this case is 0.5, which results in an average of only 2 iterations! An even more dramatic impact of ω appears when $\Delta t = 1$: Picard iteration does not convergence in 1000 iterations, but $\omega = 0.5$ again brings the average number of iterations down to 2.

Remark The simple Crank-Nicolson method with a geometric mean for the quadratic nonlinearity gives visually more accurate solutions than the Backward Euler discretization. Even with a tolerance of $\epsilon_r = 10^{-3}$, all the methods for treating the nonlinearities in the Backward Euler discretization give graphs that cannot be distinguished. So for accuracy in this problem, the time discretization is much more crucial than ϵ_r. Ideally, one should estimate the error in the time discretization, as the solution progresses, and set ϵ_r accordingly.

3.1.11   Generalization to a General Nonlinear ODE

Let us see how the various methods in the previous sections can be applied to the more generic model

$$u' = f(u, t), \tag{3.10}$$

where f is a nonlinear function of u.

Explicit time discretization Explicit ODE methods like the Forward Euler scheme, Runge-Kutta methods and Adams-Bashforth methods all evaluate f at time levels where u is already computed, so nonlinearities in f do not pose any difficulties.

Backward Euler discretization Approximating u' by a backward difference leads to a Backward Euler scheme, which can be written as

$$F(u^n) = u^n - \Delta t \, f(u^n, t_n) - u^{n-1} = 0,$$

or alternatively

$$F(u) = u - \Delta t \, f(u, t_n) - u^{(1)} = 0.$$

A simple Picard iteration, not knowing anything about the nonlinear structure of f, must approximate $f(u, t_n)$ by $f(u^-, t_n)$:

$$\hat{F}(u) = u - \Delta t \, f(u^-, t_n) - u^{(1)}.$$

The iteration starts with $u^- = u^{(1)}$ and proceeds with repeating

$$u^* = \Delta t \, f(u^-, t_n) + u^{(1)}, \quad u = \omega u^* + (1 - \omega)u^-, \quad u^- \leftarrow u,$$

until a stopping criterion is fulfilled.

Explicit vs implicit treatment of nonlinear terms

Evaluating f for a known u^- is referred to as *explicit* treatment of f, while if $f(u, t)$ has some structure, say $f(u, t) = u^3$, parts of f can involve the unknown u, as in the manual linearization $(u^-)^2 u$, and then the treatment of f is "more implicit" and "less explicit". This terminology is inspired by time discretization of $u' = f(u, t)$, where evaluating f for known u values gives explicit schemes, while treating f or parts of f implicitly, makes f contribute to the unknown terms in the equation at the new time level.

Explicit treatment of f usually means stricter conditions on Δt to achieve stability of time discretization schemes. The same applies to iteration techniques for nonlinear algebraic equations: the "less" we linearize f (i.e., the more we keep of u in the original formula), the faster the convergence may be.

We may say that $f(u,t) = u^3$ is treated explicitly if we evaluate f as $(u^-)^3$, partially implicit if we linearize as $(u^-)^2 u$ and fully implicit if we represent f by u^3. (Of course, the fully implicit representation will require further linearization, but with $f(u,t) = u^2$ a fully implicit treatment is possible if the resulting quadratic equation is solved with a formula.)

For the ODE $u' = -u^3$ with $f(u,t) = -u^3$ and coarse time resolution $\Delta t = 0.4$, Picard iteration with $(u^-)^2 u$ requires 8 iterations with $\epsilon_r = 10^{-3}$ for the first time step, while $(u^-)^3$ leads to 22 iterations. After about 10 time steps both approaches are down to about 2 iterations per time step, but this example shows a potential of treating f more implicitly.

A trick to treat f implicitly in Picard iteration is to evaluate it as $f(u^-,t)u/u^-$. For a polynomial f, $f(u,t) = u^m$, this corresponds to $(u^-)^m u/u^- = (u^-)^{m-1}u$. Sometimes this more implicit treatment has no effect, as with $f(u,t) = \exp(-u)$ and $f(u,t) = \ln(1+u)$, but with $f(u,t) = \sin(2(u+1))$, the $f(u^-,t)u/u^-$ trick leads to 7, 9, and 11 iterations during the first three steps, while $f(u^-,t)$ demands 17, 21, and 20 iterations. (Experiments can be done with the code `ODE_Picard_tricks.py`.)

Newton's method applied to a Backward Euler discretization of $u' = f(u,t)$ requires computation of the derivative

$$F'(u) = 1 - \Delta t \frac{\partial f}{\partial u}(u, t_n).$$

Starting with the solution at the previous time level, $u^- = u^{(1)}$, we can just use the standard formula

$$u = u^- - \omega \frac{F(u^-)}{F'(u^-)} = u^- - \omega \frac{u^- - \Delta t\, f(u^-, t_n) - u^{(1)}}{1 - \Delta t \frac{\partial}{\partial u} f(u^-, t_n)}. \qquad (3.11)$$

Crank-Nicolson discretization The standard Crank-Nicolson scheme with arithmetic mean approximation of f takes the form

$$\frac{u^{n+1} - u^n}{\Delta t} = \frac{1}{2}\left(f(u^{n+1}, t_{n+1}) + f(u^n, t_n) \right).$$

We can write the scheme as a nonlinear algebraic equation

$$F(u) = u - u^{(1)} - \Delta t \frac{1}{2} f(u, t_{n+1}) - \Delta t \frac{1}{2} f(u^{(1)}, t_n) = 0. \qquad (3.12)$$

A Picard iteration scheme must in general employ the linearization

$$\hat{F}(u) = u - u^{(1)} - \Delta t \frac{1}{2} f(u^-, t_{n+1}) - \Delta t \frac{1}{2} f(u^{(1)}, t_n),$$

while Newton's method can apply the general formula (3.11) with $F(u)$ given in (3.12) and

$$F'(u) = 1 - \frac{1}{2}\Delta t \frac{\partial f}{\partial u}(u, t_{n+1}).$$

3.1.12 Systems of ODEs

We may write a system of ODEs

$$\frac{d}{dt}u_0(t) = f_0(u_0(t), u_1(t), \ldots, u_N(t), t),$$

$$\frac{d}{dt}u_1(t) = f_1(u_0(t), u_1(t), \ldots, u_N(t), t),$$

$$\vdots$$

$$\frac{d}{dt}u_m(t) = f_m(u_0(t), u_1(t), \ldots, u_N(t), t),$$

as

$$u' = f(u, t), \quad u(0) = U_0, \tag{3.13}$$

if we interpret u as a vector $u = (u_0(t), u_1(t), \ldots, u_N(t))$ and f as a vector function with components $(f_0(u, t), f_1(u, t), \ldots, f_N(u, t))$.

Most solution methods for scalar ODEs, including the Forward and Backward Euler schemes and the Crank-Nicolson method, generalize in a straightforward way to systems of ODEs simply by using vector arithmetics instead of scalar arithmetics, which corresponds to applying the scalar scheme to each component of the system. For example, here is a backward difference scheme applied to each component,

$$\frac{u_0^n - u_0^{n-1}}{\Delta t} = f_0(u^n, t_n),$$

$$\frac{u_1^n - u_1^{n-1}}{\Delta t} = f_1(u^n, t_n),$$

$$\vdots$$

$$\frac{u_N^n - u_N^{n-1}}{\Delta t} = f_N(u^n, t_n),$$

which can be written more compactly in vector form as

$$\frac{u^n - u^{n-1}}{\Delta t} = f(u^n, t_n).$$

This is a *system of algebraic equations*,

$$u^n - \Delta t\, f(u^n, t_n) - u^{n-1} = 0,$$

or written out

$$u_0^n - \Delta t \, f_0(u^n, t_n) - u_0^{n-1} = 0,$$

$$\vdots$$

$$u_N^n - \Delta t \, f_N(u^n, t_n) - u_N^{n-1} = 0 \,.$$

Example We shall address the 2×2 ODE system for oscillations of a pendulum subject to gravity and air drag. The system can be written as

$$\dot{\omega} = -\sin\theta - \beta\omega|\omega|, \qquad (3.14)$$

$$\dot{\theta} = \omega, \qquad (3.15)$$

where β is a dimensionless parameter (this is the scaled, dimensionless version of the original, physical model). The unknown components of the system are the angle $\theta(t)$ and the angular velocity $\omega(t)$. We introduce $u_0 = \omega$ and $u_1 = \theta$, which leads to

$$u_0' = f_0(u, t) = -\sin u_1 - \beta u_0 |u_0|,$$

$$u_1' = f_1(u, t) = u_0 \,.$$

A Crank-Nicolson scheme reads

$$\frac{u_0^{n+1} - u_0^n}{\Delta t} = -\sin u_1^{n+\frac{1}{2}} - \beta u_0^{n+\frac{1}{2}} |u_0^{n+\frac{1}{2}}|$$

$$\approx -\sin\left(\frac{1}{2}(u_1^{n+1} + u_1 n)\right) - \beta \frac{1}{4}(u_0^{n+1} + u_0^n)|u_0^{n+1} + u_0^n|, \quad (3.16)$$

$$\frac{u_1^{n+1} - u_1^n}{\Delta t} = u_0^{n+\frac{1}{2}} \approx \frac{1}{2}(u_0^{n+1} + u_0^n) \,. \qquad (3.17)$$

This is a *coupled system* of two nonlinear algebraic equations in two unknowns u_0^{n+1} and u_1^{n+1}.

Using the notation u_0 and u_1 for the unknowns u_0^{n+1} and u_1^{n+1} in this system, writing $u_0^{(1)}$ and $u_1^{(1)}$ for the previous values u_0^n and u_1^n, multiplying by Δt and moving the terms to the left-hand sides, gives

$$u_0 - u_0^{(1)} + \Delta t \, \sin\left(\frac{1}{2}(u_1 + u_1^{(1)})\right) + \frac{1}{4}\Delta t \beta(u_0 + u_0^{(1)})|u_0 + u_0^{(1)}| = 0, \quad (3.18)$$

$$u_1 - u_1^{(1)} - \frac{1}{2}\Delta t (u_0 + u_0^{(1)}) = 0 \,. \quad (3.19)$$

Obviously, we have a need for solving systems of nonlinear algebraic equations, which is the topic of the next section.

3.2 Systems of Nonlinear Algebraic Equations

Implicit time discretization methods for a system of ODEs, or a PDE, lead to *systems* of nonlinear algebraic equations, written compactly as

$$F(u) = 0,$$

where u is a vector of unknowns $u = (u_0, \ldots, u_N)$, and F is a vector function: $F = (F_0, \ldots, F_N)$. The system at the end of Sect. 3.1.12 fits this notation with $N = 1$, $F_0(u)$ given by the left-hand side of (3.18), while $F_1(u)$ is the left-hand side of (3.19).

Sometimes the equation system has a special structure because of the underlying problem, e.g.,

$$A(u)u = b(u),$$

with $A(u)$ as an $(N + 1) \times (N + 1)$ matrix function of u and b as a vector function: $b = (b_0, \ldots, b_N)$.

We shall next explain how Picard iteration and Newton's method can be applied to systems like $F(u) = 0$ and $A(u)u = b(u)$. The exposition has a focus on ideas and practical computations. More theoretical considerations, including quite general results on convergence properties of these methods, can be found in Kelley [8].

3.2.1 Picard Iteration

We cannot apply Picard iteration to nonlinear equations unless there is some special structure. For the commonly arising case $A(u)u = b(u)$ we can linearize the product $A(u)u$ to $A(u^-)u$ and $b(u)$ as $b(u^-)$. That is, we use the most previously computed approximation in A and b to arrive at a *linear system* for u:

$$A(u^-)u = b(u^-).$$

A relaxed iteration takes the form

$$A(u^-)u^* = b(u^-), \quad u = \omega u^* + (1 - \omega)u^-.$$

In other words, we solve a system of nonlinear algebraic equations as a sequence of linear systems.

Algorithm for relaxed Picard iteration

Given $A(u)u = b(u)$ and an initial guess u^-, iterate until convergence:

1. solve $A(u^-)u^* = b(u^-)$ with respect to u^*
2. $u = \omega u^* + (1 - \omega)u^-$
3. $u^- \leftarrow u$

"Until convergence" means that the iteration is stopped when the change in the unknown, $||u - u^-||$, or the residual $||A(u)u - b||$, is sufficiently small, see Sect. 5.2.3 for more details.

3.2.2 Newton's Method

The natural starting point for Newton's method is the general nonlinear vector equation $F(u) = 0$. As for a scalar equation, the idea is to approximate F around a known value u^- by a linear function $\hat{F}$, calculated from the first two terms of a Taylor expansion of F. In the multi-variate case these two terms become

$$F(u^-) + J(u^-) \cdot (u - u^-),$$

where J is the *Jacobian* of F, defined by

$$J_{i,j} = \frac{\partial F_i}{\partial u_j}.$$

So, the original nonlinear system is approximated by

$$\hat{F}(u) = F(u^-) + J(u^-) \cdot (u - u^-) = 0,$$

which is linear in u and can be solved in a two-step procedure: first solve $J\delta u = -F(u^-)$ with respect to the vector δu and then update $u = u^- + \delta u$. A relaxation parameter can easily be incorporated:

$$u = \omega(u^- + \delta u) + (1 - \omega)u^- = u^- + \omega\delta u.$$

Algorithm for Newton's method

Given $F(u) = 0$ and an initial guess u^-, iterate until convergence:

1. solve $J\delta u = -F(u^-)$ with respect to δu
2. $u = u^- + \omega\delta u$
3. $u^- \leftarrow u$

For the special system with structure $A(u)u = b(u)$,

$$F_i = \sum_k A_{i,k}(u)u_k - b_i(u),$$

one gets

$$J_{i,j} = \sum_k \frac{\partial A_{i,k}}{\partial u_j}u_k + A_{i,j} - \frac{\partial b_i}{\partial u_j}. \tag{3.20}$$

We realize that the Jacobian needed in Newton's method consists of $A(u^-)$ as in the Picard iteration plus two additional terms arising from the differentiation. Using the notation $A'(u)$ for $\partial A/\partial u$ (a quantity with three indices: $\partial A_{i,k}/\partial u_j$), and $b'(u)$ for $\partial b/\partial u$ (a quantity with two indices: $\partial b_i/\partial u_j$), we can write the linear system to be solved as

$$(A + A'u + b')\delta u = -Au + b,$$

or

$$(A(u^-) + A'(u^-)u^- + b'(u^-))\delta u = -A(u^-)u^- + b(u^-).$$

Rearranging the terms demonstrates the difference from the system solved in each Picard iteration:

$$\underbrace{A(u^-)(u^- + \delta u) - b(u^-)}_{\text{Picard system}} + \gamma(A'(u^-)u^- + b'(u^-))\delta u = 0.$$

Here we have inserted a parameter γ such that $\gamma = 0$ gives the Picard system and $\gamma = 1$ gives the Newton system. Such a parameter can be handy in software to easily switch between the methods.

Combined algorithm for Picard and Newton iteration

Given $A(u)$, $b(u)$, and an initial guess u^-, iterate until convergence:

1. solve $(A + \gamma(A'(u^-)u^- + b'(u^-)))\delta u = -A(u^-)u^- + b(u^-)$ with respect to δu
2. $u = u^- + \omega \delta u$
3. $u^- \leftarrow u$

$\gamma = 1$ gives a Newton method while $\gamma = 0$ corresponds to Picard iteration.

3.2.3   Stopping Criteria

Let $|| \cdot ||$ be the standard Euclidean vector norm. Four termination criteria are much in use:

- Absolute change in solution: $||u - u^-|| \leq \epsilon_u$
- Relative change in solution: $||u - u^-|| \leq \epsilon_u ||u_0||$, where u_0 denotes the start value of u^- in the iteration
- Absolute residual: $||F(u)|| \leq \epsilon_r$
- Relative residual: $||F(u)|| \leq \epsilon_r ||F(u_0)||$

To prevent divergent iterations to run forever, one terminates the iterations when the current number of iterations k exceeds a maximum value k_{max}.

The relative criteria are most used since they are not sensitive to the characteristic size of u. Nevertheless, the relative criteria can be misleading when the initial start value for the iteration is very close to the solution, since an unnecessary reduction in the error measure is enforced. In such cases the absolute criteria work better. It is common to combine the absolute and relative measures of the size of the residual, as in

$$||F(u)|| \leq \epsilon_{rr} ||F(u_0)|| + \epsilon_{ra}, \tag{3.21}$$

where ϵ_{rr} is the tolerance in the relative criterion and ϵ_{ra} is the tolerance in the absolute criterion. With a very good initial guess for the iteration (typically the solution of a differential equation at the previous time level), the term $||F(u_0)||$ is small and ϵ_{ra} is the dominating tolerance. Otherwise, $\epsilon_{rr} ||F(u_0)||$ and the relative criterion dominates.

With the change in solution as criterion we can formulate a combined absolute and relative measure of the change in the solution:

$$||\delta u|| \leq \epsilon_{ur} ||u_0|| + \epsilon_{ua}. \tag{3.22}$$

The ultimate termination criterion, combining the residual and the change in solution with a test on the maximum number of iterations, can be expressed as

$$||F(u)|| \leq \epsilon_{rr} ||F(u_0)|| + \epsilon_{ra} \quad \text{or} \quad ||\delta u|| \leq \epsilon_{ur} ||u_0|| + \epsilon_{ua} \quad \text{or} \quad k > k_{\mathrm{max}}. \tag{3.23}$$

5.2.4   Example: A Nonlinear ODE Model from Epidemiology

A very simple model for the spreading of a disease, such as a flu, takes the form of a 2×2 ODE system

$$S' = -\beta SI, \tag{3.24}$$
$$I' = \beta SI - \nu I, \tag{3.25}$$

where $S(t)$ is the number of people who can get ill (susceptibles) and $I(t)$ is the number of people who are ill (infected). The constants $\beta > 0$ and $\nu > 0$ must be given along with initial conditions $S(0)$ and $I(0)$.

Implicit time discretization   A Crank-Nicolson scheme leads to a 2×2 system of nonlinear algebraic equations in the unknowns S^{n+1} and I^{n+1}:

$$\frac{S^{n+1} - S^n}{\Delta t} = -\beta[SI]^{n+\frac{1}{2}} \approx -\frac{\beta}{2}(S^n I^n + S^{n+1} I^{n+1}), \tag{3.26}$$
$$\frac{I^{n+1} - I^n}{\Delta t} = \beta[SI]^{n+\frac{1}{2}} - \nu I^{n+\frac{1}{2}} \approx \frac{\beta}{2}(S^n I^n + S^{n+1} I^{n+1}) - \frac{\nu}{2}(I^n + I^{n+1}). \tag{3.27}$$

Introducing S for S^{n+1}, $S^{(1)}$ for S^n, I for I^{n+1} and $I^{(1)}$ for I^n, we can rewrite the system as

$$F_S(S, I) = S - S^{(1)} + \frac{1}{2}\Delta t\beta(S^{(1)}I^{(1)} + SI) = 0, \tag{3.28}$$
$$F_I(S, I) = I - I^{(1)} - \frac{1}{2}\Delta t\beta(S^{(1)}I^{(1)} + SI) + \frac{1}{2}\Delta t\nu(I^{(1)} + I) = 0. \tag{3.29}$$

A Picard iteration   We assume that we have approximations S^- and I^- to S and I, respectively. A way of linearizing the only nonlinear term SI is to write I^-S in the $F_S = 0$ equation and S^-I in the $F_I = 0$ equation, which also *decouples* the equations. Solving the resulting linear equations with respect to the unknowns S and I gives

$$S = \frac{S^{(1)} - \frac{1}{2}\Delta t\beta S^{(1)}I^{(1)}}{1 + \frac{1}{2}\Delta t\beta I^-},$$
$$I = \frac{I^{(1)} + \frac{1}{2}\Delta t\beta S^{(1)}I^{(1)} - \frac{1}{2}\Delta t\nu I^{(1)}}{1 - \frac{1}{2}\Delta t\beta S^- + \frac{1}{2}\Delta t\nu}.$$

Before a new iteration, we must update $S^- \leftarrow S$ and $I^- \leftarrow I$.

Newton's method   The nonlinear system (5.28)–(5.29) can be written as $F(u) = 0$ with $F = (F_S, F_I)$ and $u = (S, I)$. The Jacobian becomes

$$J = \begin{pmatrix} \frac{\partial}{\partial S}F_S & \frac{\partial}{\partial I}F_S \\ \frac{\partial}{\partial S}F_I & \frac{\partial}{\partial I}F_I \end{pmatrix} = \begin{pmatrix} 1 + \frac{1}{2}\Delta t\beta I & \frac{1}{2}\Delta t\beta S \\ -\frac{1}{2}\Delta t\beta I & 1 - \frac{1}{2}\Delta t\beta S + \frac{1}{2}\Delta t\nu \end{pmatrix}.$$

The Newton system $J(u^-)\delta u = -F(u^-)$ to be solved in each iteration is then

$$
\begin{pmatrix}
1 + \frac{1}{2}\Delta t \beta I^- & \frac{1}{2}\Delta t \beta S^- \\
-\frac{1}{2}\Delta t \beta I^- & 1 - \frac{1}{2}\Delta t \beta S^- + \frac{1}{2}\Delta t \nu
\end{pmatrix}
\begin{pmatrix}
\delta S \\
\delta I
\end{pmatrix}
$$
$$
=
\begin{pmatrix}
S^- - S^{(1)} + \frac{1}{2}\Delta t \beta (S^{(1)}I^{(1)} + S^-I^-) \\
I^- - I^{(1)} - \frac{1}{2}\Delta t \beta (S^{(1)}I^{(1)} + S^-I^-) + \frac{1}{2}\Delta t \nu (I^{(1)} + I^-)
\end{pmatrix}.
$$

Remark For this particular system of ODEs, explicit time integration methods work very well. Even a Forward Euler scheme is fine, but (as also experienced more generally) the 4-th order Runge-Kutta method is an excellent balance between high accuracy, high efficiency, and simplicity.

3.3   Linearization at the Differential Equation Level

The attention is now turned to nonlinear partial differential equations (PDEs) and application of the techniques explained above for ODEs. The model problem is a nonlinear diffusion equation for $u(x, t)$:

$$
\frac{\partial u}{\partial t} = \nabla \cdot (\alpha(u)\nabla u) + f(u), \qquad x \in \Omega,\ t \in (0, T], \qquad (3.30)
$$

$$
-\alpha(u)\frac{\partial u}{\partial n} = g, \qquad x \in \partial\Omega_N,\ t \in (0, T], \qquad (3.31)
$$

$$
u = u_0, \qquad x \in \partial\Omega_D,\ t \in (0, T]. \qquad (3.32)
$$

In the present section, our aim is to discretize this problem in time and then present techniques for linearizing the time-discrete PDE problem "at the PDE level" such that we transform the nonlinear stationary PDE problem at each time level into a sequence of linear PDE problems, which can be solved using any method for linear PDEs. This strategy avoids the solution of systems of nonlinear algebraic equations. In Sect. 3.4 we shall take the opposite (and more common) approach: discretize the nonlinear problem in time and space first, and then solve the resulting nonlinear algebraic equations at each time level by the methods of Sect. 3.2. Very often, the two approaches are mathematically identical, so there is no preference from a computational efficiency point of view. The details of the ideas sketched above will hopefully become clear through the forthcoming examples.

3.3.1   Explicit Time Integration

The nonlinearities in the PDE are trivial to deal with if we choose an explicit time integration method for (3.30), such as the Forward Euler method:

$$
[D_t^+ u = \nabla \cdot (\alpha(u)\nabla u) + f(u)]^n,
$$

or written out,

$$
\frac{u^{n+1} - u^n}{\Delta t} = \nabla \cdot (\alpha(u^n)\nabla u^n) + f(u^n),
$$

which is a linear equation in the unknown u^{n+1} with solution

$$u^{n+1} = u^n + \Delta t \nabla \cdot (\alpha(u^n)\nabla u^n) + \Delta t f(u^n).$$

The disadvantage with this discretization is the strict stability criterion $\Delta t \leq h^2/(6 \max \alpha)$ for the case $f = 0$ and a standard 2nd-order finite difference discretization in 3D space with mesh cell sizes $h = \Delta x = \Delta y = \Delta z$.

3.3.2 Backward Euler Scheme and Picard Iteration

A Backward Euler scheme for (3.30) reads

$$[D_t^- u = \nabla \cdot (\alpha(u)\nabla u) + f(u)]^n.$$

Written out,

$$\frac{u^n - u^{n-1}}{\Delta t} = \nabla \cdot (\alpha(u^n)\nabla u^n) + f(u^n). \qquad (3.33)$$

This is a nonlinear PDE for the unknown function $u^n(x)$. Such a PDE can be viewed as a time-independent PDE where $u^{n-1}(x)$ is a known function.

We introduce a Picard iteration with k as iteration counter. A typical linearization of the $\nabla \cdot (\alpha(u^n)\nabla u^n)$ term in iteration $k+1$ is to use the previously computed $u^{n,k}$ approximation in the diffusion coefficient: $\alpha(u^{n,k})$. The nonlinear source term is treated similarly: $f(u^{n,k})$. The unknown function $u^{n,k+1}$ then fulfills the linear PDE

$$\frac{u^{n,k+1} - u^{n-1}}{\Delta t} = \nabla \cdot \left(\alpha(u^{n,k})\nabla u^{n,k+1}\right) + f(u^{n,k}). \qquad (3.34)$$

The initial guess for the Picard iteration at this time level can be taken as the solution at the previous time level: $u^{n,0} = u^{n-1}$.

We can alternatively apply the implementation-friendly notation where u corresponds to the unknown we want to solve for, i.e., $u^{n,k+1}$ above, and u^- is the most recently computed value, $u^{n,k}$ above. Moreover, $u^{(1)}$ denotes the unknown function at the previous time level, u^{n-1} above. The PDE to be solved in a Picard iteration then looks like

$$\frac{u - u^{(1)}}{\Delta t} = \nabla \cdot (\alpha(u^-)\nabla u) + f(u^-). \qquad (3.35)$$

At the beginning of the iteration we start with the value from the previous time level: $u^- = u^{(1)}$, and after each iteration, u^- is updated to u.

Remark on notation

The previous derivations of the numerical scheme for time discretizations of PDEs have, strictly speaking, a somewhat sloppy notation, but it is much used and convenient to read. A more precise notation must distinguish clearly between the exact solution of the PDE problem, here denoted $u_e(x,t)$, and the exact solution of the spatial problem, arising after time discretization at each time level, where (3.33) is an example. The latter is here represented as $u^n(x)$ and is an approximation to $u_e(x, t_n)$. Then we have another approximation $u^{n,k}(x)$ to

$u^n(x)$ when solving the nonlinear PDE problem for u^n by iteration methods, as in (3.34).

In our notation, u is a synonym for $u^{n,k+1}$ and $u^{(1)}$ is a synonym for u^{n-1}, inspired by what are natural variable names in a code. We will usually state the PDE problem in terms of u and quickly redefine the symbol u to mean the numerical approximation, while u_e is not explicitly introduced unless we need to talk about the exact solution and the approximate solution at the same time.

3.3.3 Backward Euler Scheme and Newton's Method

At time level n, we have to solve the stationary PDE (3.33). In the previous section, we saw how this can be done with Picard iterations. Another alternative is to apply the idea of Newton's method in a clever way. Normally, Newton's method is defined for systems of *algebraic equations*, but the idea of the method can be applied at the PDE level too.

Linearization via Taylor expansions Let $u^{n,k}$ be an approximation to the unknown u^n. We seek a better approximation on the form

$$u^n = u^{n,k} + \delta u. \tag{3.36}$$

The idea is to insert (3.36) in (3.33), Taylor expand the nonlinearities and keep only the terms that are linear in δu (which makes (3.36) an approximation for u^n). Then we can solve a linear PDE for the correction δu and use (3.36) to find a new approximation

$$u^{n,k+1} = u^{n,k} + \delta u$$

to u^n. Repeating this procedure gives a sequence $u^{n,k+1}, k = 0, 1, \ldots$ that hopefully converges to the goal u^n.

Let us carry out all the mathematical details for the nonlinear diffusion PDE discretized by the Backward Euler method. Inserting (3.36) in (3.33) gives

$$\frac{u^{n,k} + \delta u - u^{n-1}}{\Delta t} = \nabla \cdot (\alpha(u^{n,k} + \delta u)\nabla(u^{n,k} + \delta u)) + f(u^{n,k} + \delta u). \tag{3.37}$$

We can Taylor expand $\alpha(u^{n,k} + \delta u)$ and $f(u^{n,k} + \delta u)$:

$$\alpha(u^{n,k} + \delta u) = \alpha(u^{n,k}) + \frac{d\alpha}{du}(u^{n,k})\delta u + \mathcal{O}(\delta u^2) \approx \alpha(u^{n,k}) + \alpha'(u^{n,k})\delta u,$$

$$f(u^{n,k} + \delta u) = f(u^{n,k}) + \frac{df}{du}(u^{n,k})\delta u + \mathcal{O}(\delta u^2) \approx f(u^{n,k}) + f'(u^{n,k})\delta u.$$

Inserting the linear approximations of α and f in (3.37) results in

$$\frac{u^{n,k} + \delta u - u^{n-1}}{\Delta t} = \nabla \cdot (\alpha(u^{n,k})\nabla u^{n,k}) + f(u^{n,k})$$
$$+ \nabla \cdot (\alpha(u^{n,k})\nabla \delta u) + \nabla \cdot (\alpha'(u^{n,k})\delta u \nabla u^{n,k})$$
$$+ \nabla \cdot (\alpha'(u^{n,k})\delta u \nabla \delta u) + f'(u^{n,k})\delta u. \tag{3.38}$$

The term $\alpha'(u^{n,k})\delta u \nabla \delta u$ is of order δu^2 and therefore omitted since we expect the correction δu to be small ($\delta u \gg \delta u^2$). Reorganizing the equation gives a PDE for δu that we can write in short form as

$$\delta F(\delta u; u^{n,k}) = -F(u^{n,k}),$$

where

$$F(u^{n,k}) = \frac{u^{n,k} - u^{n-1}}{\Delta t} - \nabla \cdot (\alpha(u^{n,k})\nabla u^{n,k}) + f(u^{n,k}), \tag{3.39}$$

$$\delta F(\delta u; u^{n,k}) = -\frac{1}{\Delta t}\delta u + \nabla \cdot (\alpha(u^{n,k})\nabla \delta u)$$
$$+ \nabla \cdot (\alpha'(u^{n,k})\delta u \nabla u^{n,k}) + f'(u^{n,k})\delta u. \tag{3.40}$$

Note that δF is a linear function of δu, and F contains only terms that are known, such that the PDE for δu is indeed linear.

Observations

The notational form $\delta F = -F$ resembles the Newton system $J\delta u = -F$ for systems of algebraic equations, with δF as $J\delta u$. The unknown vector in a linear system of algebraic equations enters the system as a linear operator in terms of a matrix-vector product ($J\delta u$), while at the PDE level we have a linear differential operator instead (δF).

Similarity with Picard iteration We can rewrite the PDE for δu in a slightly different way too if we define $u^{n,k} + \delta u$ as $u^{n,k+1}$.

$$\frac{u^{n,k+1} - u^{n-1}}{\Delta t} = \nabla \cdot (\alpha(u^{n,k})\nabla u^{n,k+1}) + f(u^{n,k})$$
$$+ \nabla \cdot (\alpha'(u^{n,k})\delta u \nabla u^{n,k}) + f'(u^{n,k})\delta u. \tag{3.41}$$

Note that the first line is the same PDE as arises in the Picard iteration, while the remaining terms arise from the differentiations that are an inherent ingredient in Newton's method.

Implementation For coding we want to introduce u for u^n, u^- for $u^{n,k}$ and $u^{(1)}$ for u^{n-1}. The formulas for F and δF are then more clearly written as

$$F(u^-) = \frac{u^- - u^{(1)}}{\Delta t} - \nabla \cdot (\alpha(u^-)\nabla u^-) + f(u^-), \tag{3.42}$$

$$\delta F(\delta u; u^-) = -\frac{1}{\Delta t}\delta u + \nabla \cdot (\alpha(u^-)\nabla \delta u)$$
$$+ \nabla \cdot (\alpha'(u^-)\delta u \nabla u^-) + f'(u^-)\delta u. \tag{3.43}$$

The form that orders the PDE as the Picard iteration terms plus the Newton method's derivative terms becomes

$$\frac{u - u^{(1)}}{\Delta t} = \nabla \cdot (\alpha(u^-)\nabla u) + f(u^-)$$
$$+ \gamma(\nabla \cdot (\alpha'(u^-)(u - u^-)\nabla u^-) + f'(u^-)(u - u^-)). \tag{3.44}$$

The Picard and full Newton versions correspond to $\gamma = 0$ and $\gamma = 1$, respectively.

Derivation with alternative notation Some may prefer to derive the linearized PDE for δu using the more compact notation. We start with inserting $u^n = u^- + \delta u$ to get

$$\frac{u^- + \delta u - u^{n-1}}{\Delta t} = \nabla \cdot (\alpha(u^- + \delta u)\nabla(u^- + \delta u)) + f(u^- + \delta u).$$

Taylor expanding,

$$\alpha(u^- + \delta u) \approx \alpha(u^-) + \alpha'(u^-)\delta u,$$
$$f(u^- + \delta u) \approx f(u^-) + f'(u^-)\delta u,$$

and inserting these expressions gives a less cluttered PDE for δu:

$$\frac{u^- + \delta u - u^{n-1}}{\Delta t} = \nabla \cdot (\alpha(u^-)\nabla u^-) + f(u^-)$$
$$+ \nabla \cdot (\alpha(u^-)\nabla \delta u) + \nabla \cdot (\alpha'(u^-)\delta u \nabla u^-)$$
$$+ \nabla \cdot (\alpha'(u^-)\delta u \nabla \delta u) + f'(u^-)\delta u.$$

3.3.4 Crank-Nicolson Discretization

A Crank-Nicolson discretization of (3.30) applies a centered difference at $t_{n+\frac{1}{2}}$:

$$[D_t u = \nabla \cdot (\alpha(u)\nabla u) + f(u)]^{n+\frac{1}{2}}.$$

The standard technique is to apply an arithmetic average for quantities defined between two mesh points, e.g.,

$$u^{n+\frac{1}{2}} \approx \frac{1}{2}(u^n + u^{n+1}).$$

However, with nonlinear terms we have many choices of formulating an arithmetic mean:

$$[f(u)]^{n+\frac{1}{2}} \approx f\left(\frac{1}{2}(u^n + u^{n+1})\right) = [f(\overline{u}^t)]^{n+\frac{1}{2}}, \tag{3.45}$$

$$[f(u)]^{n+\frac{1}{2}} \approx \frac{1}{2}(f(u^n) + f(u^{n+1})) = \left[\overline{f(u)}^t\right]^{n+\frac{1}{2}}, \tag{3.46}$$

$$[\alpha(u)\nabla u]^{n+\frac{1}{2}} \approx \alpha\left(\frac{1}{2}(u^n + u^{n+1})\right)\nabla\left(\frac{1}{2}(u^n + u^{n+1})\right) = [\alpha(\overline{u}^t)\nabla\overline{u}^t]^{n+\frac{1}{2}}, \tag{3.47}$$

$$[\alpha(u)\nabla u]^{n+\frac{1}{2}} \approx \frac{1}{2}(\alpha(u^n) + \alpha(u^{n+1}))\nabla\left(\frac{1}{2}(u^n + u^{n+1})\right) = \left[\overline{\alpha(u)}^t \nabla\overline{u}^t\right]^{n+\frac{1}{2}}, \tag{3.48}$$

$$[\alpha(u)\nabla u]^{n+\frac{1}{2}} \approx \frac{1}{2}(\alpha(u^n)\nabla u^n + \alpha(u^{n+1})\nabla u^{n+1}) = \left[\overline{\alpha(u)\nabla u}^t\right]^{n+\frac{1}{2}}. \tag{3.49}$$

A big question is whether there are significant differences in accuracy between taking the products of arithmetic means or taking the arithmetic mean of products. Exercise 5.6 investigates this question, and the answer is that the approximation is $\mathcal{O}(\Delta t^2)$ in both cases.

3.4 1D Stationary Nonlinear Differential Equations

Section 3.3 presented methods for linearizing time-discrete PDEs directly prior to discretization in space. We can alternatively carry out the discretization in space of the time-discrete nonlinear PDE problem and get a system of nonlinear algebraic equations, which can be solved by Picard iteration or Newton's method as presented in Sect. 3.2. This latter approach will now be described in detail.

We shall work with the 1D problem

$$ -(\alpha(u)u')' + au = f(u), \quad x \in (0, L), \quad \alpha(u(0))u'(0) = C, \, u(L) = D. \tag{3.50} $$

The problem (3.50) arises from the stationary limit of a diffusion equation,

$$ \frac{\partial u}{\partial t} = \frac{\partial}{\partial x}\left(\alpha(u)\frac{\partial u}{\partial x}\right) - au + f(u), \tag{3.51} $$

as $t \to \infty$ and $\partial u/\partial t \to 0$. Alternatively, the problem (3.50) arises at each time level from implicit time discretization of (3.51). For example, a Backward Euler scheme for (3.51) leads to

$$ \frac{u^n - u^{n-1}}{\Delta t} = \frac{d}{dx}\left(\alpha(u^n)\frac{du^n}{dx}\right) - au^n + f(u^n). \tag{3.52} $$

Introducing $u(x)$ for $u^n(x)$, $u^{(1)}$ for u^{n-1}, and defining $f(u)$ in (3.50) to be $f(u)$ in (3.52) plus $u^{n-1}/\Delta t$, gives (3.50) with $a = 1/\Delta t$.

3.4.1 Finite Difference Discretization

The nonlinearity in the differential equation (3.50) poses no more difficulty than a variable coefficient, as in the term $(\alpha(x)u')'$. We can therefore use a standard finite difference approach when discretizing the Laplace term with a variable coefficient:

$$ [-D_x\alpha D_x u + au = f]_i. $$

Writing this out for a uniform mesh with points $x_i = i\Delta x$, $i = 0, \ldots, N_x$, leads to

$$ -\frac{1}{\Delta x^2}\left(\alpha_{i+\frac{1}{2}}(u_{i+1} - u_i) - \alpha_{i-\frac{1}{2}}(u_i - u_{i-1})\right) + au_i = f(u_i). \tag{3.53} $$

This equation is valid at all the mesh points $i = 0, 1, \ldots, N_x - 1$. At $i = N_x$ we have the Dirichlet condition $u_i = 0$. The only difference from the case with $(\alpha(x)u')'$ and $f(x)$ is that now α and f are functions of u and not only of x: $(\alpha(u(x))u')'$ and $f(u(x))$.

The quantity $\alpha_{i+\frac{1}{2}}$, evaluated between two mesh points, needs a comment. Since α depends on u and u is only known at the mesh points, we need to express $\alpha_{i+\frac{1}{2}}$ in terms of u_i and u_{i+1}. For this purpose we use an arithmetic mean, although a harmonic mean is also common in this context if α features large jumps. There are two choices of arithmetic means:

$$\alpha_{i+\frac{1}{2}} \approx \alpha\left(\frac{1}{2}(u_i + u_{i+1})\right) = [\alpha(\overline{u}^x)]^{i+\frac{1}{2}}, \tag{3.54}$$

$$\alpha_{i+\frac{1}{2}} \approx \frac{1}{2}(\alpha(u_i) + \alpha(u_{i+1})) = \left[\overline{\alpha(u)^x}\right]^{i+\frac{1}{2}}. \tag{3.55}$$

Equation (3.53) with the latter approximation then looks like

$$-\frac{1}{2\Delta x^2}\left((\alpha(u_i) + \alpha(u_{i+1}))(u_{i+1} - u_i) - (\alpha(u_{i-1}) + \alpha(u_i))(u_i - u_{i-1})\right)$$
$$+ au_i = f(u_i), \tag{3.56}$$

or written more compactly,

$$[-D_x\overline{\alpha}^x D_x u + au = f]_i .$$

At mesh point $i = 0$ we have the boundary condition $\alpha(u)u' = C$, which is discretized by

$$[\alpha(u)D_{2x}u = C]_0,$$

meaning

$$\alpha(u_0)\frac{u_1 - u_{-1}}{2\Delta x} = C . \tag{3.57}$$

The fictitious value u_{-1} can be eliminated with the aid of (3.56) for $i = 0$. Formally, (3.56) should be solved with respect to u_{i-1} and that value (for $i = 0$) should be inserted in (3.57), but it is algebraically much easier to do it the other way around. Alternatively, one can use a ghost cell $[-\Delta x, 0]$ and update the u_{-1} value in the ghost cell according to (3.57) after every Picard or Newton iteration. Such an approach means that we use a known u_{-1} value in (3.56) from the previous iteration.

3.4.2   Solution of Algebraic Equations

The structure of the equation system The nonlinear algebraic equations (3.56) are of the form $A(u)u = b(u)$ with

$$A_{i,i} = \frac{1}{2\Delta x^2}(\alpha(u_{i-1}) + 2\alpha(u_i)\alpha(u_{i+1})) + a,$$
$$A_{i,i-1} = -\frac{1}{2\Delta x^2}(\alpha(u_{i-1}) + \alpha(u_i)),$$
$$A_{i,i+1} = -\frac{1}{2\Delta x^2}(\alpha(u_i) + \alpha(u_{i+1})),$$
$$b_i = f(u_i) .$$

The matrix $A(u)$ is tridiagonal: $A_{i,j} = 0$ for $j > i + 1$ and $j < i - 1$.

The above expressions are valid for internal mesh points $1 \le i \le N_x - 1$. For $i = 0$ we need to express $u_{i-1} = u_{-1}$ in terms of u_1 using (3.57):

$$u_{-1} = u_1 - \frac{2\Delta x}{\alpha(u_0)} C \,. \tag{3.58}$$

This value must be inserted in $A_{0,0}$. The expression for $A_{i,i+1}$ applies for $i = 0$, and $A_{i,i-1}$ does not enter the system when $i = 0$.

Regarding the last equation, its form depends on whether we include the Dirichlet condition $u(L) = D$, meaning $u_{N_x} = D$, in the nonlinear algebraic equation system or not. Suppose we choose $(u_0, u_1, \ldots, u_{N_x-1})$ as unknowns, later referred to as *systems without Dirichlet conditions*. The last equation corresponds to $i = N_x - 1$. It involves the boundary value u_{N_x}, which is substituted by D. If the unknown vector includes the boundary value, $(u_0, u_1, \ldots, u_{N_x})$, later referred to as *system including Dirichlet conditions*, the equation for $i = N_x - 1$ just involves the unknown u_{N_x}, and the final equation becomes $u_{N_x} = D$, corresponding to $A_{i,i} = 1$ and $b_i = D$ for $i = N_x$.

Picard iteration The obvious Picard iteration scheme is to use previously computed values of u_i in $A(u)$ and $b(u)$, as described more in detail in Sect. 3.2. With the notation u^- for the most recently computed value of u, we have the system $F(u) \approx \hat{F}(u) = A(u^-)u - b(u^-)$, with $F = (F_0, F_1, \ldots, F_m)$, $u = (u_0, u_1, \ldots, u_m)$. The index m is N_x if the system includes the Dirichlet condition as a separate equation and $N_x - 1$ otherwise. The matrix $A(u^-)$ is tridiagonal, so the solution procedure is to fill a tridiagonal matrix data structure and the right-hand side vector with the right numbers and call a Gaussian elimination routine for tridiagonal linear systems.

Mesh with two cells It helps on the understanding of the details to write out all the mathematics in a specific case with a small mesh, say just two cells ($N_x = 2$). We use u_i^- for the i-th component in u^-.

The starting point is the basic expressions for the nonlinear equations at mesh point $i = 0$ and $i = 1$:

$$A_{0,-1}u_{-1} + A_{0,0}u_0 + A_{0,1}u_1 = b_0, \tag{3.59}$$
$$A_{1,0}u_0 + A_{1,1}u_1 + A_{1,2}u_2 = b_1 \,. \tag{3.60}$$

Equation (3.59) written out reads

$$\frac{1}{2\Delta x^2}\Big(-(\alpha(u_{-1}) + \alpha(u_0))u_{-1}$$

$$+ (\alpha(u_{-1}) + 2\alpha(u_0) + \alpha(u_1))u_0$$

$$- (\alpha(u_0) + \alpha(u_1))\Big)u_1 + au_0 = f(u_0) \,.$$

We must then replace u_{-1} by (3.58). With Picard iteration we get

$$\frac{1}{2\Delta x^2}\Big(-(\alpha(u_{-1}^-) + 2\alpha(u_0^-) + \alpha(u_1^-))u_1$$

$$+ (\alpha(u_{-1}^-) + 2\alpha(u_0^-) + \alpha(u_1^-)))\Big)u_0 + au_0$$

$$= f(u_0^-) - \frac{1}{\alpha(u_0^-)\Delta x}(\alpha(u_{-1}^-) + \alpha(u_0^-))C,$$

where

$$u_{-1}^- = u_1^- - \frac{2\Delta x}{\alpha(u_0^-)}C .$$

Equation (3.60) contains the unknown u_2 for which we have a Dirichlet condition. In case we omit the condition as a separate equation, (3.60) with Picard iteration becomes

$$\frac{1}{2\Delta x^2}\Big(-(\alpha(u_0^-) + \alpha(u_1^-))u_0$$

$$+ (\alpha(u_0^-) + 2\alpha(u_1^-) + \alpha(u_2^-))u_1$$

$$- (\alpha(u_1^-) + \alpha(u_2^-))\Big)u_2 + au_1 = f(u_1^-) .$$

We must now move the u_2 term to the right-hand side and replace all occurrences of u_2 by D:

$$\frac{1}{2\Delta x^2}\Big(-(\alpha(u_0^-) + \alpha(u_1^-))u_0$$

$$+ (\alpha(u_0^-) + 2\alpha(u_1^-) + \alpha(D)))\Big)u_1 + au_1$$

$$= f(u_1^-) + \frac{1}{2\Delta x^2}(\alpha(u_1^-) + \alpha(D))D .$$

The two equations can be written as a 2×2 system:

$$\begin{pmatrix} B_{0,0} & B_{0,1} \\ B_{1,0} & B_{1,1} \end{pmatrix} \begin{pmatrix} u_0 \\ u_1 \end{pmatrix} = \begin{pmatrix} d_0 \\ d_1 \end{pmatrix},$$

where

$$B_{0,0} = \frac{1}{2\Delta x^2}(\alpha(u_{-1}^-) + 2\alpha(u_0^-) + \alpha(u_1^-)) + a, \tag{3.61}$$

$$B_{0,1} = -\frac{1}{2\Delta x^2}(\alpha(u_{-1}^-) + 2\alpha(u_0^-) + \alpha(u_1^-)), \tag{3.62}$$

$$B_{1,0} = -\frac{1}{2\Delta x^2}(\alpha(u_0^-) + \alpha(u_1^-)), \tag{3.63}$$

$$B_{1,1} = \frac{1}{2\Delta x^2}(\alpha(u_0^-) + 2\alpha(u_1^-) + \alpha(D)) + a, \tag{3.64}$$

$$d_0 = f(u_0^-) - \frac{1}{\alpha(u_0^-)\Delta x}(\alpha(u_{-1}^-) + \alpha(u_0^-))C, \qquad (3.65)$$

$$d_1 = f(u_1^-) + \frac{1}{2\Delta x^2}(\alpha(u_1^-) + \alpha(D))D. \qquad (3.66)$$

The system with the Dirichlet condition becomes

$$\begin{pmatrix} B_{0,0} & B_{0,1} & 0 \\ B_{1,0} & B_{1,1} & B_{1,2} \\ 0 & 0 & 1 \end{pmatrix} \begin{pmatrix} u_0 \\ u_1 \\ u_2 \end{pmatrix} = \begin{pmatrix} d_0 \\ d_1 \\ D \end{pmatrix},$$

with

$$B_{1,1} = \frac{1}{2\Delta x^2}(\alpha(u_0^-) + 2\alpha(u_1^-) + \alpha(u_2)) + a, \qquad (3.67)$$

$$B_{1,2} = -\frac{1}{2\Delta x^2}(\alpha(u_1^-) + \alpha(u_2))), \qquad (3.68)$$

$$d_1 = f(u_1^-). \qquad (3.69)$$

Other entries are as in the 2×2 system.

Newton's method The Jacobian must be derived in order to use Newton's method. Here it means that we need to differentiate $F(u) = A(u)u - b(u)$ with respect to the unknown parameters $u_0, u_1, \ldots, u_m$ ($m = N_x$ or $m = N_x - 1$, depending on whether the Dirichlet condition is included in the nonlinear system $F(u) = 0$ or not). Nonlinear equation number i has the structure

$$F_i = A_{i,i-1}(u_{i-1}, u_i)u_{i-1} + A_{i,i}(u_{i-1}, u_i, u_{i+1})u_i + A_{i,i+1}(u_i, u_{i+1})u_{i+1} - b_i(u_i).$$

Computing the Jacobian requires careful differentiation. For example,

$$\begin{aligned}
\frac{\partial}{\partial u_i}(A_{i,i}(u_{i-1}, u_i, u_{i+1})u_i) &= \frac{\partial A_{i,i}}{\partial u_i}u_i + A_{i,i}\frac{\partial u_i}{\partial u_i} \\
&= \frac{\partial}{\partial u_i}\left(\frac{1}{2\Delta x^2}(\alpha(u_{i-1}) + 2\alpha(u_i) + \alpha(u_{i+1})) + a\right)u_i \\
&\quad + \frac{1}{2\Delta x^2}(\alpha(u_{i-1}) + 2\alpha(u_i) + \alpha(u_{i+1})) + a \\
&= \frac{1}{2\Delta x^2}(2\alpha'(u_i)u_i + \alpha(u_{i-1}) + 2\alpha(u_i) + \alpha(u_{i+1})) \\
&\quad + a.
\end{aligned}$$

The complete Jacobian becomes

$$
\begin{aligned}
J_{i,i} &= \frac{\partial F_i}{\partial u_i} = \frac{\partial A_{i,i-1}}{\partial u_i} u_{i-1} + \frac{\partial A_{i,i}}{\partial u_i} u_i + A_{i,i} + \frac{\partial A_{i,i+1}}{\partial u_i} u_{i+1} - \frac{\partial b_i}{\partial u_i} \\
&= \frac{1}{2\Delta x^2} (-\alpha'(u_i) u_{i-1} + 2\alpha'(u_i) u_i + \alpha(u_{i-1}) + 2\alpha(u_i) + \alpha(u_{i+1})) \\
&\quad + a - \frac{1}{2\Delta x^2} \alpha'(u_i) u_{i+1} - b'(u_i),
\end{aligned}
$$

$$
\begin{aligned}
J_{i,i-1} &= \frac{\partial F_i}{\partial u_{i-1}} = \frac{\partial A_{i,i-1}}{\partial u_{i-1}} u_{i-1} + A_{i-1,i} + \frac{\partial A_{i,i}}{\partial u_{i-1}} u_i - \frac{\partial b_i}{\partial u_{i-1}} \\
&= \frac{1}{2\Delta x^2} (-\alpha'(u_{i-1}) u_{i-1} - (\alpha(u_{i-1}) + \alpha(u_i)) + \alpha'(u_{i-1}) u_i),
\end{aligned}
$$

$$
\begin{aligned}
J_{i,i+1} &= \frac{\partial A_{i,i+1}}{\partial u_{i-1}} u_{i+1} + A_{i+1,i} + \frac{\partial A_{i,i}}{\partial u_{i+1}} u_i - \frac{\partial b_i}{\partial u_{i+1}} \\
&= \frac{1}{2\Delta x^2} (-\alpha'(u_{i+1}) u_{i+1} - (\alpha(u_i) + \alpha(u_{i+1})) + \alpha'(u_{i+1}) u_i).
\end{aligned}
$$

The explicit expression for nonlinear equation number i, $F_i(u_0, u_1, \ldots)$, arises from moving the $f(u_i)$ term in (3.56) to the left-hand side:

$$
\begin{aligned}
F_i = &- \frac{1}{2\Delta x^2} ((\alpha(u_i) + \alpha(u_{i+1}))(u_{i+1} - u_i) - (\alpha(u_{i-1}) + \alpha(u_i))(u_i - u_{i-1})) \\
&+ au_i - f(u_i) = 0.
\end{aligned}
$$

$$(3.70)$$

At the boundary point $i = 0$, u_{-1} must be replaced using the formula (3.58). When the Dirichlet condition at $i = N_x$ is not a part of the equation system, the last equation $F_m = 0$ for $m = N_x - 1$ involves the quantity u_{N_x-1} which must be replaced by D. If u_{N_x} is treated as an unknown in the system, the last equation $F_m = 0$ has $m = N_x$ and reads

$$
F_{N_x}(u_0, \ldots, u_{N_x}) = u_{N_x} - D = 0.
$$

Similar replacement of u_{-1} and u_{N_x} must be done in the Jacobian for the first and last row. When u_{N_x} is included as an unknown, the last row in the Jacobian must help implement the condition $\delta u_{N_x} = 0$, since we assume that u contains the right Dirichlet value at the beginning of the iteration ($u_{N_x} = D$), and then the Newton update should be zero for $i = 0$, i.e., $\delta u_{N_x} = 0$. This also forces the right-hand side to be $b_i = 0$, $i = N_x$.

We have seen, and can see from the present example, that the linear system in Newton's method contains all the terms present in the system that arises in the Picard iteration method. The extra terms in Newton's method can be multiplied by a factor such that it is easy to program one linear system and set this factor to 0 or 1 to generate the Picard or Newton system.

3.5 Multi-Dimensional Nonlinear PDE Problems

The fundamental ideas in the derivation of F_i and $J_{i,j}$ in the 1D model problem are easily generalized to multi-dimensional problems. Nevertheless, the expressions involved are slightly different, with derivatives in x replaced by ∇, so we present some examples below in detail.

3.5.1 Finite Difference Discretization

A typical diffusion equation

$$u_t = \nabla \cdot (\alpha(u)\nabla u) + f(u),$$

can be discretized by (e.g.) a Backward Euler scheme, which in 2D can be written

$$\left[D_t^- u = D_x \overline{\alpha(u)}^x D_x u + D_y \overline{\alpha(u)}^y D_y u + f(u) \right]_{i,j}^n.$$

We do not dive into the details of handling boundary conditions now. Dirichlet and Neumann conditions are handled as in corresponding linear, variable-coefficient diffusion problems.

Writing the scheme out, putting the unknown values on the left-hand side and known values on the right-hand side, and introducing $\Delta x = \Delta y = h$ to save some writing, one gets

$$
\begin{aligned}
u_{i,j}^n - \frac{\Delta t}{h^2} \Bigg(&\frac{1}{2}(\alpha(u_{i,j}^n) + \alpha(u_{i+1,j}^n))(u_{i+1,j}^n - u_{i,j}^n) \\
&- \frac{1}{2}(\alpha(u_{i-1,j}^n) + \alpha(u_{i,j}^n))(u_{i,j}^n - u_{i-1,j}^n) \\
&+ \frac{1}{2}(\alpha(u_{i,j}^n) + \alpha(u_{i,j+1}^n))(u_{i,j+1}^n - u_{i,j}^n) \\
&- \frac{1}{2}(\alpha(u_{i,j-1}^n) + \alpha(u_{i,j}^n))(u_{i,j}^n - u_{i-1,j-1}^n) \Bigg) - \Delta t f(u_{i,j}^n) = u_{i,j}^{n-1}.
\end{aligned}
$$

This defines a nonlinear algebraic system on the form $A(u)u = b(u)$.

Picard iteration The most recently computed values u^- of u^n can be used in α and f for a Picard iteration, or equivalently, we solve $A(u^-)u = b(u^-)$. The result is a linear system of the same type as arising from $u_t = \nabla \cdot (\alpha(x)\nabla u) + f(x,t)$.

The Picard iteration scheme can also be expressed in operator notation:

$$\left[D_t^- u = D_x \overline{\alpha(u^-)}^x D_x u + D_y \overline{\alpha(u^-)}^y D_y u + f(u^-) \right]_{i,j}^n.$$

Newton's method As always, Newton's method is technically more involved than Picard iteration. We first define the nonlinear algebraic equations to be solved, drop the superscript n (use u for u^n), and introduce $u^{(1)}$ for u^{n-1}:

$$F_{i,j} = u_{i,j} - \frac{\Delta t}{h^2}\left(\frac{1}{2}(\alpha(u_{i,j}) + \alpha(u_{i+1,j}))(u_{i+1,j} - u_{i,j})\right.$$
$$-\frac{1}{2}(\alpha(u_{i-1,j}) + \alpha(u_{i,j}))(u_{i,j} - u_{i-1,j})$$
$$+\frac{1}{2}(\alpha(u_{i,j}) + \alpha(u_{i,j+1}))(u_{i,j+1} - u_{i,j})$$
$$\left.-\frac{1}{2}(\alpha(u_{i,j-1}) + \alpha(u_{i,j}))(u_{i,j} - u_{i-1,j-1})\right)$$
$$-\Delta t\, f(u_{i,j}) - u_{i,j}^{(1)} = 0.$$

It is convenient to work with two indices i and j in 2D finite difference discretizations, but it complicates the derivation of the Jacobian, which then gets four indices. (Make sure you really understand the 1D version of this problem as treated in Sect. 3.4.1.) The left-hand expression of an equation $F_{i,j} = 0$ is to be differentiated with respect to each of the unknowns $u_{r,s}$ (recall that this is short notation for $u_{r,s}^n$), $r \in \mathcal{I}_x, s \in \mathcal{I}_y$:

$$J_{i,j,r,s} = \frac{\partial F_{i,j}}{\partial u_{r,s}}.$$

The Newton system to be solved in each iteration can be written as

$$\sum_{r\in\mathcal{I}_x}\sum_{s\in\mathcal{I}_y} J_{i,j,r,s}\delta u_{r,s} = -F_{i,j}, \quad i \in \mathcal{I}_x,\ j \in \mathcal{I}_y.$$

Given i and j, only a few r and s indices give nonzero contribution to the Jacobian since $F_{i,j}$ contains $u_{i\pm1,j}$, $u_{i,j\pm1}$, and $u_{i,j}$. This means that $J_{i,j,r,s}$ has nonzero contributions only if $r = i \pm 1$, $s = j \pm 1$, as well as $r = i$ and $s = j$. The corresponding terms in $J_{i,j,r,s}$ are $J_{i,j,i-1,j}$, $J_{i,j,i+1,j}$, $J_{i,j,i,j-1}$, $J_{i,j,i,j+1}$ and $J_{i,j,i,j}$. Therefore, the left-hand side of the Newton system, $\sum_r \sum_s J_{i,j,r,s}\delta u_{r,s}$ collapses to

$$J_{i,j,r,s}\delta u_{r,s} = J_{i,j,i,j}\delta u_{i,j} + J_{i,j,i-1,j}\delta u_{i-1,j} + J_{i,j,i+1,j}\delta u_{i+1,j} + J_{i,j,i,j-1}\delta u_{i,j-1}$$
$$+ J_{i,j,i,j+1}\delta u_{i,j+1}.$$

The specific derivatives become

$$J_{i,j,i-1,j} = \frac{\partial F_{i,j}}{\partial u_{i-1,j}}$$
$$= \frac{\Delta t}{h^2}(\alpha'(u_{i-1,j})(u_{i,j} - u_{i-1,j}) + \alpha(u_{i-1,j})(-1)),$$
$$J_{i,j,i+1,j} = \frac{\partial F_{i,j}}{\partial u_{i+1,j}}$$
$$= \frac{\Delta t}{h^2}(-\alpha'(u_{i+1,j})(u_{i+1,j} - u_{i,j}) - \alpha(u_{i-1,j})),$$

$$J_{i,j,i,j-1} = \frac{\partial F_{i,j}}{\partial u_{i,j-1}}$$

$$= \frac{\Delta t}{h^2}(\alpha'(u_{i,j-1})(u_{i,j} - u_{i,j-1}) + \alpha(u_{i,j-1})(-1)),$$

$$J_{i,j,i,j+1} = \frac{\partial F_{i,j}}{\partial u_{i,j+1}}$$

$$= \frac{\Delta t}{h^2}(-\alpha'(u_{i,j+1})(u_{i,j+1} - u_{i,j}) - \alpha(u_{i,j-1})).$$

The $J_{i,j,i,j}$ entry has a few more terms and is left as an exercise. Inserting the most recent approximation u^- for u in the J and F formulas and then forming $J\delta u = -F$ gives the linear system to be solved in each Newton iteration. Boundary conditions will affect the formulas when any of the indices coincide with a boundary value of an index.

3.5.2 Continuation Methods

Picard iteration or Newton's method may diverge when solving PDEs with severe nonlinearities. Relaxation with $\omega < 1$ may help, but in highly nonlinear problems it can be necessary to introduce a *continuation parameter* Λ in the problem: $\Lambda = 0$ gives a version of the problem that is easy to solve, while $\Lambda = 1$ is the target problem. The idea is then to increase Λ in steps, $\Lambda_0 = 0, \Lambda_1 < \cdots < \Lambda_n = 1$, and use the solution from the problem with Λ_{i-1} as initial guess for the iterations in the problem corresponding to Λ_i.

The continuation method is easiest to understand through an example. Suppose we intend to solve

$$-\nabla \cdot (||\nabla u||^q \nabla u) = f,$$

which is an equation modeling the flow of a non-Newtonian fluid through a channel or pipe. For $q = 0$ we have the Poisson equation (corresponding to a Newtonian fluid) and the problem is linear. A typical value for pseudo-plastic fluids may be $q_n = -0.8$. We can introduce the continuation parameter $\Lambda \in [0, 1]$ such that $q = q_n \Lambda$. Let $\{\Lambda_\ell\}_{\ell=0}^n$ be the sequence of Λ values in $[0, 1]$, with corresponding q values $\{q_\ell\}_{\ell=0}^n$. We can then solve a sequence of problems

$$-\nabla \cdot (||\nabla u^\ell||_\ell^q \nabla u^\ell) = f, \quad \ell = 0, \ldots, n,$$

where the initial guess for iterating on u^ℓ is the previously computed solution $u^{\ell-1}$. If a particular Λ_ℓ leads to convergence problems, one may try a smaller increase in Λ: $\Lambda_* = \frac{1}{2}(\Lambda_{\ell-1} + \Lambda_\ell)$, and repeat halving the step in Λ until convergence is reestablished.

3.6   Operator Splitting Methods

Operator splitting is a natural and old idea. When a PDE or system of PDEs contains different terms expressing different physics, it is natural to use different numerical methods for different physical processes. This can optimize and simplify the overall solution process. The idea was especially popularized in the context of the Navier-Stokes equations and reaction-diffusion PDEs. Common names for the technique are *operator splitting*, *fractional step* methods, and *split-step* methods. We shall stick to the former name. In the context of nonlinear differential equations, operator splitting can be used to isolate nonlinear terms and simplify the solution methods.

A related technique, often known as dimensional splitting or alternating direction implicit (ADI) methods, is to split the spatial dimensions and solve a 2D or 3D problem as two or three consecutive 1D problems, but this type of splitting is not to be further considered here.

3.6.1   Ordinary Operator Splitting for ODEs

Consider first an ODE where the right-hand side is split into two terms:

$$u' = f_0(u) + f_1(u) \,. \tag{3.71}$$

In case f_0 and f_1 are linear functions of u, $f_0 = au$ and $f_1 = bu$, we have $u(t) = Ie^{(a+b)t}$, if $u(0) = I$. When going one time step of length Δt from t_n to t_{n+1}, we have

$$u(t_{n+1}) = u(t_n)e^{(a+b)\Delta t} \,.$$

This expression can be also be written as

$$u(t_{n+1}) = u(t_n)e^{a\Delta t}e^{b\Delta t},$$

or

$$u^* = u(t_n)e^{a\Delta t}, \tag{3.72}$$

$$u(t_{n+1}) = u^*e^{b\Delta t} \,. \tag{3.73}$$

The first step (3.72) means solving $u' = f_0$ over a time interval Δt with $u(t_n)$ as start value. The second step (3.73) means solving $u' = f_1$ over a time interval Δt with the value at the end of the first step as start value. That is, we progress the solution in two steps and solve two ODEs $u' = f_0$ and $u' = f_1$. The order of the equations is not important. From the derivation above we see that solving $u' = f_1$ prior to $u' = f_0$ can equally well be done.

The technique is exact if the ODEs are linear. For nonlinear ODEs it is only an approximate method with error Δt. The technique can be extended to an arbitrary number of steps; i.e., we may split the PDE system into any number of subsystems. Examples will illuminate this principle.

3.6.2   Strang Splitting for ODEs

The accuracy of the splitting method in Sect. 3.6.1 can be improved from $\mathcal{O}(\Delta t)$ to $\mathcal{O}(\Delta t^2)$ using so-called *Strang splitting*, where we take half a step with the f_0 operator, a full step with the f_1 operator, and finally half another step with the f_0 operator. During a time interval Δt the algorithm can be written as follows.

$$\frac{du^*}{dt} = f_0(u^*), \quad u^*(t_n) = u(t_n), \quad t \in \left[t_n, t_n + \frac{1}{2}\Delta t\right],$$

$$\frac{du^{***}}{dt} = f_1(u^{***}), \quad u^{***}(t_n) = u^*\left(t_{n+\frac{1}{2}}\right), \quad t \in [t_n, t_n + \Delta t],$$

$$\frac{du^{**}}{dt} = f_0(u^{**}), \quad u^{**}\left(t_{n+\frac{1}{2}}\right) = u^{***}(t_{n+1}), \quad t \in \left[t_n + \frac{1}{2}\Delta t, t_n + \Delta t\right].$$

The global solution is set as $u(t_{n+1}) = u^{**}(t_{n+1})$.

There is no use in combining higher-order methods with ordinary splitting since the error due to splitting is $\mathcal{O}(\Delta t)$, but for Strang splitting it makes sense to use schemes of order $\mathcal{O}(\Delta t^2)$.

With the notation introduced for Strang splitting, we may express ordinary first-order splitting as

$$\frac{du^*}{dt} = f_0(u^*), \quad u^*(t_n) = u(t_n), \quad t \in [t_n, t_n + \Delta t],$$

$$\frac{du^{**}}{dt} = f_1(u^{**}), \quad u^{**}(t_n) = u^*(t_{n+1}), \quad t \in [t_n, t_n + \Delta t],$$

with global solution set as $u(t_{n+1}) = u^{**}(t_{n+1})$.

3.6.3   Example: Logistic Growth

Let us split the (scaled) logistic equation

$$u' = u(1 - u), \quad u(0) = 0.1,$$

with solution $u = (9e^{-t} + 1)^{-1}$, into

$$u' = u - u^2 = f_0(u) + f_1(u), \quad f_0(u) = u, \quad f_1(u) = -u^2.$$

We solve $u' = f_0(u)$ and $u' = f_1(u)$ by a Forward Euler step. In addition, we add a method where we solve $u' = f_0(u)$ analytically, since the equation is actually $u' = u$ with solution e^t. The software that accompanies the following methods is the file `split_logistic.py`.

Splitting techniques   Ordinary splitting takes a Forward Euler step for each of the ODEs according to

$$\frac{u^{*,n+1} - u^{*,n}}{\Delta t} = f_0(u^{*,n}), \quad u^{*,n} = u(t_n), \quad t \in [t_n, t_n + \Delta t], \tag{3.74}$$

$$\frac{u^{**,n+1} - u^{**,n}}{\Delta t} = f_1(u^{**,n}), \quad u^{**,n} = u^{*,n+1}, \quad t \in [t_n, t_n + \Delta t], \tag{3.75}$$

with $u(t_{n+1}) = u^{**,n+1}$.

Strang splitting takes the form

$$\frac{u^{*,n+\frac{1}{2}} - u^{*,n}}{\frac{1}{2}\Delta t} = f_0(u^{*,n}), \quad u^{*,n} = u(t_n), \ t \in \left[t_n, t_n + \frac{1}{2}\Delta t \right], \quad (3.76)$$

$$\frac{u^{***,n+1} - u^{***,n}}{\Delta t} = f_1(u^{***,n}), \quad u^{***,n} = u^{*,n+\frac{1}{2}}, \ t \in [t_n, t_n + \Delta t], \quad (3.77)$$

$$\frac{u^{**,n+1} - u^{**,n+\frac{1}{2}}}{\frac{1}{2}\Delta t} = f_0\left(u^{**,n+\frac{1}{2}} \right), \quad u^{**,n+\frac{1}{2}} = u^{***,n+1},$$

$$t \in \left[t_n + \frac{1}{2}\Delta t, t_n + \Delta t \right]. \quad (3.78)$$

Verbose implementation The following function computes four solutions arising from the Forward Euler method, ordinary splitting, Strang splitting, as well as Strang splitting with exact treatment of $u' = f_0(u)$:

```
import numpy as np

def solver(dt, T, f, f_0, f_1):
    """
    Solve u'=f by the Forward Euler method and by ordinary and
    Strang splitting: f(u) = f_0(u) + f_1(u).
    """
    Nt = int(round(T/float(dt)))
    t = np.linspace(0, Nt*dt, Nt+1)
    u_FE = np.zeros(len(t))
    u_split1 = np.zeros(len(t))  # 1st-order splitting
    u_split2 = np.zeros(len(t))  # 2nd-order splitting
    u_split3 = np.zeros(len(t))  # 2nd-order splitting w/exact f_0

    # Set initial values
    u_FE[0] = 0.1
    u_split1[0] = 0.1
    u_split2[0] = 0.1
    u_split3[0] = 0.1

    for n in range(len(t)-1):
        # Forward Euler method
        u_FE[n+1] = u_FE[n] + dt*f(u_FE[n])

        # --- Ordinary splitting ---
        # First step
        u_s_n = u_split1[n]
        u_s = u_s_n + dt*f_0(u_s_n)
        # Second step
        u_ss_n = u_s
        u_ss = u_ss_n + dt*f_1(u_ss_n)
        u_split1[n+1] = u_ss

        # --- Strang splitting ---
        # First step
        u_s_n = u_split2[n]
        u_s = u_s_n + dt/2.*f_0(u_s_n)
```

```
                # Second step
                u_sss_n = u_s
                u_sss = u_sss_n + dt*f_1(u_sss_n)
                # Third step
                u_ss_n = u_sss
                u_ss = u_ss_n + dt/2.*f_0(u_ss_n)
                u_split2[n+1] = u_ss

                # --- Strang splitting using exact integrator for u'=f_0 ---
                # First step
                u_s_n = u_split3[n]
                u_s = u_s_n*np.exp(dt/2.)  # exact
                # Second step
                u_sss_n = u_s
                u_sss = u_sss_n + dt*f_1(u_sss_n)
                # Third step
                u_ss_n = u_sss
                u_ss = u_ss_n*np.exp(dt/2.)  # exact
                u_split3[n+1] = u_ss

        return u_FE, u_split1, u_split2, u_split3, t
```

Compact implementation We have used quite many lines for the steps in the splitting methods. Many will prefer to condense the code a bit, as done here:

```
# Ordinary splitting
u_s = u_split1[n] + dt*f_0(u_split1[n])
u_split1[n+1] = u_s + dt*f_1(u_s)
# Strang splitting
u_s = u_split2[n] + dt/2.*f_0(u_split2[n])
u_sss = u_s + dt*f_1(u_s)
u_split2[n+1] = u_sss + dt/2.*f_0(u_sss)
# Strang splitting using exact integrator for u'=f_0
u_s = u_split3[n]*np.exp(dt/2.)  # exact
u_ss = u_s + dt*f_1(u_s)
u_split3[n+1] = u_ss*np.exp(dt/2.)
```

Results Figure 3.3 shows that the impact of splitting is significant. Interestingly, however, the Forward Euler method applied to the entire problem directly is much more accurate than any of the splitting schemes. We also see that Strang splitting is definitely more accurate than ordinary splitting and that it helps a bit to use an exact solution of $u' = f_0(u)$. With a large time step ($\Delta t = 0.2$, left plot in Fig. 3.3), the asymptotic values are off by 20–30 %. A more reasonable time step ($\Delta t = 0.05$, right plot in Fig. 3.3) gives better results, but still the asymptotic values are up to 10 % wrong.

As technique for solving nonlinear ODEs, we realize that the present case study is not particularly promising, as the Forward Euler method both linearizes the original problem and provides a solution that is much more accurate than any of the splitting techniques. In complicated multi-physics settings, on the other hand, splitting may be the only feasible way to go, and sometimes you really need to apply different numerics to different parts of a PDE problem. But in very simple problems, like the logistic ODE, splitting is just an inferior technique. Still, the logistic

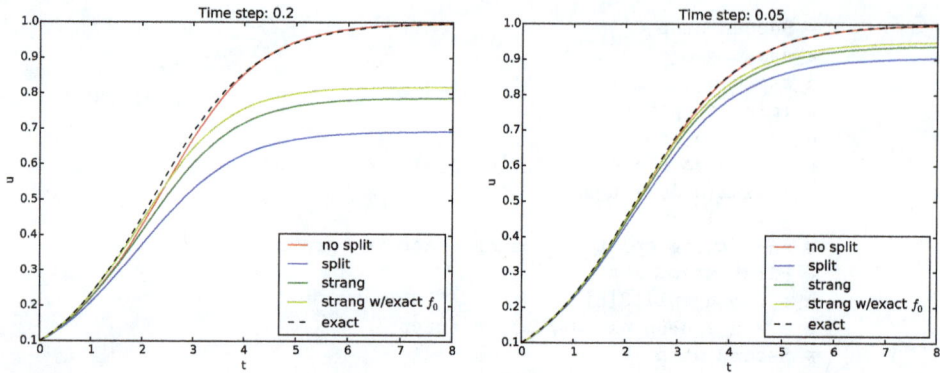

Fig. 3.3 Effect of ordinary and Strang splitting for the logistic equation

ODE is ideal for introducing all the mathematical details and for investigating the behavior.

3.6.4 Reaction-Diffusion Equation

Consider a diffusion equation coupled to chemical reactions modeled by a nonlinear term $f(u)$:

$$\frac{\partial u}{\partial t} = \alpha \nabla^2 u + f(u).$$

This is a physical process composed of two individual processes: u is the concentration of a substance that is locally generated by a chemical reaction $f(u)$, while u is spreading in space because of diffusion. There are obviously two time scales: one for the chemical reaction and one for diffusion. Typically, fast chemical reactions require much finer time stepping than slower diffusion processes. It could therefore be advantageous to split the two physical effects in separate models and use different numerical methods for the two.

A natural spitting in the present case is

$$\frac{\partial u^*}{\partial t} = \alpha \nabla^2 u^*, \tag{3.79}$$

$$\frac{\partial u^{**}}{\partial t} = f(u^{**}). \tag{3.80}$$

Looking at these familiar problems, we may apply a θ rule (implicit) scheme for (3.79) over one time step and avoid dealing with nonlinearities by applying an explicit scheme for (3.80) over the same time step.

Suppose we have some solution u at time level t_n. For flexibility, we define a θ method for the diffusion part (3.79) by

$$[D_t u^* = \alpha(D_x D_x u^* + D_y D_y u^*)]^{n+\theta}.$$

We use u^n as initial condition for u^*.

The reaction part, which is defined at each mesh point (without coupling values in different mesh points), can employ any scheme for an ODE. Here we use an Adams-Bashforth method of order 2. Recall that the overall accuracy of the splitting method is maximum $\mathcal{O}(\Delta t^2)$ for Strang splitting, otherwise it is just $\mathcal{O}(\Delta t)$. Higher-order methods for ODEs will therefore be a waste of work. The 2nd-order Adams-Bashforth method reads

$$u_{i,j}^{**,n+1} = u_{i,j}^{**,n} + \frac{1}{2}\Delta t \left(3f(u_{i,j}^{**,n}, t_n) - f(u_{i,j}^{**,n-1}, t_{n-1})\right) . \tag{3.81}$$

We can use a Forward Euler step to start the method, i.e, compute $u_{i,j}^{**,1}$.

The algorithm goes like this:

1. Solve the diffusion problem for one time step as usual.
2. Solve the reaction ODEs at each mesh point in $[t_n, t_n + \Delta t]$, using the diffusion solution in 1. as initial condition. The solution of the ODEs constitutes the solution of the original problem at the end of each time step.

We may use a much smaller time step when solving the reaction part, adapted to the dynamics of the problem $u' = f(u)$. This gives great flexibility in splitting methods.

3.6.5   Example: Reaction-Diffusion with Linear Reaction Term

The methods above may be explored in detail through a specific computational example in which we compute the convergence rates associated with four different solution approaches for the reaction-diffusion equation with a linear reaction term, i.e. $f(u) = -bu$. The methods comprise solving without splitting (just straight Forward Euler), ordinary splitting, first order Strang splitting, and second order Strang splitting. In all four methods, a standard centered difference approximation is used for the spatial second derivative. The methods share the error model $E = Ch^r$, while differing in the step h (being either Δx^2 or Δx) and the convergence rate r (being either 1 or 2).

All code commented below is found in the file split_diffu_react.py. When executed, a function convergence_rates is called, from which all convergence rate computations are handled:

```
def convergence_rates(scheme='diffusion'):

    F = 0.5       # Upper limit for FE (stability). For CN, this
                  # limit does not apply, but for simplicity, we
                  # choose F = 0.5 as the initial F value.
    T = 1.2
    a = 3.5
    b = 1
    L = 1.5
    k = np.pi/L
```

```
def exact(x, t):
    '''exact sol. to: du/dt = a*d^2u/dx^2 - b*u'''
    return np.exp(-(a*k**2 + b)*t) * np.sin(k*x)

def f(u, t):
    return -b*u

def I(x):
    return exact(x, 0)

global error    # error computed in the user action function
error = 0

# Convergence study
def action(u, x, t, n):
    global error
    if n == 1:        # New simulation, - reset error
        error = 0
    else:
        error = max(error, np.abs(u - exact(x, t[n])).max())

E = []
h = []
Nx_values = [10, 20, 40, 80]    # i.e., dx halved each time
for Nx in Nx_values:
    dx = L/Nx
    if scheme == 'Strang_splitting_2ndOrder':
        print 'Strang splitting with 2nd order schemes...'
        # In this case, E = C*h**r (with r = 2)  and  since
        # h = dx = K*dt, the ratio dt/dx must be constant.
        # To fulfill this demand, we must let F change
        # when dx changes. From F = a*dt/dx**2, it follows
        # that halving dx AND doubling F assures dt/dx const.
        # Initially, we simply choose F = 0.5.


        dt = F/a*dx**2
        #print 'dt/dx:', dt/dx
        Nt = int(round(T/float(dt)))
        t = np.linspace(0, Nt*dt, Nt+1)    # global time
        Strang_splitting_2ndOrder(I=I, a=a, b=b, f=f, L=L, dt=dt,
                                  dt_Rfactor=1, F=F, t=t, T=T,
                                  user_action=action)
        h.append(dx)
        # prepare for next iteration (make F match dx/2)
        F = F*2        # assures dt/dx const. when dx = dx/2
    else:
        # In these cases, E = C*h**r (with r = 1) and since
        # h = dx**2 = K*dt, the ratio dt/dx**2 must be constant.
        # This is fulfilled by choosing F = 0.5 (for FE stability)
        # and make sure that F, dx and dt comply to F = a*dt/dx**2.
        dt = F/a*dx**2
        Nt = int(round(T/float(dt)))
        t = np.linspace(0, Nt*dt, Nt+1)    # global time
```

```
            if scheme == 'diffusion':
                print 'FE on whole eqn...'
                diffusion_theta(I, a, f, L, dt, F, t, T,
                                step_no=0, theta=0,
                                u_L=0, u_R=0, user_action=action)
                h.append(dx**2)
            elif scheme == 'ordinary_splitting':
                print 'Ordinary splitting...'
                ordinary_splitting(I=I, a=a, b=b, f=f, L=L, dt=dt,
                                   dt_Rfactor=1, F=F, t=t, T=T,
                                   user_action=action)
                h.append(dx**2)
            elif scheme == 'Strang_splitting_1stOrder':
                print 'Strang splitting with 1st order schemes...'
                Strang_splitting_1stOrder(I=I, a=a, b=b, f=f, L=L, dt=dt,
                                          dt_Rfactor=1, F=F, t=t, T=T,
                                          user_action=action)
                h.append(dx**2)
            else:
                print 'Unknown scheme requested!'
                sys.exit(0)

            #print 'dt/dx**2:', dt/dx**2


        E.append(error)
        Nx *= 2            # Nx doubled gives dx/2

    print 'E:', E
    print 'h:', h

    # Convergence rates
    r = [np.log(E[i]/E[i-1])/np.log(h[i]/h[i-1])
         for i in range(1,len(Nx_values))]
    print 'Computed rates:', r


if __name__ == '__main__':

    schemes = ['diffusion',
               'ordinary_splitting',
               'Strang_splitting_1stOrder',
               'Strang_splitting_2ndOrder']

    for scheme in schemes:
        convergence_rates(scheme=scheme)
```

Now, with respect to the error ($E = Ch^r$), the Forward Euler scheme, the ordinary splitting scheme and first order Strang splitting scheme are all first order ($r = 1$), with a step $h = \Delta x^2 = K^{-1}\Delta t$, where K is some constant. This implies that the *ratio* $\frac{\Delta t}{\Delta x^2}$ must be held constant during convergence rate calculations. Furthermore, the Fourier number $F = \frac{\alpha\Delta t}{\Delta x^2}$ is upwards limited to $F = 0.5$, being the stability limit with explicit schemes. Thus, in these cases, we use the fixed value of F and a given (but changing) spatial resolution Δx to compute the corresponding value of Δt according to the expression for F. This assures that $\frac{\Delta t}{\Delta x^2}$ is kept constant. The loop in `convergence_rates` runs over a chosen set of grid points

(Nx_values) which gives a doubling of spatial resolution with each iteration (Δx is halved).

For the second order Strang splitting scheme, we have $r = 2$ and a step $h = \Delta x = K^{-1}\Delta t$, where K again is some constant. In this case, it is thus the ratio $\frac{\Delta t}{\Delta x}$ that must be held constant during the convergence rate calculations. From the expression for F, it is clear then that F must change with each halving of Δx. In fact, if F is doubled each time Δx is halved, the ratio $\frac{\Delta t}{\Delta x}$ will be constant (this follows, e.g., from the expression for F). This is utilized in our code.

A solver diffusion_theta is used in each of the four solution approaches:

```
def diffusion_theta(I, a, f, L, dt, F, t, T, step_no, theta=0.5,
                    u_L=0, u_R=0, user_action=None):
    """
    Full solver for the model problem using the theta-rule
    difference approximation in time (no restriction on F,
    i.e., the time step when theta >= 0.5).      Vectorized
    implementation and sparse (tridiagonal) coefficient matrix.
    Note that t always covers the whole global time interval, whether
    splitting is the case or not. T, on the other hand, is
    the end of the global time interval if there is no split,
    but if splitting, we use T=dt. When splitting, step_no
    keeps track of the time step number (for lookup in t).
    """

    Nt = int(round(T/float(dt)))
    dx = np.sqrt(a*dt/F)
    Nx = int(round(L/dx))
    x = np.linspace(0, L, Nx+1)        # Mesh points in space
    # Make sure dx and dt are compatible with x and t
    dx = x[1] - x[0]
    dt = t[1] - t[0]

    u   = np.zeros(Nx+1)    # solution array at t[n+1]
    u_1 = np.zeros(Nx+1)    # solution at t[n]

    # Representation of sparse matrix and right-hand side
    diagonal = np.zeros(Nx+1)
    lower    = np.zeros(Nx)
    upper    = np.zeros(Nx)
    b        = np.zeros(Nx+1)

    # Precompute sparse matrix (scipy format)
    Fl = F*theta
    Fr = F*(1-theta)
    diagonal[:] = 1 + 2*Fl
    lower[:] = -Fl  #1
    upper[:] = -Fl  #1
    # Insert boundary conditions
    diagonal[0] = 1
    upper[0] = 0
    diagonal[Nx] = 1
    lower[-1] = 0
```

```
diags = [0, -1, 1]
A = scipy.sparse.diags(
    diagonals=[diagonal, lower, upper],
    offsets=[0, -1, 1], shape=(Nx+1, Nx+1),
    format='csr')
#print A.todense()

# Allow f to be None or 0
if f is None or f == 0:
    f = lambda x, t: np.zeros((x.size)) \
        if isinstance(x, np.ndarray) else 0

# Set initial condition
if isinstance(I, np.ndarray):    # I is an array
    u_1 = np.copy(I)
else:                            # I is a function
    for i in range(0, Nx+1):
        u_1[i] = I(x[i])

if user_action is not None:
    user_action(u_1, x, t, step_no+0)


# Time loop
for n in range(0, Nt):
    b[1:-1] = u_1[1:-1] + \
                Fr*(u_1[:-2] - 2*u_1[1:-1] + u_1[2:]) + \
                dt*theta*f(u_1[1:-1], t[step_no+n+1]) + \
                dt*(1-theta)*f(u_1[1:-1], t[step_no+n])
    b[0] = u_L; b[-1] = u_R  # boundary conditions
    u[:] = scipy.sparse.linalg.spsolve(A, b)

    if user_action is not None:
        user_action(u, x, t, step_no+(n+1))

    # Update u_1 before next step
    u_1, u = u, u_1


# u is now contained in u_1 (swapping)
return u_1
```

For the no splitting approach with Forward Euler in time, this solver handles both the diffusion and the reaction term. When splitting, `diffusion_theta` takes care of the diffusion term only, while the reaction term is handled either by a Forward Euler scheme in `reaction_FE`, or by a second order Adams-Bashforth scheme from Odespy. The `reaction_FE` function covers one complete time step dt during ordinary splitting, while Strang splitting (both first and second order) applies it with $dt/2$ twice during each time step dt. Since the reaction term typically represents a much faster process than the diffusion term, a further refinement of the time step is made possible in `reaction_FE`. It was implemented as

```
def reaction_FE(I, f, L, Nx, dt, dt_Rfactor, t, step_no,
                user_action=None):
    """Reaction solver, Forward Euler method.
    Note the at t covers the whole global time interval.
    dt is either one complete,or one half, of the step in the
    diffusion part, i.e. there is a local time interval
    [0, dt] or [0, dt/2] that the reaction_FE
    deals with each time it is called. step_no keeps
    track of the (global) time step number (required
    for lookup in t).
    """

    u = np.copy(I)
    dt_local = dt/float(dt_Rfactor)
    Nt_local = int(round(dt/float(dt_local)))
    x = np.linspace(0, L, Nx+1)

    for n in range(Nt_local):
        time = t[step_no] + n*dt_local
        u[1:Nx] = u[1:Nx] + dt_local*f(u[1:Nx], time)

    # BC already inserted in diffusion step, i.e. no action here
    return u
```

With the ordinary splitting approach, each time step dt is covered twice. First computing the impact of the reaction term, then the contribution from the diffusion term:

```
def ordinary_splitting(I, a, b, f, L, dt,
                       dt_Rfactor, F, t, T,
                       user_action=None):
    '''1st order scheme, i.e. Forward Euler is enough for both
    the diffusion and the reaction part. The time step dt is
    given for the diffusion step, while the time step for the
    reaction part is found as dt/dt_Rfactor, where dt_Rfactor >= 1.
    '''
    Nt = int(round(T/float(dt)))
    dx = np.sqrt(a*dt/F)
    Nx = int(round(L/dx))
    x = np.linspace(0, L, Nx+1)        # Mesh points in space
    u = np.zeros(Nx+1)

    # Set initial condition u(x,0) = I(x)
    for i in range(0, Nx+1):
        u[i] = I(x[i])

    # In the following loop, each time step is "covered twice",
    # first for reaction, then for diffusion
    for n in range(0, Nt):
        # Reaction step (potentially many smaller steps within dt)
        u_s = reaction_FE(I=u, f=f, L=L, Nx=Nx,
                          dt=dt, dt_Rfactor=dt_Rfactor,
                          t=t, step_no=n,
                          user_action=None)
```

```
        u = diffusion_theta(I=u_s, a=a, f=0, L=L, dt=dt, F=F,
                            t=t, T=dt, step_no=n, theta=0,
                            u_L=0, u_R=0, user_action=None)

    if user_action is not None:
        user_action(u, x, t, n+1)

return
```

For the two Strang splitting approaches, each time step `dt` is handled by first computing the reaction step for (the first) `dt/2`, followed by a diffusion step `dt`, before the reaction step is treated once again for (the remaining) `dt/2`. Since first order Strang splitting is no better than first order accurate, both the reaction and diffusion steps are computed explicitly. The solver was implemented as

```
def Strang_splitting_1stOrder(I, a, b, f, L, dt, dt_Rfactor,
                              F, t, T, user_action=None):
    '''Strang splitting while still using FE for the reaction
    step and for the diffusion step. Gives 1st order scheme.
    The time step dt is given for the diffusion step, while
    the time step for the reaction part is found as
    0.5*dt/dt_Rfactor, where dt_Rfactor >= 1. Introduce an
    extra time mesh t2 for the reaction part, since it steps dt/2.
    '''
    Nt = int(round(T/float(dt)))
    t2 = np.linspace(0, Nt*dt, (Nt+1)+Nt)   # Mesh points in diff
    dx = np.sqrt(a*dt/F)
    Nx = int(round(L/dx))
    x = np.linspace(0, L, Nx+1)
    u = np.zeros(Nx+1)

    # Set initial condition u(x,0) = I(x)
    for i in range(0, Nx+1):
        u[i] = I(x[i])

    for n in range(0, Nt):
        # Reaction step (1/2 dt: from t_n to t_n+1/2)
        # (potentially many smaller steps within dt/2)
        u_s = reaction_FE(I=u, f=f, L=L, Nx=Nx,
                          dt=dt/2.0, dt_Rfactor=dt_Rfactor,
                          t=t2, step_no=2*n,
                          user_action=None)
        # Diffusion step (1 dt: from t_n to t_n+1)
        u_sss = diffusion_theta(I=u_s, a=a, f=0, L=L, dt=dt, F=F,
                                t=t, T=dt, step_no=n, theta=0,
                                u_L=0, u_R=0, user_action=None)
        # Reaction step (1/2 dt: from t_n+1/2 to t_n+1)
        # (potentially many smaller steps within dt/2)
        u = reaction_FE(I=u_sss, f=f, L=L, Nx=Nx,
                        dt=dt/2.0, dt_Rfactor=dt_Rfactor,
                        t=t2, step_no=2*n+1,
                        user_action=None)

        if user_action is not None:
            user_action(u, x, t, n+1)

    return
```

The second order version of the Strang splitting approach utilizes a second order Adams-Bashforth solver for the reaction part and a Crank-Nicolson scheme for the diffusion part. The solver has the same structure as the one for first order Strang splitting and was implemented as

```python
def Strang_splitting_2ndOrder(I, a, b, f, L, dt, dt_Rfactor,
                              F, t, T, user_action=None):
    '''Strang splitting using Crank-Nicolson for the diffusion
    step (theta-rule) and Adams-Bashforth 2 for the reaction step.
    Gives 2nd order scheme. Introduce an extra time mesh t2 for
    the reaction part, since it steps dt/2.
    '''
    import odespy
    Nt = int(round(T/float(dt)))
    t2 = np.linspace(0, Nt*dt, (Nt+1)+Nt)   # Mesh points in diff
    dx = np.sqrt(a*dt/F)
    Nx = int(round(L/dx))
    x = np.linspace(0, L, Nx+1)
    u = np.zeros(Nx+1)

    # Set initial condition u(x,0) = I(x)
    for i in range(0, Nx+1):
        u[i] = I(x[i])

    reaction_solver = odespy.AdamsBashforth2(f)

    for n in range(0, Nt):
        # Reaction step (1/2 dt: from t_n to t_n+1/2)
        # (potentially many smaller steps within dt/2)
        reaction_solver.set_initial_condition(u)
        t_points = np.linspace(0, dt/2.0, dt_Rfactor+1)
        u_AB2, t_ = reaction_solver.solve(t_points) # t_ not needed
        u_s = u_AB2[-1,:]  # pick sol at last point in time

        # Diffusion step (1 dt: from t_n to t_n+1)
        u_sss = diffusion_theta(I=u_s, a=a, f=0, L=L, dt=dt, F=F,
                                t=t, T=dt, step_no=n, theta=0.5,
                                u_L=0, u_R=0, user_action=None)
        # Reaction step (1/2 dt: from t_n+1/2 to t_n+1)
        # (potentially many smaller steps within dt/2)
        reaction_solver.set_initial_condition(u_sss)
        t_points = np.linspace(0, dt/2.0, dt_Rfactor+1)
        u_AB2, t_ = reaction_solver.solve(t_points) # t_ not needed
        u = u_AB2[-1,:]  # pick sol at last point in time

        if user_action is not None:
            user_action(u, x, t, n+1)

    return
```

When executing split_diffu_react.py, we find that the estimated convergence rates are as expected. The second order Strang splitting gives the least error (about $4e^{-5}$) and has second order convergence ($r = 2$), while the remaining three approaches have first order convergence ($r = 1$).

3.6.6 Analysis of the Splitting Method

Let us address a linear PDE problem for which we can develop analytical solutions of the discrete equations, with and without splitting, and discuss these. Choosing $f(u) = -\beta u$ for a constant β gives a linear problem. We use the Forward Euler method for both the PDE and ODE problems.

We seek a 1D Fourier wave component solution of the problem, assuming homogeneous Dirichlet conditions at $x = 0$ and $x = L$:

$$u = e^{-\alpha k^2 t - \beta t} \sin kx, \quad k = \frac{\pi}{L}.$$

This component fits the 1D PDE problem ($f = 0$). On complex form we can write

$$u = e^{-\alpha k^2 t - \beta t + ikx},$$

where $i = \sqrt{-1}$ and the imaginary part is taken as the physical solution.

We refer to Sect. 3.3 and to the book [9] for a discussion of exact numerical solutions to diffusion and decay problems, respectively. The key idea is to search for solutions $A^n e^{ikx}$ and determine A. For the diffusion problem solved by a Forward Euler method one has

$$A = 1 - 4F \sin^p,$$

where $F = \alpha \Delta t / \Delta x^2$ is the mesh Fourier number and $p = k\Delta x/2$ is a dimensionless number reflecting the spatial resolution (number of points per wave length in space). For the decay problem $u' = -\beta u$, we have $A = 1 - q$, where q is a dimensionless parameter for the resolution in the decay problem: $q = \beta \Delta t$.

The original model problem can also be discretized by a Forward Euler scheme,

$$[D_t^+ u = \alpha D_x D_x u - \beta u]_i^n.$$

Assuming $A^n e^{ikx}$ we find that

$$u_i^n = (1 - 4F \sin^p - q)^n \sin kx.$$

We are particularly interested in what happens at one time step. That is,

$$u_i^n = (1 - 4F \sin^2 p)u_i^{n-1}.$$

In the two stage algorithm, we first compute the diffusion step

$$u_i^{*,n+1} = (1 - 4F \sin^2 p)u_i^{n-1}.$$

Then we use this as input to the decay algorithm and arrive at

$$u^{**,n+1} = (1 - q)u^{*,n+1} = (1 - q)(1 - 4F \sin^2 p)u_i^{n-1}.$$

The splitting approximation over one step is therefore

$$E = 1 - 4F \sin^p - q - (1 - q)(1 - 4F \sin^2 p) = -q(2 - F \sin^2 p).$$

3.7 Exercises

Problem 5.1: Determine if equations are nonlinear or not
Classify each term in the following equations as linear or nonlinear. Assume that u, $\boldsymbol{u}$, and p are unknown functions and that all other symbols are known quantities.

1. $mu'' + \beta|u'|u' + cu = F(t)$
2. $u_t = \alpha u_{xx}$
3. $u_{tt} = c^2 \nabla^2 u$
4. $u_t = \nabla \cdot (\alpha(u)\nabla u) + f(x, y)$
5. $u_t + f(u)_x = 0$
6. $\boldsymbol{u}_t + \boldsymbol{u} \cdot \nabla \boldsymbol{u} = -\nabla p + r\nabla^2 \boldsymbol{u}, \ \nabla \cdot \boldsymbol{u} = 0$ ($\boldsymbol{u}$ is a vector field)
7. $u' = f(u, t)$
8. $\nabla^2 u = \lambda e^u$

Filename: `nonlinear_vs_linear`.

Problem 3.2: Derive and investigate a generalized logistic model
The logistic model for population growth is derived by assuming a nonlinear growth rate,

$$u' = a(u)u, \quad u(0) = I, \tag{3.82}$$

and the logistic model arises from the simplest possible choice of $a(u)$: $r(u) = \varrho(1-u/M)$, where M is the maximum value of u that the environment can sustain, and ϱ is the growth under unlimited access to resources (as in the beginning when u is small). The idea is that $a(u) \sim \varrho$ when u is small and that $a(t) \to 0$ as $u \to M$.

An $a(u)$ that generalizes the linear choice is the polynomial form

$$a(u) = \varrho(1 - u/M)^p, \tag{3.83}$$

where $p > 0$ is some real number.

a) Formulate a Forward Euler, Backward Euler, and a Crank-Nicolson scheme for (3.82).

Hint Use a geometric mean approximation in the Crank-Nicolson scheme: $[a(u)u]^{n+1/2} \approx a(u^n)u^{n+1}$.

b) Formulate Picard and Newton iteration for the Backward Euler scheme in a).
c) Implement the numerical solution methods from a) and b). Use `logistic.py` to compare the case $p = 1$ and the choice (3.83).
d) Implement unit tests that check the asymptotic limit of the solutions: $u \to M$ as $t \to \infty$.

Hint You need to experiment to find what "infinite time" is (increases substantially with p) and what the appropriate tolerance is for testing the asymptotic limit.

e) Perform experiments with Newton and Picard iteration for the model (3.83). See how sensitive the number of iterations is to Δt and p.

Filename: `logistic_p`.

Problem 3.3: Experience the behavior of Newton's method

The program `Newton_demo.py` illustrates graphically each step in Newton's method and is run like

---- Terminal ----

```
Terminal> python Newton_demo.py f dfdx x0 xmin xmax
```

Use this program to investigate potential problems with Newton's method when solving $e^{-0.5x^2} \cos(\pi x) = 0$. Try a starting point $x_0 = 0.8$ and $x_0 = 0.85$ and watch the different behavior. Just run

---- Terminal ----

```
Terminal>   python Newton_demo.py '0.2 + exp(-0.5*x**2)*cos(pi*x)' \
            '-x*exp(-x**2)*cos(pi*x) - pi*exp(-x**2)*sin(pi*x)' \
            0.85 -3 3
```

---

and repeat with 0.85 replaced by 0.8.

Exercise 3.4: Compute the Jacobian of a 2 × 2 system

Write up the system (3.18)–(3.19) in the form $F(u) = 0$, $F = (F_0, F_1)$, $u = (u_0, u_1)$, and compute the Jacobian $J_{i,j} = \partial F_i / \partial u_j$.

Problem 3.5: Solve nonlinear equations arising from a vibration ODE

Consider a nonlinear vibration problem

$$mu'' + bu'|u'| + s(u) = F(t), \tag{3.84}$$

where $m > 0$ is a constant, $b \geq 0$ is a constant, $s(u)$ a possibly nonlinear function of u, and $F(t)$ is a prescribed function. Such models arise from Newton's second law of motion in mechanical vibration problems where $s(u)$ is a spring or restoring force, mu'' is mass times acceleration, and $bu'|u'|$ models water or air drag.

a) Rewrite the equation for u as a system of two first-order ODEs, and discretize this system by a Crank-Nicolson (centered difference) method. With $v = u'$, we get a nonlinear term $v^{n+\frac{1}{2}}|v^{n+\frac{1}{2}}|$. Use a geometric average for $v^{n+\frac{1}{2}}$.
b) Formulate a Picard iteration method to solve the system of nonlinear algebraic equations.
c) Explain how to apply Newton's method to solve the nonlinear equations at each time level. Derive expressions for the Jacobian and the right-hand side in each Newton iteration.

Filename: `nonlin_vib`.

Exercise 3.6: Find the truncation error of arithmetic mean of products

In Sect. 3.3.4 we introduce alternative arithmetic means of a product. Say the product is $P(t)Q(t)$ evaluated at $t = t_{n+\frac{1}{2}}$. The exact value is

$$[PQ]^{n+\frac{1}{2}} = P^{n+\frac{1}{2}} Q^{n+\frac{1}{2}} .$$

There are two obvious candidates for evaluating $[PQ]^{n+\frac{1}{2}}$ as a mean of values of P and Q at t_n and t_{n+1}. Either we can take the arithmetic mean of each factor P and Q,

$$[PQ]^{n+\frac{1}{2}} \approx \frac{1}{2}(P^n + P^{n+1})\frac{1}{2}(Q^n + Q^{n+1}), \tag{3.85}$$

or we can take the arithmetic mean of the product PQ:

$$[PQ]^{n+\frac{1}{2}} \approx \frac{1}{2}(P^n Q^n + P^{n+1} Q^{n+1}). \tag{3.86}$$

The arithmetic average of $P(t_{n+\frac{1}{2}})$ is $\mathcal{O}(\Delta t^2)$:

$$P\left(t_{n+\frac{1}{2}}\right) = \frac{1}{2}(P^n + P^{n+1}) + \mathcal{O}(\Delta t^2).$$

A fundamental question is whether (3.85) and (3.86) have different orders of accuracy in $\Delta t = t_{n+1} - t_n$. To investigate this question, expand quantities at t_{n+1} and t_n in Taylor series around $t_{n+\frac{1}{2}}$, and subtract the true value $[PQ]^{n+\frac{1}{2}}$ from the approximations (3.85) and (3.86) to see what the order of the error terms are.

Hint You may explore sympy for carrying out the tedious calculations. A general Taylor series expansion of $P(t + \frac{1}{2}\Delta t)$ around t involving just a general function $P(t)$ can be created as follows:

```
>>> from sympy import *
>>> t, dt = symbols('t dt')
>>> P = symbols('P', cls=Function)
>>> P(t).series(t, 0, 4)
P(0) + t*Subs(Derivative(P(_x), _x), (_x,), (0,)) +
t**2*Subs(Derivative(P(_x), _x, _x), (_x,), (0,))/2 +
t**3*Subs(Derivative(P(_x), _x, _x, _x), (_x,), (0,))/6 + O(t**4)
>>> P_p = P(t).series(t, 0, 4).subs(t, dt/2)
>>> P_p
P(0) + dt*Subs(Derivative(P(_x), _x), (_x,), (0,))/2 +
dt**2*Subs(Derivative(P(_x), _x, _x), (_x,), (0,))/8 +
dt**3*Subs(Derivative(P(_x), _x, _x, _x), (_x,), (0,))/48 + O(dt**4)
```

The error of the arithmetic mean, $\frac{1}{2}(P(-\frac{1}{2}\Delta t) + P(-\frac{1}{2}\Delta t))$ for $t = 0$ is then

```
>>> P_m = P(t).series(t, 0, 4).subs(t, -dt/2)
>>> mean = Rational(1,2)*(P_m + P_p)
>>> error = simplify(expand(mean) - P(0))
>>> error
dt**2*Subs(Derivative(P(_x), _x, _x), (_x,), (0,))/8 + O(dt**4)
```

Use these examples to investigate the error of (3.85) and (3.86) for $n = 0$. (Choosing $n = 0$ is necessary for not making the expressions too complicated for sympy, but there is of course no lack of generality by using $n = 0$ rather than an arbitrary n - the main point is the product and addition of Taylor series.)
Filename: `product_arith_mean`.

Problem 5.7: Newton's method for linear problems

Suppose we have a linear system $F(u) = Au - b = 0$. Apply Newton's method to this system, and show that the method converges in one iteration.
Filename: `Newton_linear`.

Problem 3.8: Discretize a 1D problem with a nonlinear coefficient

We consider the problem

$$((1 + u^2)u')' = 1, \quad x \in (0, 1), \quad u(0) = u(1) = 0. \qquad (3.87)$$

Discretize (3.87) by a centered finite difference method on a uniform mesh.
Filename: `nonlin_1D_coeff_discretize`.

Problem 5.9: Linearize a 1D problem with a nonlinear coefficient

We have a two-point boundary value problem

$$((1 + u^2)u')' = 1, \quad x \in (0, 1), \quad u(0) = u(1) = 0. \qquad (3.88)$$

a) Construct a Picard iteration method for (3.88) without discretizing in space.
b) Apply Newton's method to (3.88) without discretizing in space.
c) Discretize (3.88) by a centered finite difference scheme. Construct a Picard method for the resulting system of nonlinear algebraic equations.
d) Discretize (3.88) by a centered finite difference scheme. Define the system of nonlinear algebraic equations, calculate the Jacobian, and set up Newton's method for solving the system.

Filename: `nonlin_1D_coeff_linearize`.

Problem 3.10: Finite differences for the 1D Bratu problem

We address the so-called Bratu problem

$$u'' + \lambda e^u = 0, \quad x \in (0, 1), \quad u(0) = u(1) = 0, \qquad (3.89)$$

where λ is a given parameter and u is a function of x. This is a widely used model problem for studying numerical methods for nonlinear differential equations. The problem (3.89) has an exact solution

$$u_e(x) = -2 \ln \left(\frac{\cosh((x - \frac{1}{2})\theta/2)}{\cosh(\theta/4)} \right),$$

where θ solves

$$\theta = \sqrt{2\lambda} \cosh(\theta/4).$$

There are two solutions of (3.89) for $0 < \lambda < \lambda_c$ and no solution for $\lambda > \lambda_c$. For $\lambda = \lambda_c$ there is one unique solution. The critical value λ_c solves

$$1 = \sqrt{2\lambda_c} \frac{1}{4} \sinh(\theta(\lambda_c)/4).$$

A numerical value is $\lambda_c = 3.513830719$.

a) Discretize (3.89) by a centered finite difference method.
b) Set up the nonlinear equations $F_i(u_0, u_1, \ldots, u_{N_x}) = 0$ from a). Calculate the associated Jacobian.
c) Implement a solver that can compute $u(x)$ using Newton's method. Plot the error as a function of x in each iteration.
d) Investigate whether Newton's method gives second-order convergence by computing $||u_e - u||/||u_e - u^-||^2$ in each iteration, where u is solution in the current iteration and u^- is the solution in the previous iteration.

Filename: `nonlin_1D_Bratu_fd`.

Problem 3.11: Discretize a nonlinear 1D heat conduction PDE by finite differences

We address the 1D heat conduction PDE

$$\varrho c(T)T_t = (k(T)T_x)_x,$$

for $x \in [0, L]$, where ϱ is the density of the solid material, $c(T)$ is the heat capacity, T is the temperature, and $k(T)$ is the heat conduction coefficient. $T(x, 0) = I(x)$, and ends are subject to a cooling law:

$$k(T)T_x|_{x=0} = h(T)(T - T_s), \quad -k(T)T_x|_{x=L} = h(T)(T - T_s),$$

where $h(T)$ is a heat transfer coefficient and T_s is the given surrounding temperature.

a) Discretize this PDE in time using either a Backward Euler or Crank-Nicolson scheme.
b) Formulate a Picard iteration method for the time-discrete problem (i.e., an iteration method before discretizing in space).
c) Formulate a Newton method for the time-discrete problem in b).
d) Discretize the PDE by a finite difference method in space. Derive the matrix and right-hand side of a Picard iteration method applied to the space-time discretized PDE.
e) Derive the matrix and right-hand side of a Newton method applied to the discretized PDE in d).

Filename: `nonlin_1D_heat_FD`.

Problem 3.12: Differentiate a highly nonlinear term

The operator $\nabla \cdot (\alpha(u)\nabla u)$ with $\alpha(u) = |\nabla u|^q$ appears in several physical problems, especially flow of Non-Newtonian fluids. The expression $|\nabla u|$ is defined as the Euclidean norm of a vector: $|\nabla u|^2 = \nabla u \cdot \nabla u$. In a Newton method one has to carry out the differentiation $\partial \alpha(u)/\partial c_j$, for $u = \sum_k c_k \psi_k$. Show that

$$\frac{\partial}{\partial u_j}|\nabla u|^q = q|\nabla u|^{q-2}\nabla u \cdot \nabla \psi_j.$$

Filename: `nonlin_differentiate`.

Exercise 3.13: Crank-Nicolson for a nonlinear 3D diffusion equation

Redo Sect. 3.5.1 when a Crank-Nicolson scheme is used to discretize the equations in time and the problem is formulated for three spatial dimensions.

Hint Express the Jacobian as $J_{i,j,k,r,s,t} = \partial F_{i,j,k}/\partial u_{r,s,t}$ and observe, as in the 2D case, that $J_{i,j,k,r,s,t}$ is very sparse: $J_{i,j,k,r,s,t} \neq 0$ only for $r = i \pm i$, $s = j \pm 1$, and $t = k \pm 1$ as well as $r = i$, $s = j$, and $t = k$.
Filename: `nonlin_heat_FD_CN_2D`.

Problem 3.14: Find the sparsity of the Jacobian

Consider a typical nonlinear Laplace term like $\nabla \cdot \alpha(u)\nabla u$ discretized by centered finite differences. Explain why the Jacobian corresponding to this term has the same sparsity pattern as the matrix associated with the corresponding linear term $\alpha\nabla^2 u$.

Hint Set up the unknowns that enter the difference equation at a point (i, j) in 2D or (i, j, k) in 3D, and identify the nonzero entries of the Jacobian that can arise from such a type of difference equation.
Filename: `nonlin_sparsity_Jacobian`.

Problem 3.15: Investigate a 1D problem with a continuation method

Flow of a pseudo-plastic power-law fluid between two flat plates can be modeled by

$$\frac{d}{dx}\left(\mu_0 \left|\frac{du}{dx}\right|^{n-1}\frac{du}{dx}\right) = -\beta, \quad u'(0) = 0, \ u(H) = 0,$$

where $\beta > 0$ and $\mu_0 > 0$ are constants. A target value of n may be $n = 0.2$.

a) Formulate a Picard iteration method directly for the differential equation problem.

b) Perform a finite difference discretization of the problem in each Picard iteration. Implement a solver that can compute u on a mesh. Verify that the solver gives an exact solution for $n = 1$ on a uniform mesh regardless of the cell size.

c) Given a sequence of decreasing n values, solve the problem for each n using the solution for the previous n as initial guess for the Picard iteration. This is called a continuation method. Experiment with $n = (1, 0.6, 0.2)$ and $n = (1, 0.9, 0.8, \ldots, 0.2)$ and make a table of the number of Picard iterations versus n.

d) Derive a Newton method at the differential equation level and discretize the resulting linear equations in each Newton iteration with the finite difference method.

e) Investigate if Newton's method has better convergence properties than Picard iteration, both in combination with a continuation method.

A

Important Formulas

A.1 Finite Difference Operator Notation

$$u'(t_n) \approx [D_t u]^n = \frac{u^{n+\frac{1}{2}} - u^{n-\frac{1}{2}}}{\Delta t} \tag{A.1}$$

$$u'(t_n) \approx [D_{2t} u]^n = \frac{u^{n+1} - u^{n-1}}{2\Delta t} \tag{A.2}$$

$$u'(t_n) = [D_t^- u]^n = \frac{u^n - u^{n-1}}{\Delta t} \tag{A.3}$$

$$u'(t_n) \approx [D_t^+ u]^n = \frac{u^{n+1} - u^n}{\Delta t} \tag{A.4}$$

$$u'(t_{n+\theta}) = [\bar{D}_t u]^{n+\theta} = \frac{u^{n+1} - u^n}{\Delta t} \tag{A.5}$$

$$u'(t_n) \approx [D_t^{2-} u]^n = \frac{3u^n - 4u^{n-1} + u^{n-2}}{2\Delta t} \tag{A.6}$$

$$u''(t_n) \approx [D_t D_t u]^n = \frac{u^{n+1} - 2u^n + u^{n-1}}{\Delta t^2} \tag{A.7}$$

$$u\left(t_{n+\frac{1}{2}}\right) \approx [\bar{u}^t]^{n+\frac{1}{2}} = \frac{1}{2}(u^{n+1} + u^n) \tag{A.8}$$

$$u\left(t_{n+\frac{1}{2}}\right)^2 \approx [\overline{u^2}^{t,g}]^{n+\frac{1}{2}} = u^{n+1} u^n \tag{A.9}$$

$$u\left(t_{n+\frac{1}{2}}\right) \approx [\bar{u}^{t,h}]^{n+\frac{1}{2}} = \frac{2}{\frac{1}{u^{n+1}} + \frac{1}{u^n}} \tag{A.10}$$

$$u(t_{n+\theta}) \approx [\bar{u}^{t,\theta}]^{n+\theta} = \theta u^{n+1} + (1-\theta)u^n, \tag{A.11}$$

$$t_{n+\theta} = \theta t_{n+1} + (1-\theta)t_{n-1} \tag{A.12}$$

Some may wonder why θ is absent on the right-hand side of (A.5). The fraction is an approximation to the derivative at the point $t_{n+\theta} = \theta t_{n+1} + (1 - \theta)t_n$.

A.2   Truncation Errors of Finite Difference Approximations

$$u_e'(t_n) = [D_t u_e]^n + R^n = \frac{u_e^{n+\frac{1}{2}} - u_e^{n-\frac{1}{2}}}{\Delta t} + R^n,$$

$$R^n = -\frac{1}{24} u_e'''(t_n)\Delta t^2 + \mathcal{O}(\Delta t^4) \tag{A.13}$$

$$u_e'(t_n) = [D_{2t} u_e]^n + R^n = \frac{u_e^{n+1} - u_e^{n-1}}{2\Delta t} + R^n,$$

$$R^n = -\frac{1}{6} u_e'''(t_n)\Delta t^2 + \mathcal{O}(\Delta t^4) \tag{A.14}$$

$$u_e'(t_n) = [D_t^- u_e]^n + R^n = \frac{u_e^n - u_e^{n-1}}{\Delta t} + R^n,$$

$$R^n = -\frac{1}{2} u_e''(t_n)\Delta t + \mathcal{O}(\Delta t^2) \tag{A.15}$$

$$u_e'(t_n) = [D_t^+ u_e]^n + R^n = \frac{u_e^{n+1} - u_e^n}{\Delta t} + R^n,$$

$$R^n = \frac{1}{2} u_e''(t_n)\Delta t + \mathcal{O}(\Delta t^2) \tag{A.16}$$

$$u_e'(t_{n+\theta}) = [\bar{D}_t u_e]^{n+\theta} + R^{n+\theta} = \frac{u_e^{n+1} - u_e^n}{\Delta t} + R^{n+\theta},$$

$$R^{n+\theta} = -\frac{1}{2}(1-2\theta)u_e''(t_{n+\theta})\Delta t + \frac{1}{6}((1-\theta)^3 - \theta^3)u_e'''(t_{n+\theta})\Delta t^2$$
$$+ \mathcal{O}(\Delta t^3) \tag{A.17}$$

$$u_e'(t_n) = [D_t^{2-} u_e]^n + R^n = \frac{3u_e^n - 4u_e^{n-1} + u_e^{n-2}}{2\Delta t} + R^n,$$

$$R^n = \frac{1}{3} u_e'''(t_n)\Delta t^2 + \mathcal{O}(\Delta t^3) \tag{A.18}$$

$$u_e''(t_n) = [D_t D_t u_e]^n + R^n = \frac{u_e^{n+1} - 2u_e^n + u_e^{n-1}}{\Delta t^2} + R^n,$$

$$R^n = -\frac{1}{12} u_e''''(t_n)\Delta t^2 + \mathcal{O}(\Delta t^4) \tag{A.19}$$

$$u_e(t_{n+\theta}) = [\overline{u_e^{t,\theta}}]^{n+\theta} + R^{n+\theta} = \theta u_e^{n+1} + (1-\theta)u_e^n + R^{n+\theta},$$

$$R^{n+\theta} = -\frac{1}{2} u_e''(t_{n+\theta})\Delta t^2 \theta(1-\theta) + \mathcal{O}(\Delta t^3). \tag{A.20}$$

A.3   Finite Differences of Exponential Functions

Complex exponentials Let $u^n = \exp\left(i\omega n\Delta t\right) = e^{i\omega t_n}$.

$$[D_t D_t u]^n = u^n \frac{2}{\Delta t}(\cos\omega\Delta t - 1) = -\frac{4}{\Delta t}\sin^2\left(\frac{\omega\Delta t}{2}\right), \qquad (\text{A.21})$$

$$[D_t^+ u]^n = u^n \frac{1}{\Delta t}(\exp\left(i\omega\Delta t\right) - 1), \qquad (\text{A.22})$$

$$[D_t^- u]^n = u^n \frac{1}{\Delta t}(1 - \exp\left(-i\omega\Delta t\right)), \qquad (\text{A.23})$$

$$[D_t u]^n = u^n \frac{2}{\Delta t}i\,\sin\left(\frac{\omega\Delta t}{2}\right), \qquad (\text{A.24})$$

$$[D_{2t} u]^n = u^n \frac{1}{\Delta t}i\,\sin\left(\omega\Delta t\right). \qquad (\text{A.25})$$

Real exponentials Let $u^n = \exp\left(\omega n\Delta t\right) = e^{\omega t_n}$.

$$[D_t D_t u]^n = u^n \frac{2}{\Delta t}(\cos\omega\Delta t - 1) = -\frac{4}{\Delta t}\sin^2\left(\frac{\omega\Delta t}{2}\right), \qquad (\text{A.26})$$

$$[D_t^+ u]^n = u^n \frac{1}{\Delta t}(\exp\left(i\omega\Delta t\right) - 1), \qquad (\text{A.27})$$

$$[D_t^- u]^n = u^n \frac{1}{\Delta t}(1 - \exp\left(-i\omega\Delta t\right)), \qquad (\text{A.28})$$

$$[D_t u]^n = u^n \frac{2}{\Delta t}i\,\sin\left(\frac{\omega\Delta t}{2}\right), \qquad (\text{A.29})$$

$$[D_{2t} u]^n = u^n \frac{1}{\Delta t}i\,\sin\left(\omega\Delta t\right). \qquad (\text{A.30})$$

A.4   Finite Differences of t^n

The following results are useful when checking if a polynomial term in a solution fulfills the discrete equation for the numerical method.

$$[D_t^+ t]^n = 1, \qquad (\text{A.31})$$
$$[D_t^- t]^n = 1, \qquad (\text{A.32})$$
$$[D_t t]^n = 1, \qquad (\text{A.33})$$
$$[D_{2t} t]^n = 1, \qquad (\text{A.34})$$
$$[D_t D_t t]^n = 0. \qquad (\text{A.35})$$

The next formulas concern the action of difference operators on a t^2 term.

$$[D_t^+ t^2]^n = (2n + 1)\Delta t, \tag{A.36}$$

$$[D_t^- t^2]^n = (2n - 1)\Delta t, \tag{A.37}$$

$$[D_t t^2]^n = 2n\Delta t, \tag{A.38}$$

$$[D_{2t} t^2]^n = 2n\Delta t, \tag{A.39}$$

$$[D_t D_t t^2]^n = 2. \tag{A.40}$$

Finally, we present formulas for a t^3 term:

$$[D_t^+ t^3]^n = 3(n\Delta t)^2 + 3n\Delta t^2 + \Delta t^2, \tag{A.41}$$

$$[D_t^- t^3]^n = 3(n\Delta t)^2 - 3n\Delta t^2 + \Delta t^2, \tag{A.42}$$

$$[D_t t^3]^n = 3(n\Delta t)^2 + \frac{1}{4}\Delta t^2, \tag{A.43}$$

$$[D_{2t} t^3]^n = 3(n\Delta t)^2 + \Delta t^2, \tag{A.44}$$

$$[D_t D_t t^3]^n = 6n\Delta t. \tag{A.45}$$

A.4.1 Software

Application of finite difference operators to polynomials and exponential functions, resulting in the formulas above, can easily be computed by some sympy code (from the file lib.py):

```
from sympy import *
t, dt, n, w = symbols('t dt n w', real=True)

# Finite difference operators

def D_t_forward(u):
    return (u(t + dt) - u(t))/dt

def D_t_backward(u):
    return (u(t) - u(t-dt))/dt

def D_t_centered(u):
    return (u(t + dt/2) - u(t-dt/2))/dt

def D_2t_centered(u):
    return (u(t + dt) - u(t-dt))/(2*dt)

def D_t_D_t(u):
    return (u(t + dt) - 2*u(t) + u(t-dt))/(dt**2)


op_list = [D_t_forward, D_t_backward,
           D_t_centered, D_2t_centered, D_t_D_t]
```

```
def ft1(t):
    return t

def ft2(t):
    return t**2

def ft3(t):
    return t**3

def f_expiwt(t):
    return exp(I*w*t)

def f_expwt(t):
    return exp(w*t)

func_list = [ft1, ft2, ft3, f_expiwt, f_expwt]
```

To see the results, one can now make a simple loop over the different types of functions and the various operators associated with them:

```
for func in func_list:
    for op in op_list:
        f = func
        e = op(f)
        e = simplify(expand(e))
        print e
        if func in [f_expiwt, f_expwt]:
            e = e/f(t)
        e = e.subs(t, n*dt)
        print expand(e)
        print factor(simplify(expand(e)))
```

Analysis of Truncation Error

Truncation error analysis provides a widely applicable framework for analyzing the accuracy of finite difference schemes. This type of analysis can also be used for finite element and finite volume methods if the discrete equations are written in finite difference form. The result of the analysis is an asymptotic estimate of the error in the scheme on the form $C h^r$, where h is a discretization parameter (Δt, Δx, etc.), r is a number, known as the convergence rate, and C is a constant, typically dependent on the derivatives of the exact solution.

Knowing r gives understanding of the accuracy of the scheme. But maybe even more important, a powerful verification method for computer codes is to check that the empirically observed convergence rates in experiments coincide with the theoretical value of r found from truncation error analysis.

The analysis can be carried out by hand, by symbolic software, and also numerically. All three methods will be illustrated. From examining the symbolic expressions of the truncation error we can add correction terms to the differential equations in order to increase the numerical accuracy.

In general, the term truncation error refers to the discrepancy that arises from performing a finite number of steps to approximate a process with infinitely many steps. The term is used in a number of contexts, including truncation of infinite series, finite precision arithmetic, finite differences, and differential equations. We shall be concerned with computing truncation errors arising in finite difference formulas and in finite difference discretizations of differential equations.

B.1 Overview of Truncation Error Analysis

B.1.1 Abstract Problem Setting

Consider an abstract differential equation

$$\mathcal{L}(u) = 0,$$

where $\mathcal{L}(u)$ is some formula involving the unknown u and its derivatives. One example is $\mathcal{L}(u) = u'(t) + a(t)u(t) - b(t)$, where a and b are constants or functions of time. We can discretize the differential equation and obtain a corresponding discrete model, here written as

$$\mathcal{L}_\Delta(u) = 0.$$

The solution u of this equation is the *numerical solution*. To distinguish the numerical solution from the exact solution of the differential equation problem, we denote the latter by u_e and write the differential equation and its discrete counterpart as

$$\mathcal{L}(u_e) = 0,$$
$$\mathcal{L}_\Delta(u) = 0.$$

Initial and/or boundary conditions can usually be left out of the truncation error analysis and are omitted in the following.

   The numerical solution u is, in a finite difference method, computed at a collection of mesh points. The discrete equations represented by the abstract equation $\mathcal{L}_\Delta(u) = 0$ are usually algebraic equations involving u at some neighboring mesh points.

B.1.2 Error Measures

A key issue is how accurate the numerical solution is. The ultimate way of addressing this issue would be to compute the error $u_e - u$ at the mesh points. This is usually extremely demanding. In very simplified problem settings we may, however, manage to derive formulas for the numerical solution u, and therefore closed form expressions for the error $u_e - u$. Such special cases can provide considerable insight regarding accuracy and stability, but the results are established for special problems.

   The error $u_e - u$ can be computed empirically in special cases where we know u_e. Such cases can be constructed by the method of manufactured solutions, where we choose some exact solution $u_e = v$ and fit a source term f in the governing differential equation $\mathcal{L}(u_e) = f$ such that $u_e = v$ is a solution (i.e., $f = \mathcal{L}(v)$). Assuming an error model of the form Ch^r, where h is the discretization parameter, such as Δt or Δx, one can estimate the convergence rate r. This is a widely applicable procedure, but the validity of the results is, strictly speaking, tied to the chosen test problems.

   Another error measure arises by asking to what extent the exact solution u_e fits the discrete equations. Clearly, u_e is in general not a solution of $\mathcal{L}_\Delta(u) = 0$, but we can define the residual

$$R = \mathcal{L}_\Delta(u_e),$$

and investigate how close R is to zero. A small R means intuitively that the discrete equations are close to the differential equation, and then we are tempted to think that u^n must also be close to $u_e(t_n)$.

   The residual R is known as the truncation error of the finite difference scheme $\mathcal{L}_\Delta(u) = 0$. It appears that the truncation error is relatively straightforward to compute by hand or symbolic software *without specializing the differential equation and the discrete model to a special case*. The resulting R is found as a power series in the discretization parameters. The leading-order terms in the series provide an asymptotic measure of the accuracy of the numerical solution method (as the discretization parameters tend to zero). An advantage of truncation error analysis, compared to empirical estimation of convergence rates, or detailed analysis of a special problem with a mathematical expression for the numerical solution, is that the truncation error analysis reveals the accuracy of the various building blocks in the numerical method and how each building block impacts the overall accuracy. The analysis can therefore be used to detect building blocks with lower accuracy than the others.

Knowing the truncation error or other error measures is important for verification of programs by empirically establishing convergence rates. The forthcoming text will provide many examples on how to compute truncation errors for finite difference discretizations of ODEs and PDEs.

B.2   Truncation Errors in Finite Difference Formulas

The accuracy of a finite difference formula is a fundamental issue when discretizing differential equations. We shall first go through a particular example in detail and thereafter list the truncation error in the most common finite difference approximation formulas.

B.2.1   Example: The Backward Difference for $u'(t)$

Consider a backward finite difference approximation of the first-order derivative u':

$$[D_t^- u]^n = \frac{u^n - u^{n-1}}{\Delta t} \approx u'(t_n) .$$ (B.1)

Here, u^n means the value of some function $u(t)$ at a point t_n, and $[D_t^- u]^n$ is the *discrete derivative* of $u(t)$ at $t = t_n$. The discrete derivative computed by a finite difference is, in general, not exactly equal to the derivative $u'(t_n)$. The error in the approximation is

$$R^n = [D_t^- u]^n - u'(t_n) .$$ (B.2)

The common way of calculating R^n is to

1. expand $u(t)$ in a Taylor series around the point where the derivative is evaluated, here t_n,
2. insert this Taylor series in (B.2), and
3. collect terms that cancel and simplify the expression.

The result is an expression for R^n in terms of a power series in Δt. The error R^n is commonly referred to as the *truncation error* of the finite difference formula.

The Taylor series formula often found in calculus books takes the form

$$f(x + h) = \sum_{i=0}^{\infty} \frac{1}{i!} \frac{d^i f}{dx^i}(x) h^i .$$

In our application, we expand the Taylor series around the point where the finite difference formula approximates the derivative. The Taylor series of u^n at t_n is simply $u(t_n)$, while the Taylor series of u^{n-1} at t_n must employ the general formula,

$$u(t_{n-1}) = u(t - \Delta t) = \sum_{i=0}^{\infty} \frac{1}{i!} \frac{d^i u}{dt^i}(t_n)(-\Delta t)^i$$

$$= u(t_n) - u'(t_n)\Delta t + \frac{1}{2}u''(t_n)\Delta t^2 + \mathcal{O}(\Delta t^3),$$

where $\mathcal{O}(\Delta t^3)$ means a power-series in Δt where the lowest power is Δt^3. We assume that Δt is small such that $\Delta t^p \gg \Delta t^q$ if p is smaller than q. The details of higher-order terms in Δt are therefore not of much interest. Inserting the Taylor series above in the right-hand side of (B.2) gives rise to some algebra:

$$[D_t^- u]^n - u'(t_n) = \frac{u(t_n) - u(t_{n-1})}{\Delta t} - u'(t_n)$$

$$= \frac{u(t_n) - (u(t_n) - u'(t_n)\Delta t + \frac{1}{2}u''(t_n)\Delta t^2 + \mathcal{O}(\Delta t^3))}{\Delta t} - u'(t_n)$$

$$= -\frac{1}{2}u''(t_n)\Delta t + \mathcal{O}(\Delta t^2),$$

which is, according to (B.2), the truncation error:

$$R^n = -\frac{1}{2}u''(t_n)\Delta t + \mathcal{O}(\Delta t^2). \tag{B.3}$$

The dominating term for small Δt is $-\frac{1}{2}u''(t_n)\Delta t$, which is proportional to Δt, and we say that the truncation error is of *first order* in Δt.

B.2.2 Example: The Forward Difference for $u'(t)$

We can analyze the approximation error in the forward difference

$$u'(t_n) \approx [D_t^+ u]^n = \frac{u^{n+1} - u^n}{\Delta t},$$

by writing

$$R^n = [D_t^+ u]^n - u'(t_n),$$

and expanding u^{n+1} in a Taylor series around t_n,

$$u(t_{n+1}) = u(t_n) + u'(t_n)\Delta t + \frac{1}{2}u''(t_n)\Delta t^2 + \mathcal{O}(\Delta t^3).$$

The result becomes

$$R = \frac{1}{2}u''(t_n)\Delta t + \mathcal{O}(\Delta t^2),$$

showing that also the forward difference is of first order.

B.2.3   Example: The Central Difference for $u'(t)$

For the central difference approximation,

$$u'(t_n) \approx [D_t u]^n, \quad [D_t u]^n = \frac{u^{n+\frac{1}{2}} - u^{n-\frac{1}{2}}}{\Delta t},$$

we write

$$R^n = [D_t u]^n - u'(t_n),$$

and expand $u(t_{n+\frac{1}{2}})$ and $u(t_{n-\frac{1}{2}})$ in Taylor series around the point t_n where the derivative is evaluated. We have

$$u\left(t_{n+\frac{1}{2}}\right) = u(t_n) + u'(t_n)\frac{1}{2}\Delta t + \frac{1}{2}u''(t_n)\left(\frac{1}{2}\Delta t\right)^2 +$$

$$\frac{1}{6}u'''(t_n)\left(\frac{1}{2}\Delta t\right)^3 + \frac{1}{24}u''''(t_n)\left(\frac{1}{2}\Delta t\right)^4 +$$

$$\frac{1}{120}u'''''(t_n)\left(\frac{1}{2}\Delta t\right)^5 + \mathcal{O}(\Delta t^6),$$

$$u\left(t_{n-\frac{1}{2}}\right) = u(t_n) - u'(t_n)\frac{1}{2}\Delta t + \frac{1}{2}u''(t_n)\left(\frac{1}{2}\Delta t\right)^2 -$$

$$\frac{1}{6}u'''(t_n)\left(\frac{1}{2}\Delta t\right)^3 + \frac{1}{24}u''''(t_n)\left(\frac{1}{2}\Delta t\right)^4 -$$

$$\frac{1}{120}u'''''(t_n)\left(\frac{1}{2}\Delta t\right)^5 + \mathcal{O}(\Delta t^6).$$

Now,

$$u\left(t_{n+\frac{1}{2}}\right) - u\left(t_{n-\frac{1}{2}}\right) = u'(t_n)\Delta t + \frac{1}{24}u'''(t_n)\Delta t^3 + \frac{1}{960}u'''''(t_n)\Delta t^5 + \mathcal{O}(\Delta t^7).$$

By collecting terms in $[D_t u]^n - u'(t_n)$ we find the truncation error to be

$$R^n = \frac{1}{24}u'''(t_n)\Delta t^2 + \mathcal{O}(\Delta t^4), \tag{B.4}$$

with only even powers of Δt. Since $R \sim \Delta t^2$ we say the centered difference is of *second order* in Δt.

B.2.4 Overview of Leading-Order Error Terms in Finite Difference Formulas

Here we list the leading-order terms of the truncation errors associated with several common finite difference formulas for the first and second derivatives.

$$[D_t u]^n = \frac{u^{n+\frac{1}{2}} - u^{n-\frac{1}{2}}}{\Delta t} = u'(t_n) + R^n, \tag{B.5}$$

$$R^n = \frac{1}{24} u'''(t_n) \Delta t^2 + \mathcal{O}(\Delta t^4) \tag{B.6}$$

$$[D_{2t} u]^n = \frac{u^{n+1} - u^{n-1}}{2\Delta t} = u'(t_n) + R^n, \tag{B.7}$$

$$R^n = \frac{1}{6} u'''(t_n) \Delta t^2 + \mathcal{O}(\Delta t^4) \tag{B.8}$$

$$[D_t^- u]^n = \frac{u^n - u^{n-1}}{\Delta t} = u'(t_n) + R^n, \tag{B.9}$$

$$R^n = -\frac{1}{2} u''(t_n) \Delta t + \mathcal{O}(\Delta t^2) \tag{B.10}$$

$$[D_t^+ u]^n = \frac{u^{n+1} - u^n}{\Delta t} = u'(t_n) + R^n, \tag{B.11}$$

$$R^n = \frac{1}{2} u''(t_n) \Delta t + \mathcal{O}(\Delta t^2) \tag{B.12}$$

$$[\bar{D}_t u]^{n+\theta} = \frac{u^{n+1} - u^n}{\Delta t} = u'(t_{n+\theta}) + R^{n+\theta}, \tag{B.13}$$

$$R^{n+\theta} = \frac{1}{2}(1 - 2\theta) u''(t_{n+\theta}) \Delta t - \frac{1}{6}((1-\theta)^3 - \theta^3) u'''(t_{n+\theta}) \Delta t^2 + \mathcal{O}(\Delta t^3) \tag{B.14}$$

$$[D_t^{2-} u]^n = \frac{3u^n - 4u^{n-1} + u^{n-2}}{2\Delta t} = u'(t_n) + R^n, \tag{B.15}$$

$$R^n = -\frac{1}{3} u'''(t_n) \Delta t^2 + \mathcal{O}(\Delta t^3) \tag{B.16}$$

$$[D_t D_t u]^n = \frac{u^{n+1} - 2u^n + u^{n-1}}{\Delta t^2} = u''(t_n) + R^n, \tag{B.17}$$

$$R^n = \frac{1}{12} u''''(t_n) \Delta t^2 + \mathcal{O}(\Delta t^4) \tag{B.18}$$

It will also be convenient to have the truncation errors for various means or averages. The weighted arithmetic mean leads to

$$[\bar{u}^{t,\theta}]^{n+\theta} = \theta u^{n+1} + (1 - \theta) u^n = u(t_{n+\theta}) + R^{n+\theta}, \tag{B.19}$$

$$R^{n+\theta} = \frac{1}{2} u''(t_{n+\theta}) \Delta t^2 \theta(1 - \theta) + \mathcal{O}(\Delta t^3). \tag{B.20}$$

The standard arithmetic mean follows from this formula when $\theta = \frac{1}{2}$. Expressed at point t_n we get

$$[\overline{u^t}]^n = \frac{1}{2}\left(u^{n-\frac{1}{2}} + u^{n+\frac{1}{2}}\right) = u(t_n) + R^n, \tag{B.21}$$

$$R^n = \frac{1}{8}u''(t_n)\Delta t^2 + \frac{1}{384}u''''(t_n)\Delta t^4 + \mathcal{O}(\Delta t^6). \tag{B.22}$$

The geometric mean also has an error $\mathcal{O}(\Delta t^2)$:

$$\left[\overline{u^{2}}^{t,g}\right]^n = u^{n-\frac{1}{2}}u^{n+\frac{1}{2}} = (u^n)^2 + R^n, \tag{B.23}$$

$$R^n = -\frac{1}{4}u'(t_n)^2\Delta t^2 + \frac{1}{4}u(t_n)u''(t_n)\Delta t^2 + \mathcal{O}(\Delta t^4). \tag{B.24}$$

The harmonic mean is also second-order accurate:

$$[\overline{u^{t,h}}]^n = u^n = \frac{2}{\frac{1}{u^{n-\frac{1}{2}}} + \frac{1}{u^{n+\frac{1}{2}}}} + R^{n+\frac{1}{2}}, \tag{B.25}$$

$$R^n = -\frac{u'(t_n)^2}{4u(t_n)}\Delta t^2 + \frac{1}{8}u''(t_n)\Delta t^2. \tag{B.26}$$

B.2.5   Software for Computing Truncation Errors

We can use sympy to aid calculations with Taylor series. The derivatives can be defined as symbols, say D3f for the 3rd derivative of some function f. A truncated Taylor series can then be written as f + D1f*h + D2f*h**2/2. The following class takes some symbol f for the function in question and makes a list of symbols for the derivatives. The __call__ method computes the symbolic form of the series truncated at num_terms terms.

```
import sympy as sym

class TaylorSeries:
    """Class for symbolic Taylor series."""
    def __init__(self, f, num_terms=4):
        self.f = f
        self.N = num_terms
        # Introduce symbols for the derivatives
        self.df = [f]
        for i in range(1, self.N+1):
            self.df.append(sym.Symbol('D%d%s' % (i, f.name)))

    def __call__(self, h):
        """Return the truncated Taylor series at x+h."""
        terms = self.f
        for i in range(1, self.N+1):
            terms += sym.Rational(1, sym.factorial(i))*self.df[i]*h**i
        return terms
```

We may, for example, use this class to compute the truncation error of the Forward Euler finite difference formula:

```
>>> from truncation_errors import TaylorSeries
>>> from sympy import *
>>> u, dt = symbols('u dt')
>>> u_Taylor = TaylorSeries(u, 4)
>>> u_Taylor(dt)
D1u*dt + D2u*dt**2/2 + D3u*dt**3/6 + D4u*dt**4/24 + u
>>> FE = (u_Taylor(dt) - u)/dt
>>> FE
(D1u*dt + D2u*dt**2/2 + D3u*dt**3/6 + D4u*dt**4/24)/dt
>>> simplify(FE)
D1u + D2u*dt/2 + D3u*dt**2/6 + D4u*dt**3/24
```

The truncation error consists of the terms after the first one (u').

The module file `trunc/truncation_errors.py` contains another class `DiffOp` with symbolic expressions for most of the truncation errors listed in the previous section. For example:

```
>>> from truncation_errors import DiffOp
>>> from sympy import *
>>> u = Symbol('u')
>>> diffop = DiffOp(u, independent_variable='t')
>>> diffop['geometric_mean']
-D1u**2*dt**2/4 - D1u*D3u*dt**4/48 + D2u**2*dt**4/64 + ...
>>> diffop['Dtm']
D1u + D2u*dt/2 + D3u*dt**2/6 + D4u*dt**3/24
>>> >>> diffop.operator_names()
['geometric_mean', 'harmonic_mean', 'Dtm', 'D2t', 'DtDt',
 'weighted_arithmetic_mean', 'Dtp', 'Dt']
```

The indexing of `diffop` applies names that correspond to the operators: `Dtp` for D_t^+, `Dtm` for D_t^-, `Dt` for D_t, `D2t` for D_{2t}, `DtDt` for $D_t D_t$.

B.3 Exponential Decay ODEs

We shall now compute the truncation error of a finite difference scheme for a differential equation. Our first problem involves the following linear ODE that models exponential decay,

$$u'(t) = -au(t). \tag{B.27}$$

B.3.1 Forward Euler Scheme

We begin with the Forward Euler scheme for discretizing (B.27):

$$[D_t^+ u = -au]^n. \tag{B.28}$$

The idea behind the truncation error computation is to insert the exact solution u_e of the differential equation problem (B.27) in the discrete equations (B.28) and find the residual that arises because u_e does not solve the discrete equations. Instead, u_e solves the discrete equations with a residual R^n:

$$[D_t^+ u_e + a u_e = R]^n .$$ (B.29)

From (B.11)–(B.12) it follows that

$$[D_t^+ u_e]^n = u_e'(t_n) + \frac{1}{2} u_e''(t_n) \Delta t + \mathcal{O}(\Delta t^2),$$

which inserted in (B.29) results in

$$u_e'(t_n) + \frac{1}{2} u_e''(t_n) \Delta t + \mathcal{O}(\Delta t^2) + a u_e(t_n) = R^n .$$

Now, $u_e'(t_n) + a u_e^n = 0$ since u_e solves the differential equation. The remaining terms constitute the residual:

$$R^n = \frac{1}{2} u_e''(t_n) \Delta t + \mathcal{O}(\Delta t^2) .$$ (B.30)

This is the truncation error R^n of the Forward Euler scheme.

Because R^n is proportional to Δt, we say that the Forward Euler scheme is of first order in Δt. However, the truncation error is just one error measure, and it is not equal to the true error $u_e^n - u^n$. For this simple model problem we can compute a range of different error measures for the Forward Euler scheme, including the true error $u_e^n - u^n$, and all of them have dominating terms proportional to Δt.

B.3.2 Crank-Nicolson Scheme

For the Crank-Nicolson scheme,

$$[D_t u = -au]^{n+\frac{1}{2}},$$ (B.31)

we compute the truncation error by inserting the exact solution of the ODE and adding a residual R,

$$[D_t u_e + a \overline{u_e}^t = R]^{n+\frac{1}{2}} .$$ (B.32)

The term $[D_t u_e]^{n+\frac{1}{2}}$ is easily computed from (B.5)–(B.6) by replacing n with $n + \frac{1}{2}$ in the formula,

$$[D_t u_e]^{n+\frac{1}{2}} = u_e'\left(t_{n+\frac{1}{2}}\right) + \frac{1}{24} u_e'''\left(t_{n+\frac{1}{2}}\right) \Delta t^2 + \mathcal{O}(\Delta t^4) .$$

The arithmetic mean is related to $u(t_{n+\frac{1}{2}})$ by (B.21)–(B.22) so

$$[a \overline{u_e}^t]^{n+\frac{1}{2}} = u_e\left(t_{n+\frac{1}{2}}\right) + \frac{1}{8} u_e''(t_n) \Delta t^2 + \mathcal{O}(\Delta t^4) .$$

Inserting these expressions in (B.32) and observing that $u'_e(t_{n+\frac{1}{2}}) + au_e^{n+\frac{1}{2}} = 0$, because $u_e(t)$ solves the ODE $u'(t) = -au(t)$ at any point t, we find that

$$R^{n+\frac{1}{2}} = \left(\frac{1}{24}u'''_e\left(t_{n+\frac{1}{2}}\right) + \frac{1}{8}u''_e(t_n)\right)\Delta t^2 + \mathcal{O}(\Delta t^4). \tag{B.33}$$

Here, the truncation error is of second order because the leading term in R is proportional to Δt^2.

At this point it is wise to redo some of the computations above to establish the truncation error of the Backward Euler scheme, see Exercise B.4.

B.3.3 The θ-Rule

We may also compute the truncation error of the θ-rule,

$$[\bar{D}_t u = -a\overline{u}^{t,\theta}]^{n+\theta}.$$

Our computational task is to find $R^{n+\theta}$ in

$$[\bar{D}_t u_e + a\overline{u_e}^{t,\theta} = R]^{n+\theta}.$$

From (B.13)–(B.14) and (B.19)–(B.20) we get expressions for the terms with u_e. Using that $u'_e(t_{n+\theta}) + au_e(t_{n+\theta}) = 0$, we end up with

$$R^{n+\theta} = \left(\frac{1}{2} - \theta\right)u''_e(t_{n+\theta})\Delta t + \frac{1}{2}\theta(1-\theta)u''_e(t_{n+\theta})\Delta t^2$$
$$+ \frac{1}{2}(\theta^2 - \theta + 3)u'''_e(t_{n+\theta})\Delta t^2 + \mathcal{O}(\Delta t^3). \tag{B.34}$$

For $\theta = \frac{1}{2}$ the first-order term vanishes and the scheme is of second order, while for $\theta \neq \frac{1}{2}$ we only have a first-order scheme.

B.3.4 Using Symbolic Software

The previously mentioned `truncation_error` module can be used to automate the Taylor series expansions and the process of collecting terms. Here is an example on possible use:

```
from truncation_error import DiffOp
from sympy import *

def decay():
    u, a = symbols('u a')
    diffop = DiffOp(u, independent_variable='t',
                    num_terms_Taylor_series=3)
    D1u = diffop.D(1)    # symbol for du/dt
    ODE = D1u + a*u      # define ODE
```

```
# Define schemes
FE = diffop['Dtp'] + a*u
CN = diffop['Dt' ] + a*u
BE = diffop['Dtm'] + a*u
theta = diffop['barDt'] + a*diffop['weighted_arithmetic_mean']
theta = sm.simplify(sm.expand(theta))
# Residuals (truncation errors)
R = {'FE': FE-ODE, 'BE': BE-ODE, 'CN': CN-ODE,
     'theta': theta-ODE}
return R
```

The returned dictionary becomes

```
decay: {
 'BE': D2u*dt/2 + D3u*dt**2/6,
 'FE': -D2u*dt/2 + D3u*dt**2/6,
 'CN': D3u*dt**2/24,
 'theta': -D2u*a*dt**2*theta**2/2 + D2u*a*dt**2*theta/2 -
          D2u*dt*theta + D2u*dt/2 + D3u*a*dt**3*theta**3/3 -
          D3u*a*dt**3*theta**2/2 + D3u*a*dt**3*theta/6 +
          D3u*dt**2*theta**2/2 - D3u*dt**2*theta/2 + D3u*dt**2/6,
}
```

The results are in correspondence with our hand-derived expressions.

B.3.5 Empirical Verification of the Truncation Error

The task of this section is to demonstrate how we can compute the truncation error R numerically. For example, the truncation error of the Forward Euler scheme applied to the decay ODE $u' = -ua$ is

$$R^n = [D_t^+ u_e + a u_e]^n . \tag{B.35}$$

If we happen to know the exact solution $u_e(t)$, we can easily evaluate R^n from the above formula.

To estimate how R varies with the discretization parameter Δt, which has been our focus in the previous mathematical derivations, we first make the assumption that $R = C\Delta t^r$ for appropriate constants C and r and small enough Δt. The rate r can be estimated from a series of experiments where Δt is varied. Suppose we have m experiments $(\Delta t_i, R_i)$, $i = 0, \ldots, m-1$. For two consecutive experiments $(\Delta t_{i-1}, R_{i-1})$ and $(\Delta t_i, R_i)$, a corresponding r_{i-1} can be estimated by

$$r_{i-1} = \frac{\ln(R_{i-1}/R_i)}{\ln(\Delta t_{i-1}/\Delta t_i)}, \tag{B.36}$$

for $i = 1, \ldots, m-1$. Note that the truncation error R_i varies through the mesh, so (B.36) is to be applied pointwise. A complicating issue is that R_i and R_{i-1} refer to different meshes. Pointwise comparisons of the truncation error at a certain point in all meshes therefore requires any computed R to be restricted to the *coarsest mesh*

and that all finer meshes contain all the points in the coarsest mesh. Suppose we
have N_0 intervals in the coarsest mesh. Inserting a superscript n in (B.36), where n
counts mesh points in the coarsest mesh, $n = 0, \ldots, N_0$, leads to the formula

$$r_{i-1}^n = \frac{\ln(R_{i-1}^n/R_i^n)}{\ln(\Delta t_{i-1}/\Delta t_i)} . \tag{B.37}$$

Experiments are most conveniently defined by N_0 and a number of refinements m.
Suppose each mesh has twice as many cells N_i as the previous one:

$$N_i = 2^i N_0, \quad \Delta t_i = T N_i^{-1},$$

where $[0, T]$ is the total time interval for the computations. Suppose the computed
R_i values on the mesh with N_i intervals are stored in an array R[i] (R being a list of
arrays, one for each mesh). Restricting this R_i function to the coarsest mesh means
extracting every N_i/N_0 point and is done as follows:

```
stride = N[i]/N_0
R[i] = R[i][::stride]
```

The quantity R[i][n] now corresponds to R_i^n.
    In addition to estimating r for the pointwise values of $R = C\Delta t^r$, we may also
consider an integrated quantity on mesh i,

$$R_{I,i} = \left(\Delta t_i \sum_{n=0}^{N_i} (R_i^n)^2 \right)^{\frac{1}{2}} \approx \int_0^T R_i(t) dt . \tag{B.38}$$

The sequence $R_{I,i}, i = 0, \ldots, m-1$, is also expected to behave as $C\Delta t^r$, with the
same r as for the pointwise quantity R, as $\Delta t \to 0$.
    The function below computes the R_i and $R_{I,i}$ quantities, plots them and com-
pares with the theoretically derived truncation error (R_a) if available.

```
import numpy as np
import scitools.std as plt

def estimate(truncation_error, T, N_0, m, makeplot=True):
    """
    Compute the truncation error in a problem with one independent
    variable, using m meshes, and estimate the convergence
    rate of the truncation error.

    The user-supplied function truncation_error(dt, N) computes
    the truncation error on a uniform mesh with N intervals of
    length dt::

      R, t, R_a = truncation_error(dt, N)

    where R holds the truncation error at points in the array t,
    and R_a are the corresponding theoretical truncation error
    values (None if not available).
```

```
The truncation_error function is run on a series of meshes
with 2**i*N_0 intervals, i=0,1,...,m-1.
The values of R and R_a are restricted to the coarsest mesh.
and based on these data, the convergence rate of R (pointwise)
and time-integrated R can be estimated empirically.
"""
N = [2**i*N_0 for i in range(m)]


R_I = np.zeros(m) # time-integrated R values on various meshes
R   = [None]*m    # time series of R restricted to coarsest mesh
R_a = [None]*m    # time series of R_a restricted to coarsest mesh
dt = np.zeros(m)
legends_R = [];  legends_R_a = []   # all legends of curves

for i in range(m):
    dt[i] = T/float(N[i])
    R[i], t, R_a[i] = truncation_error(dt[i], N[i])

    R_I[i] = np.sqrt(dt[i]*np.sum(R[i]**2))

    if i == 0:
        t_coarse = t              # the coarsest mesh

    stride = N[i]/N_0
    R[i] = R[i][::stride]         # restrict to coarsest mesh
    R_a[i] = R_a[i][::stride]

    if makeplot:
        plt.figure(1)
        plt.plot(t_coarse, R[i], log='y')
        legends_R.append('N=%d' % N[i])
        plt.hold('on')

        plt.figure(2)
        plt.plot(t_coarse, R_a[i] - R[i], log='y')
        plt.hold('on')
        legends_R_a.append('N=%d' % N[i])


if makeplot:
    plt.figure(1)
    plt.xlabel('time')
    plt.ylabel('pointwise truncation error')
    plt.legend(legends_R)
    plt.savefig('R_series.png')
    plt.savefig('R_series.pdf')
    plt.figure(2)
    plt.xlabel('time')
    plt.ylabel('pointwise error in estimated truncation error')
    plt.legend(legends_R_a)
    plt.savefig('R_error.png')
    plt.savefig('R_error.pdf')
```

```
# Convergence rates
r_R_I = convergence_rates(dt, R_I)
print 'R integrated in time; r:',
print ' '.join(['%.1f' % r for r in r_R_I])
R = np.array(R)   # two-dim. numpy array
r_R = [convergence_rates(dt, R[:,n])[-1]
       for n in range(len(t_coarse))]
```

The first `makeplot` block demonstrates how to build up two figures in parallel, using `plt.figure(i)` to create and switch to figure number `i`. Figure numbers start at 1. A logarithmic scale is used on the y axis since we expect that R as a function of time (or mesh points) is exponential. The reason is that the theoretical estimate (B.30) contains u_e'', which for the present model goes like e^{-at}. Taking the logarithm makes a straight line.

The code follows closely the previously stated mathematical formulas, but the statements for computing the convergence rates might deserve an explanation. The generic help function `convergence_rate(h, E)` computes and returns r_{i-1}, $i = 1, \ldots, m-1$ from (B.37), given Δt_i in h and R_i^n in E:

```
def convergence_rates(h, E):
    from math import log
    r = [log(E[i]/E[i-1])/log(h[i]/h[i-1])
         for i in range(1, len(h))]
    return r
```

Calling `r_R_I = convergence_rates(dt, R_I)` computes the sequence of rates $r_0, r_1, \ldots, r_{m-2}$ for the model $R_I \sim \Delta t^r$, while the statements

```
R = np.array(R)   # two-dim. numpy array
r_R = [convergence_rates(dt, R[:,n])[-1]
       for n in range(len(t_coarse))]
```

compute the final rate r_{m-2} for $R^n \sim \Delta t^r$ at each mesh point t_n in the coarsest mesh. This latter computation deserves more explanation. Since `R[i][n]` holds the estimated truncation error R_i^n on mesh i, at point t_n in the coarsest mesh, `R[:,n]` picks out the sequence R_i^n for $i = 0, \ldots, m-1$. The `convergence_rate` function computes the rates at t_n, and by indexing `[-1]` on the returned array from `convergence_rate`, we pick the rate r_{m-2}, which we believe is the best estimation since it is based on the two finest meshes.

The `estimate` function is available in a module `trunc_empir.py`. Let us apply this function to estimate the truncation error of the Forward Euler scheme. We need a function `decay_FE(dt, N)` that can compute (B.35) at the points in a mesh with time step `dt` and N intervals:

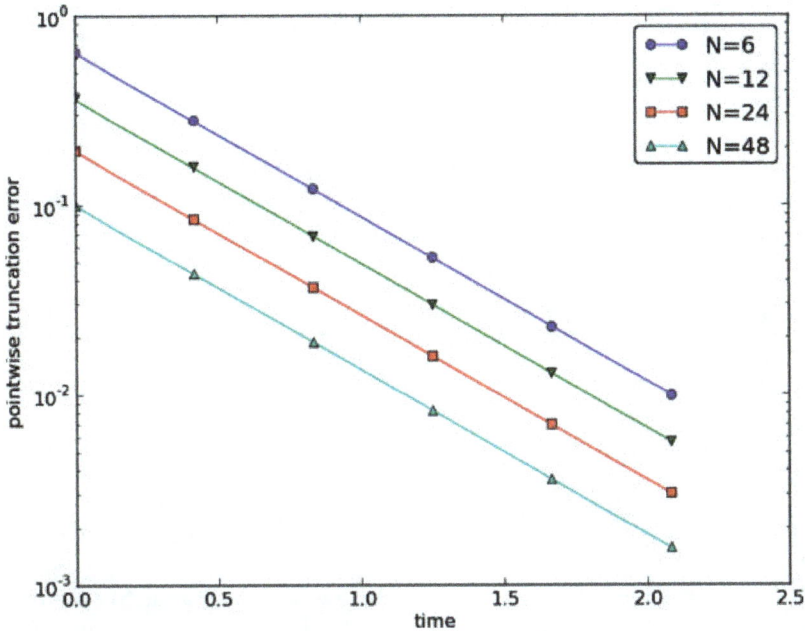

Fig. B.1 Estimated truncation error at mesh points for different meshes

```
import numpy as np
import trunc_empir

def decay_FE(dt, N):
    dt = float(dt)
    t = np.linspace(0, N*dt, N+1)
    u_e = I*np.exp(-a*t)  # exact solution, I and a are global
    u = u_e  # naming convention when writing up the scheme
    R = np.zeros(N)

    for n in range(0, N):
        R[n] = (u[n+1] - u[n])/dt + a*u[n]

    # Theoretical expression for the trunction error
    R_a = 0.5*I*(-a)**2*np.exp(-a*t)*dt

    return R, t[:-1], R_a[:-1]

if __name__ == '__main__':
    I = 1; a = 2  # global variables needed in decay_FE
    trunc_empir.estimate(decay_FE, T=2.5, N_0=6, m=4, makeplot=True)
```

The estimated rates for the integrated truncation error R_I become 1.1, 1.0, and 1.0 for this sequence of four meshes. All the rates for R^n, computed as r_R, are also very close to 1 at all mesh points. The agreement between the theoretical formula (B.30) and the computed quantity (ref(B.35)) is very good, as illustrated in Fig. B.1 and B.2. The program trunc_decay_FE.py was used to perform the simulations and it can easily be modified to test other schemes (see also Exercise B.5).

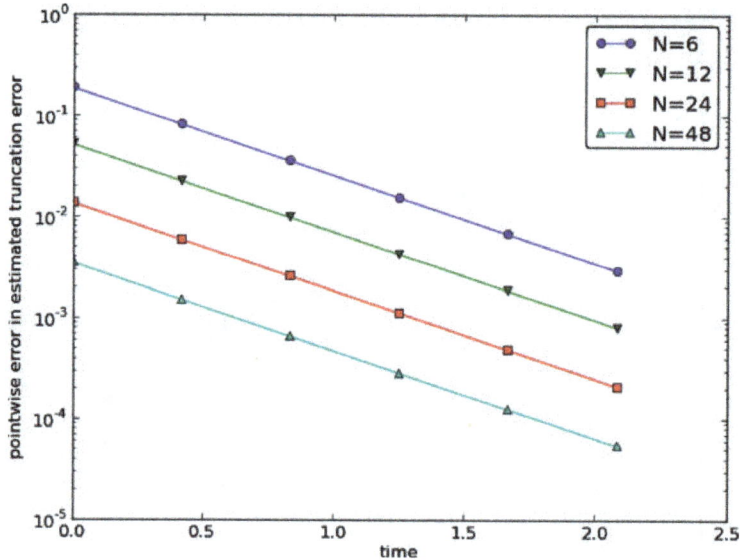

Fig. B.2 Difference between theoretical and estimated truncation error at mesh points for different meshes

B.3.6   Increasing the Accuracy by Adding Correction Terms

Now we ask the question: can we add terms in the differential equation that can help increase the order of the truncation error? To be precise, let us revisit the Forward Euler scheme for $u' = -au$, insert the exact solution u_e, include a residual R, but also include new terms C:

$$[D_t^+ u_e + au_e = C + R]^n . \tag{B.39}$$

Inserting the Taylor expansions for $[D_t^+ u_e]^n$ and keeping terms up to 3rd order in Δt gives the equation

$$\frac{1}{2}u_e''(t_n)\Delta t - \frac{1}{6}u_e'''(t_n)\Delta t^2 + \frac{1}{24}u_e''''(t_n)\Delta t^3 + \mathcal{O}(\Delta t^4) = C^n + R^n .$$

Can we find C^n such that R^n is $\mathcal{O}(\Delta t^2)$? Yes, by setting

$$C^n = \frac{1}{2}u_e''(t_n)\Delta t,$$

we manage to cancel the first-order term and

$$R^n = \frac{1}{6}u_e'''(t_n)\Delta t^2 + \mathcal{O}(\Delta t^3) .$$

The correction term C^n introduces $\frac{1}{2}\Delta t u''$ in the discrete equation, and we have to get rid of the derivative u''. One idea is to approximate u'' by a second-order accurate finite difference formula, $u'' \approx (u^{n+1} - 2u^n + u^{n-1})/\Delta t^2$, but this introduces an additional time level with u^{n-1}. Another approach is to rewrite u'' in terms of u' or u using the ODE:

$$u' = -au \quad \Rightarrow \quad u'' = -au' = -a(-au) = a^2 u \, .$$

This means that we can simply set $C^n = \frac{1}{2}a^2 \Delta t u^n$. We can then either solve the discrete equation

$$\left[D_t^+ u = -au + \frac{1}{2}a^2 \Delta t u \right]^n , \tag{B.40}$$

or we can equivalently discretize the perturbed ODE

$$u' = -\hat{a}u, \quad \hat{a} = a\left(1 - \frac{1}{2}a\Delta t \right), \tag{B.41}$$

by a Forward Euler method. That is, we replace the original coefficient a by the perturbed coefficient $\hat{a}$. Observe that $\hat{a} \to a$ as $\Delta t \to 0$.

The Forward Euler method applied to (B.41) results in

$$\left[D_t^+ u = -a\left(1 - \frac{1}{2}a\Delta t \right) u \right]^n .$$

We can control our computations and verify that the truncation error of the scheme above is indeed $\mathcal{O}(\Delta t^2)$.

Another way of revealing the fact that the perturbed ODE leads to a more accurate solution is to look at the amplification factor. Our scheme can be written as

$$u^{n+1} = Au^n, \quad A = 1 - \hat{a}\Delta t = 1 - p + \frac{1}{2}p^2, \quad p = a\Delta t,$$

The amplification factor A as a function of $p = a\Delta t$ is seen to be the first three terms of the Taylor series for the exact amplification factor e^{-p}. The Forward Euler scheme for $u = -au$ gives only the first two terms $1 - p$ of the Taylor series for e^{-p}. That is, using $\hat{a}$ increases the order of the accuracy in the amplification factor.

Instead of replacing u'' by $a^2 u$, we use the relation $u'' = -au'$ and add a term $-\frac{1}{2}a\Delta t u'$ in the ODE:

$$u' = -au - \frac{1}{2}a\Delta t u' \quad \Rightarrow \quad \left(1 + \frac{1}{2}a\Delta t \right) u' = -au \, .$$

Using a Forward Euler method results in

$$\left(1 + \frac{1}{2}a\Delta t \right) \frac{u^{n+1} - u^n}{\Delta t} = -au^n,$$

which after some algebra can be written as

$$u^{n+1} = \frac{1 - \frac{1}{2}a\Delta t}{1 + \frac{1}{2}a\Delta t} u^n \, .$$

This is the same formula as the one arising from a Crank-Nicolson scheme applied to $u' = -au$! It is now recommended to do Exercise B.6 and repeat the above steps to see what kind of correction term is needed in the Backward Euler scheme to make it second order.

The Crank-Nicolson scheme is a bit more challenging to analyze, but the ideas and techniques are the same. The discrete equation reads

$$[D_t u = -au]^{n+\frac{1}{2}},$$

and the truncation error is defined through

$$[D_t u_e + a\overline{u_e}^t = C + R]^{n+\frac{1}{2}},$$

where we have added a correction term. We need to Taylor expand both the discrete derivative and the arithmetic mean with aid of (B.5)–(B.6) and (B.21)–(B.22), respectively. The result is

$$\frac{1}{24}u_e'''\left(t_{n+\frac{1}{2}}\right)\Delta t^2 + \mathcal{O}(\Delta t^4) + \frac{a}{8}u_e''\left(t_{n+\frac{1}{2}}\right)\Delta t^2 + \mathcal{O}(\Delta t^4) = C^{n+\frac{1}{2}} + R^{n+\frac{1}{2}}.$$

The goal now is to make $C^{n+\frac{1}{2}}$ cancel the Δt^2 terms:

$$C^{n+\frac{1}{2}} = \frac{1}{24}u_e'''\left(t_{n+\frac{1}{2}}\right)\Delta t^2 + \frac{a}{8}u_e''(t_n)\Delta t^2.$$

Using $u' = -au$, we have that $u'' = a^2 u$, and we find that $u''' = -a^3 u$. We can therefore solve the perturbed ODE problem

$$u' = -\hat{a}u, \quad \hat{a} = a\left(1 - \frac{1}{12}a^2\Delta t^2\right),$$

by the Crank-Nicolson scheme and obtain a method that is of fourth order in Δt. Exercise B.7 encourages you to implement these correction terms and calculate empirical convergence rates to verify that higher-order accuracy is indeed obtained in real computations.

B.3.7   Extension to Variable Coefficients

Let us address the decay ODE with variable coefficients,

$$u'(t) = -a(t)u(t) + b(t),$$

discretized by the Forward Euler scheme,

$$[D_t^+ u = -au + b]^n. \tag{B.42}$$

The truncation error R is as always found by inserting the exact solution $u_e(t)$ in the discrete scheme:

$$[D_t^+ u_e + au_e - b = R]^n. \tag{B.43}$$

Using (B.11)–(B.12),

$$u'_e(t_n) - \frac{1}{2}u''_e(t_n)\Delta t + \mathcal{O}(\Delta t^2) + a(t_n)u_e(t_n) - b(t_n) = R^n \,.$$

Because of the ODE,

$$u'_e(t_n) + a(t_n)u_e(t_n) - b(t_n) = 0,$$

we are left with the result

$$R^n = -\frac{1}{2}u''_e(t_n)\Delta t + \mathcal{O}(\Delta t^2)\,. \tag{B.44}$$

We see that the variable coefficients do not pose any additional difficulties in this case. Exercise B.8 takes the analysis above one step further to the Crank-Nicolson scheme.

B.3.8 Exact Solutions of the Finite Difference Equations

Having a mathematical expression for the numerical solution is very valuable in program verification, since we then know the exact numbers that the program should produce. Looking at the various formulas for the truncation errors in (B.5)–(B.6) and (B.25)–(B.26) in Sect. B.2.4, we see that all but two of the R expressions contain a second or higher order derivative of u_e. The exceptions are the geometric and harmonic means where the truncation error involves u'_e and even u_e in case of the harmonic mean. So, apart from these two means, choosing u_e to be a linear function of t, $u_e = ct + d$ for constants c and d, will make the truncation error vanish since $u''_e = 0$. Consequently, the truncation error of a finite difference scheme will be zero since the various approximations used will all be exact. This means that the linear solution is an exact solution of the discrete equations.

In a particular differential equation problem, the reasoning above can be used to determine if we expect a linear u_e to fulfill the discrete equations. To actually prove that this is true, we can either compute the truncation error and see that it vanishes, or we can simply insert $u_e(t) = ct + d$ in the scheme and see that it fulfills the equations. The latter method is usually the simplest. It will often be necessary to add some source term to the ODE in order to allow a linear solution.

Many ODEs are discretized by centered differences. From Sect. B.2.4 we see that all the centered difference formulas have truncation errors involving u'''_e or higher-order derivatives. A quadratic solution, e.g., $u_e(t) = t^2 + ct + d$, will then make the truncation errors vanish. This observation can be used to test if a quadratic solution will fulfill the discrete equations. Note that a quadratic solution will not obey the equations for a Crank-Nicolson scheme for $u' = -au + b$ because the approximation applies an arithmetic mean, which involves a truncation error with u''_e.

B.3.9   Computing Truncation Errors in Nonlinear Problems

The general nonlinear ODE

$$u' = f(u, t),$$ (B.45)

can be solved by a Crank-Nicolson scheme

$$[D_t u = \overline{f}^t]^{n+\frac{1}{2}} .$$ (B.46)

The truncation error is as always defined as the residual arising when inserting the exact solution u_e in the scheme:

$$[D_t u_e - \overline{f}^t = R]^{n+\frac{1}{2}} .$$ (B.47)

Using (B.21)–(B.22) for $\overline{f}^t$ results in

$$
\begin{aligned}
[\overline{f}^t]^{n+\frac{1}{2}} &= \frac{1}{2}(f(u_e^n, t_n) + f(u_e^{n+1}, t_{n+1})) \\
&= f\left(u_e^{n+\frac{1}{2}}, t_{n+\frac{1}{2}}\right) + \frac{1}{8}u_e''\left(t_{n+\frac{1}{2}}\right)\Delta t^2 + \mathcal{O}(\Delta t^4) .
\end{aligned}
$$

With (B.5)–(B.6) the discrete equations (B.47) lead to

$$
\begin{aligned}
u_e'\left(t_{n+\frac{1}{2}}\right) + \frac{1}{24}u_e'''\left(t_{n+\frac{1}{2}}\right)\Delta t^2 - f\left(u_e^{n+\frac{1}{2}}, t_{n+\frac{1}{2}}\right) \\
- \frac{1}{8}u_e''\left(t_{n+\frac{1}{2}}\right)\Delta t^2 + \mathcal{O}(\Delta t^4) = R^{n+\frac{1}{2}} .
\end{aligned}
$$

Since $u_e'(t_{n+\frac{1}{2}}) - f(u_e^{n+\frac{1}{2}}, t_{n+\frac{1}{2}}) = 0$, the truncation error becomes

$$R^{n+\frac{1}{2}} = \left(\frac{1}{24}u_e'''\left(t_{n+\frac{1}{2}}\right) - \frac{1}{8}u_e''\left(t_{n+\frac{1}{2}}\right)\right)\Delta t^2 .$$

The computational techniques worked well even for this nonlinear ODE.

B.4   Vibration ODEs

B.4.1   Linear Model Without Damping

The next example on computing the truncation error involves the following ODE for vibration problems:

$$u''(t) + \omega^2 u(t) = 0 .$$ (B.48)

Here, ω is a given constant.

The truncation error of a centered finite difference scheme Using a standard, second-ordered, central difference for the second-order derivative in time, we have the scheme

$$[D_t D_t u + \omega^2 u = 0]^n .$$ (B.49)

Inserting the exact solution u_e in this equation and adding a residual R so that u_e can fulfill the equation results in

$$[D_t D_t u_e + \omega^2 u_e = R]^n .$$ (B.50)

To calculate the truncation error R^n, we use (B.17)–(B.18), i.e.,

$$[D_t D_t u_e]^n = u_e''(t_n) + \frac{1}{12} u_e''''(t_n) \Delta t^2 + \mathcal{O}(\Delta t^4),$$

and the fact that $u_e''(t) + \omega^2 u_e(t) = 0$. The result is

$$R^n = \frac{1}{12} u_e''''(t_n) \Delta t^2 + \mathcal{O}(\Delta t^4) .$$ (B.51)

The truncation error of approximating $u'(0)$ The initial conditions for (B.48) are $u(0) = I$ and $u'(0) = V$. The latter involves a finite difference approximation. The standard choice

$$[D_{2t} u = V]^0,$$

where u^{-1} is eliminated with the aid of the discretized ODE for $n = 0$, involves a centered difference with an $\mathcal{O}(\Delta t^2)$ truncation error given by (B.7)–(B.8). The simpler choice

$$[D_t^+ u = V]^0,$$

is based on a forward difference with a truncation error $\mathcal{O}(\Delta t)$. A central question is if this initial error will impact the order of the scheme throughout the simulation. Exercise B.11 asks you to perform an experiment to investigate this question.

Truncation error of the equation for the first step We have shown that the truncation error of the difference used to approximate the initial condition $u'(0) = 0$ is $\mathcal{O}(\Delta t^2)$, but we can also investigate the difference equation used for the first step. In a truncation error setting, the right way to view this equation is not to use the initial condition $[D_{2t} u = V]^0$ to express $u^{-1} = u^1 - 2\Delta t V$ in order to eliminate u^{-1} from the discretized differential equation, but the other way around: the fundamental equation is the discretized initial condition $[D_{2t} u = V]^0$ and we use the discretized ODE $[D_t D_t + \omega^2 u = 0]^0$ to eliminate u^{-1} in the discretized initial condition. From $[D_t D_t + \omega^2 u = 0]^0$ we have

$$u^{-1} = 2u^0 - u^1 - \Delta t^2 \omega^2 u^0,$$

which inserted in $[D_{2t} u = V]^0$ gives

$$\frac{u^1 - u^0}{\Delta t} + \frac{1}{2} \omega^2 \Delta t u^0 = V .$$ (B.52)

The first term can be recognized as a forward difference such that the equation can be written in operator notation as

$$\left[D_t^+ u + \frac{1}{2}\omega^2 \Delta t u = V \right]^0 .$$

The truncation error is defined as

$$\left[D_t^+ u_e + \frac{1}{2}\omega^2 \Delta t u_e - V = R \right]^0 .$$

Using (B.11)–(B.12) with one more term in the Taylor series, we get that

$$u_e'(0) + \frac{1}{2}u_e''(0)\Delta t + \frac{1}{6}u_e'''(0)\Delta t^2 + \mathcal{O}(\Delta t^3) + \frac{1}{2}\omega^2 \Delta t u_e(0) - V = R^n .$$

Now, $u_e'(0) = V$ and $u_e''(0) = -\omega^2 u_e(0)$ so we get

$$R^n = \frac{1}{6}u_e'''(0)\Delta t^2 + \mathcal{O}(\Delta t^3) .$$

There is another way of analyzing the discrete initial condition, because eliminating u^{-1} via the discretized ODE can be expressed as

$$[D_{2t}u + \Delta t(D_t D_t u - \omega^2 u) = V]^0 . \qquad (B.53)$$

Writing out (B.53) shows that the equation is equivalent to (B.52). The truncation error is defined by

$$[D_{2t}u_e + \Delta t(D_t D_t u_e - \omega^2 u_e) = V + R]^0 .$$

Replacing the difference via (B.7)–(B.8) and (B.17)–(B.18), as well as using $u_e'(0) = V$ and $u_e''(0) = -\omega^2 u_e(0)$, gives

$$R^n = \frac{1}{6}u_e'''(0)\Delta t^2 + \mathcal{O}(\Delta t^3) .$$

Computing correction terms The idea of using correction terms to increase the order of R^n can be applied as described in Sect. B.3.6. We look at

$$[D_t D_t u_e + \omega^2 u_e = C + R]^n ,$$

and observe that C^n must be chosen to cancel the Δt^2 term in R^n. That is,

$$C^n = \frac{1}{12}u_e''''(t_n)\Delta t^2 .$$

To get rid of the 4th-order derivative we can use the differential equation: $u'' = -\omega^2 u$, which implies $u'''' = \omega^4 u$. Adding the correction term to the ODE results in

$$u'' + \omega^2 \left(1 - \frac{1}{12}\omega^2 \Delta t^2 \right) u = 0 . \qquad (B.54)$$

Solving this equation by the standard scheme

$$\left[D_t D_t u + \omega^2 \left(1 - \frac{1}{12}\omega^2 \Delta t^2 \right) u = 0 \right]^n ,$$

will result in a scheme with truncation error $\mathcal{O}(\Delta t^4)$.

We can use another set of arguments to justify that (B.54) leads to a higher-order method. Mathematical analysis of the scheme (B.49) reveals that the numerical frequency $\tilde{\omega}$ is (approximately as $\Delta t \to 0$)

$$\tilde{\omega} = \omega \left(1 + \frac{1}{24}\omega^2 \Delta t^2 \right) .$$

One can therefore attempt to replace ω in the ODE by a slightly smaller ω since the numerics will make it larger:

$$\left[u'' + \left(\omega \left(1 - \frac{1}{24}\omega^2 \Delta t^2 \right) \right)^2 u = 0 \right]^n .$$

Expanding the squared term and omitting the higher-order term Δt^4 gives exactly the ODE (B.54). Experiments show that u^n is computed to 4th order in Δt. You can confirm this by running a little program in the vib directory:

```
from vib_undamped import convergence_rates, solver_adjust_w

r = convergence_rates(
      m=5, solver_function=solver_adjust_w, num_periods=8)
```

One will see that the rates r lie around 4.

B.4.2 Model with Damping and Nonlinearity

The model (B.48) can be extended to include damping $\beta u'$, a nonlinear restoring (spring) force $s(u)$, and some known excitation force $F(t)$:

$$mu'' + \beta u' + s(u) = F(t) . \tag{B.55}$$

The coefficient m usually represents the mass of the system. This governing equation can be discretized by centered differences:

$$[mD_t D_t u + \beta D_{2t} u + s(u) = F]^n . \tag{B.56}$$

The exact solution u_e fulfills the discrete equations with a residual term:

$$[mD_t D_t u_e + \beta D_{2t} u_e + s(u_e) = F + R]^n . \tag{B.57}$$

Using (B.17)–(B.18) and (B.7)–(B.8) we get

$$[mD_tD_tu_e + \beta D_{2t}u_e]^n = mu_e''(t_n) + \beta u_e'(t_n)$$
$$+ \left(\frac{m}{12}u_e''''(t_n) + \frac{\beta}{6}u_e'''(t_n)\right)\Delta t^2 + \mathcal{O}(\Delta t^4).$$

Combining this with the previous equation, we can collect the terms

$$mu_e''(t_n) + \beta u_e'(t_n) + \omega^2 u_e(t_n) + s(u_e(t_n)) - F^n,$$

and set this sum to zero because u_e solves the differential equation. We are left with the truncation error

$$R^n = \left(\frac{m}{12}u_e''''(t_n) + \frac{\beta}{6}u_e'''(t_n)\right)\Delta t^2 + \mathcal{O}(\Delta t^4), \tag{B.58}$$

so the scheme is of second order.

According to (B.58), we can add correction terms

$$C^n = \left(\frac{m}{12}u_e''''(t_n) + \frac{\beta}{6}u_e'''(t_n)\right)\Delta t^2,$$

to the right-hand side of the ODE to obtain a fourth-order scheme. However, expressing u'''' and u''' in terms of lower-order derivatives is now harder because the differential equation is more complicated:

$$u''' = \frac{1}{m}(F' - \beta u'' - s'(u)u'),$$
$$u'''' = \frac{1}{m}(F'' - \beta u''' - s''(u)(u')^2 - s'(u)u''),$$
$$= \frac{1}{m}(F'' - \beta\frac{1}{m}(F' - \beta u'' - s'(u)u') - s''(u)(u')^2 - s'(u)u'').$$

It is not impossible to discretize the resulting modified ODE, but it is up to debate whether correction terms are feasible and the way to go. Computing with a smaller Δt is usually always possible in these problems to achieve the desired accuracy.

B.4.3 Extension to Quadratic Damping

Instead of the linear damping term $\beta u'$ in (B.55) we now consider quadratic damping $\beta|u'|u'$:

$$mu'' + \beta|u'|u' + s(u) = F(t). \tag{B.59}$$

A centered difference for u' gives rise to a nonlinearity, which can be linearized using a geometric mean: $[|u'|u']^n \approx |[u']^{n-\frac{1}{2}}|[u']^{n+\frac{1}{2}}$. The resulting scheme becomes

$$[mD_tD_tu]^n + \beta|[D_tu]^{n-\frac{1}{2}}|[D_tu]^{n+\frac{1}{2}} + s(u^n) = F^n. \tag{B.60}$$

The truncation error is defined through

$$[m D_t D_t u_e]^n + \beta |[D_t u_e]^{n-\frac{1}{2}}|[D_t u_e]^{n+\frac{1}{2}} + s(u_e^n) - F^n = R^n . \qquad \text{(B.61)}$$

We start with expressing the truncation error of the geometric mean. According to (B.23)–(B.24),

$$|[D_t u_e]^{n-\frac{1}{2}}|[D_t u_e]^{n+\frac{1}{2}} = [|D_t u_e| D_t u_e]^n - \frac{1}{4} u_e'(t_n)^2 \Delta t^2$$

$$+ \frac{1}{4} u_e(t_n) u_e''(t_n) \Delta t^2 + \mathcal{O}(\Delta t^4) .$$

Using (B.5)–(B.6) for the $D_t u_e$ factors results in

$$[|D_t u_e| D_t u_e]^n$$
$$= \left| u_e' + \frac{1}{24} u_e'''(t_n) \Delta t^2 + \mathcal{O}(\Delta t^4) \right| \left(u_e' + \frac{1}{24} u_e'''(t_n) \Delta t^2 + \mathcal{O}(\Delta t^4) \right) .$$

We can remove the absolute value since it essentially gives a factor 1 or -1 only. Calculating the product, we have the leading-order terms

$$[D_t u_e D_t u_e]^n = (u_e'(t_n))^2 + \frac{1}{12} u_e(t_n) u_e'''(t_n) \Delta t^2 + \mathcal{O}(\Delta t^4) .$$

With
$$m[D_t D_t u_e]^n = m u_e''(t_n) + \frac{m}{12} u_e''''(t_n) \Delta t^2 + \mathcal{O}(\Delta t^4),$$

and using the differential equation on the form $m u'' + \beta(u')^2 + s(u) = F$, we end up with

$$R^n = \left(\frac{m}{12} u_e''''(t_n) + \frac{\beta}{12} u_e(t_n) u_e'''(t_n) \right) \Delta t^2 + \mathcal{O}(\Delta t^4) .$$

This result demonstrates that we have second-order accuracy also with quadratic damping. The key elements that lead to the second-order accuracy is that the difference approximations are $\mathcal{O}(\Delta t^2)$ *and* the geometric mean approximation is also $\mathcal{O}(\Delta t^2)$.

B.4.4   The General Model Formulated as First-Order ODEs

The second-order model (B.59) can be formulated as a first-order system,

$$v' = \frac{1}{m} (F(t) - \beta |v| v - s(u)) , \qquad \text{(B.62)}$$

$$u' = v . \qquad \text{(B.63)}$$

The system (B.63)–(B.63) can be solved either by a forward-backward scheme (the Euler-Cromer method) or a centered scheme on a staggered mesh.

A centered scheme on a staggered mesh We now introduce a staggered mesh where we seek u at mesh points t_n and v at points $t_{n+\frac{1}{2}}$ in between the u points. The staggered mesh makes it easy to formulate centered differences in the system (B.63)–(B.63):

$$[D_t u = v]^{n-\frac{1}{2}}, \tag{B.64}$$

$$\left[D_t v = \frac{1}{m}(F(t) - \beta|v|v - s(u)) \right]^n . \tag{B.65}$$

The term $|v^n|v^n$ causes trouble since v^n is not computed, only $v^{n-\frac{1}{2}}$ and $v^{n+\frac{1}{2}}$. Using geometric mean, we can express $|v^n|v^n$ in terms of known quantities: $|v^n|v^n \approx |v^{n-\frac{1}{2}}|v^{n+\frac{1}{2}}$. We then have

$$[D_t u]^{n-\frac{1}{2}} = v^{n-\frac{1}{2}}, \tag{B.66}$$

$$[D_t v]^n = \frac{1}{m}\left(F(t_n) - \beta \left| v^{n-\frac{1}{2}} \right| v^{n+\frac{1}{2}} - s(u^n) \right) . \tag{B.67}$$

The truncation error in each equation fulfills

$$[D_t u_e]^{n-\frac{1}{2}} = v_e\left(t_{n-\frac{1}{2}} \right) + R_u^{n-\frac{1}{2}},$$

$$[D_t v_e]^n = \frac{1}{m}\left(F(t_n) - \beta \left| v_e\left(t_{n-\frac{1}{2}} \right) \right| v_e\left(t_{n+\frac{1}{2}} \right) - s(u^n) \right) + R_v^n .$$

The truncation error of the centered differences is given by (B.5)–(B.6), and the geometric mean approximation analysis can be taken from (B.23)–(B.24). These results lead to

$$u_e'\left(t_{n-\frac{1}{2}} \right) + \frac{1}{24} u_e'''\left(t_{n-\frac{1}{2}} \right) \Delta t^2 + \mathcal{O}(\Delta t^4) = v_e\left(t_{n-\frac{1}{2}} \right) + R_u^{n-\frac{1}{2}},$$

and

$$v_e'(t_n) = \frac{1}{m}(F(t_n) - \beta|v_e(t_n)|v_e(t_n) + \mathcal{O}(\Delta t^2) - s(u^n)) + R_v^n .$$

The ODEs fulfilled by u_e and v_e are evident in these equations, and we achieve second-order accuracy for the truncation error in both equations:

$$R_u^{n-\frac{1}{2}} = \mathcal{O}(\Delta t^2), \quad R_v^n = \mathcal{O}(\Delta t^2) .$$

B.5 Wave Equations

B.5.1 Linear Wave Equation in 1D

The standard, linear wave equation in 1D for a function $u(x, t)$ reads

$$\frac{\partial^2 u}{\partial t^2} = c^2 \frac{\partial^2 u}{\partial x^2} + f(x,t), \quad x \in (0, L), \ t \in (0, T], \tag{B.68}$$

where c is the constant wave velocity of the physical medium in $[0, L]$. The equation can also be more compactly written as

$$u_{tt} = c^2 u_{xx} + f, \quad x \in (0, L),\, t \in (0, T]. \tag{B.69}$$

Centered, second-order finite differences are a natural choice for discretizing the derivatives, leading to

$$[D_t D_t u = c^2 D_x D_x u + f]_i^n. \tag{B.70}$$

Inserting the exact solution $u_e(x, t)$ in (B.70) makes this function fulfill the equation if we add the term R:

$$[D_t D_t u_e = c^2 D_x D_x u_e + f + R]_i^n. \tag{B.71}$$

Our purpose is to calculate the truncation error R. From (B.17)–(B.18) we have that

$$[D_t D_t u_e]_i^n = u_{e,tt}(x_i, t_n) + \frac{1}{12} u_{e,tttt}(x_i, t_n)\Delta t^2 + \mathcal{O}(\Delta t^4),$$

when we use a notation taking into account that u_e is a function of two variables and that derivatives must be partial derivatives. The notation $u_{e,tt}$ means $\partial^2 u_e / \partial t^2$.
The same formula may also be applied to the x-derivative term:

$$[D_x D_x u_e]_i^n = u_{e,xx}(x_i, t_n) + \frac{1}{12} u_{e,xxxx}(x_i, t_n)\Delta x^2 + \mathcal{O}(\Delta x^4).$$

Equation (B.71) now becomes

$$u_{e,tt} + \frac{1}{12} u_{e,tttt}(x_i, t_n)\Delta t^2 = c^2 u_{e,xx} + c^2 \frac{1}{12} u_{e,xxxx}(x_i, t_n)\Delta x^2 + f(x_i, t_n)$$
$$+ \mathcal{O}(\Delta t^4, \Delta x^4) + R_i^n.$$

Because u_e fulfills the partial differential equation (PDE) (B.69), the first, third, and fifth term cancel out, and we are left with

$$R_i^n = \frac{1}{12} u_{e,tttt}(x_i, t_n)\Delta t^2 - c^2 \frac{1}{12} u_{e,xxxx}(x_i, t_n)\Delta x^2 + \mathcal{O}(\Delta t^4, \Delta x^4), \tag{B.72}$$

showing that the scheme (B.70) is of second order in the time and space mesh spacing.

B.5.2 Finding Correction Terms

Can we add correction terms to the PDE and increase the order of R_i^n in (B.72)? The starting point is

$$[D_t D_t u_e = c^2 D_x D_x u_e + f + C + R]_i^n. \tag{B.73}$$

From the previous analysis we simply get (B.72) again, but now with C:

$$R_i^n + C_i^n = \frac{1}{12}u_{e,tttt}(x_i,t_n)\Delta t^2 - c^2\frac{1}{12}u_{e,xxxx}(x_i,t_n)\Delta x^2 + \mathcal{O}(\Delta t^4, \Delta x^4).$$

(B.74)

The idea is to let C_i^n cancel the Δt^2 and Δx^2 terms to make $R_i^n = \mathcal{O}(\Delta t^4, \Delta x^4)$:

$$C_i^n = \frac{1}{12}u_{e,tttt}(x_i,t_n)\Delta t^2 - c^2\frac{1}{12}u_{e,xxxx}(x_i,t_n)\Delta x^2.$$

Essentially, it means that we add a new term

$$C = \frac{1}{12}\left(u_{tttt}\Delta t^2 - c^2 u_{xxxx}\Delta x^2\right),$$

to the right-hand side of the PDE. We must either discretize these 4th-order derivatives directly or rewrite them in terms of lower-order derivatives with the aid of the PDE. The latter approach is more feasible. From the PDE we have the operator equality

$$\frac{\partial^2}{\partial t^2} = c^2\frac{\partial^2}{\partial x^2},$$

so

$$u_{tttt} = c^2 u_{xxtt}, \quad u_{xxxx} = c^{-2}u_{ttxx}.$$

Assuming u is smooth enough, so that $u_{xxtt} = u_{ttxx}$, these relations lead to

$$C = \frac{1}{12}((c^2\Delta t^2 - \Delta x^2)u_{xx})_{tt}.$$

A natural discretization is

$$C_i^n = \frac{1}{12}((c^2\Delta t^2 - \Delta x^2)[D_x D_x D_t D_t u]_i^n.$$

Writing out $[D_x D_x D_t D_t u]_i^n$ as $[D_x D_x (D_t D_t u)]_i^n$ gives

$$\frac{1}{\Delta t^2}\left(\frac{u_{i+1}^{n+1} - 2u_{i+1}^n + u_{i+1}^{n-1}}{\Delta x^2}\right.$$
$$\left. - 2\frac{u_i^{n+1} - 2u_i^n + u_i^{n-1}}{\Delta x^2} + \frac{u_{i-1}^{n+1} - 2u_{i-1}^n + u_{i-1}^{n-1}}{\Delta x^2}\right).$$

Now the unknown values u_{i+1}^{n+1}, u_i^{n+1}, and u_{i-1}^{n+1} are *coupled*, and we must solve a tridiagonal system to find them. This is in principle straightforward, but it results in an implicit finite difference scheme, while we had a convenient explicit scheme without the correction terms.

B.5.3 Extension to Variable Coefficients

Now we address the variable coefficient version of the linear 1D wave equation,

$$\frac{\partial^2 u}{\partial t^2} = \frac{\partial}{\partial x}\left(\lambda(x)\frac{\partial u}{\partial x}\right),$$

or written more compactly as

$$u_{tt} = (\lambda u_x)_x .$$ (B.75)

The discrete counterpart to this equation, using arithmetic mean for λ and centered differences, reads

$$\left[D_t D_t u = D_x \overline{\lambda}^x D_x u \right]_i^n .$$ (B.76)

The truncation error is the residual R in the equation

$$\left[D_t D_t u_e = D_x \overline{\lambda}^x D_x u_e + R \right]_i^n .$$ (B.77)

The difficulty with (B.77) is how to compute the truncation error of the term $[D_x \overline{\lambda}^x D_x u_e]_i^n$.

We start by writing out the outer operator:

$$\left[D_x \overline{\lambda}^x D_x u_e \right]_i^n = \frac{1}{\Delta x} \left(\left[\overline{\lambda}^x D_x u_e \right]_{i+\frac{1}{2}}^n - \left[\overline{\lambda}^x D_x u_e \right]_{i-\frac{1}{2}}^n \right).$$ (B.78)

With the aid of (B.5)–(B.6) and (B.21)–(B.22) we have

$$[D_x u_e]_{i+\frac{1}{2}}^n = u_{e,x}\left(x_{i+\frac{1}{2}}, t_n\right) + \frac{1}{24} u_{e,xxx}\left(x_{i+\frac{1}{2}}, t_n\right) \Delta x^2 + \mathcal{O}(\Delta x^4),$$

$$\left[\overline{\lambda}^x\right]_{i+\frac{1}{2}} = \lambda\left(x_{i+\frac{1}{2}}\right) + \frac{1}{8} \lambda''\left(x_{i+\frac{1}{2}}\right) \Delta x^2 + \mathcal{O}(\Delta x^4),$$

$$\left[\overline{\lambda}^x D_x u_e\right]_{i+\frac{1}{2}}^n = \left(\lambda\left(x_{i+\frac{1}{2}}\right) + \frac{1}{8} \lambda''\left(x_{i+\frac{1}{2}}\right) \Delta x^2 + \mathcal{O}(\Delta x^4)\right)$$

$$\times \left(u_{e,x}\left(x_{i+\frac{1}{2}}, t_n\right) + \frac{1}{24} u_{e,xxx}\left(x_{i+\frac{1}{2}}, t_n\right) \Delta x^2 + \mathcal{O}(\Delta x^4)\right)$$

$$= \lambda\left(x_{i+\frac{1}{2}}\right) u_{e,x}\left(x_{i+\frac{1}{2}}, t_n\right) + \lambda\left(x_{i+\frac{1}{2}}\right) \frac{1}{24} u_{e,xxx}\left(x_{i+\frac{1}{2}}, t_n\right) \Delta x^2$$

$$+ u_{e,x}\left(x_{i+\frac{1}{2}}, t_n\right) \frac{1}{8} \lambda''\left(x_{i+\frac{1}{2}}\right) \Delta x^2 + \mathcal{O}(\Delta x^4)$$

$$= [\lambda u_{e,x}]_{i+\frac{1}{2}}^n + G_{i+\frac{1}{2}}^n \Delta x^2 + \mathcal{O}(\Delta x^4),$$

where we have introduced the short form

$$G_{i+\frac{1}{2}}^n = \frac{1}{24} u_{e,xxx}\left(x_{i+\frac{1}{2}}, t_n\right) \lambda\left(x_{i+\frac{1}{2}}\right) + u_{e,x}\left(x_{i+\frac{1}{2}}, t_n\right) \frac{1}{8} \lambda''\left(x_{i+\frac{1}{2}}\right).$$

Similarly, we find that

$$\left[\overline{\lambda}^x D_x u_e\right]_{i-\frac{1}{2}}^n = [\lambda u_{e,x}]_{i-\frac{1}{2}}^n + G_{i-\frac{1}{2}}^n \Delta x^2 + \mathcal{O}(\Delta x^4).$$

Inserting these expressions in the outer operator (B.78) results in

$$\left[D_x \overline{\lambda}^x D_x u_e \right]_i^n = \frac{1}{\Delta x} \left(\left[\overline{\lambda}^x D_x u_e \right]_{i+\frac{1}{2}}^n - \left[\overline{\lambda}^x D_x u_e \right]_{i-\frac{1}{2}}^n \right)$$

$$= \frac{1}{\Delta x} \left([\lambda u_{e,x}]_{i+\frac{1}{2}}^n + G_{i+\frac{1}{2}}^n \Delta x^2 - [\lambda u_{e,x}]_{i-\frac{1}{2}}^n - G_{i-\frac{1}{2}}^n \Delta x^2 + \mathcal{O}(\Delta x^4) \right)$$

$$= [D_x \lambda u_{e,x}]_i^n + [D_x G]_i^n \Delta x^2 + \mathcal{O}(\Delta x^4).$$

The reason for $\mathcal{O}(\Delta x^4)$ in the remainder is that there are coefficients in front of this term, say $H\Delta x^4$, and the subtraction and division by Δx results in $[D_x H]_i^n \Delta x^4$.

We can now use (B.5)–(B.6) to express the D_x operator in $[D_x \lambda u_{e,x}]_i^n$ as a derivative and a truncation error:

$$[D_x \lambda u_{e,x}]_i^n = \frac{\partial}{\partial x}\lambda(x_i)u_{e,x}(x_i, t_n) + \frac{1}{24}(\lambda u_{e,x})_{xxx}(x_i, t_n)\Delta x^2 + \mathcal{O}(\Delta x^4).$$

Expressions like $[D_x G]_i^n \Delta x^2$ can be treated in an identical way,

$$[D_x G]_i^n \Delta x^2 = G_x(x_i, t_n)\Delta x^2 + \frac{1}{24}G_{xxx}(x_i, t_n)\Delta x^4 + \mathcal{O}(\Delta x^4).$$

There will be a number of terms with the Δx^2 factor. We lump these now into $\mathcal{O}(\Delta x^2)$. The result of the truncation error analysis of the spatial derivative is therefore summarized as

$$\left[D_x \overline{\lambda}^x D_x u_e\right]_i^n = \frac{\partial}{\partial x}\lambda(x_i)u_{e,x}(x_i, t_n) + \mathcal{O}(\Delta x^2).$$

After having treated the $[D_t D_t u_e]_i^n$ term as well, we achieve

$$R_i^n = \mathcal{O}(\Delta x^2) + \frac{1}{12}u_{e,tttt}(x_i, t_n)\Delta t^2.$$

The main conclusion is that the scheme is of second-order in time and space also in this variable coefficient case. The key ingredients for second order are the centered differences and the arithmetic mean for λ: all those building blocks feature second-order accuracy.

B.5.4 Linear Wave Equation in 2D/3D

The two-dimensional extension of (B.68) takes the form

$$\frac{\partial^2 u}{\partial t^2} = c^2\left(\frac{\partial^2 u}{\partial x^2} + \frac{\partial^2 u}{\partial y^2}\right) + f(x, y, t), \quad (x, y) \in (0, L) \times (0, H), \ t \in (0, T],$$
(B.79)

where now $c(x, y)$ is the constant wave velocity of the physical medium $[0, L] \times [0, H]$. In compact notation, the PDE (B.79) can be written

$$u_{tt} = c^2(u_{xx} + u_{yy}) + f(x, y, t), \quad (x, y) \in (0, L) \times (0, H), \ t \in (0, T], \quad \text{(B.80)}$$

in 2D, while the 3D version reads

$$u_{tt} = c^2(u_{xx} + u_{yy} + u_{zz}) + f(x, y, z, t), \tag{B.81}$$

for $(x, y, z) \in (0, L) \times (0, H) \times (0, B)$ and $t \in (0, T]$.

Approximating the second-order derivatives by the standard formulas (B.17)–(B.18) yields the scheme

$$[D_t D_t u = c^2(D_x D_x u + D_y D_y u + D_z D_z u) + f]_{i,j,k}^n. \tag{B.82}$$

The truncation error is found from

$$[D_t D_t u_e = c^2 (D_x D_x u_e + D_y D_y u_e + D_z D_z u_e) + f + R]_{i,j,k}^n. \quad \text{(B.83)}$$

The calculations from the 1D case can be repeated with the terms in the y and z directions. Collecting terms that fulfill the PDE, we end up with

$$R_{i,j,k}^n = \left[\frac{1}{12} u_{e,tttt} \Delta t^2 - c^2 \frac{1}{12} \left(u_{e,xxxx} \Delta x^2 + u_{e,yyyy} \Delta x^2 + u_{e,zzzz} \Delta z^2 \right) \right]_{i,j,k}^n$$
$$+ \mathcal{O}(\Delta t^4, \Delta x^4, \Delta y^4, \Delta z^4).$$
$$\text{(B.84)}$$

B.6 Diffusion Equations

B.6.1 Linear Diffusion Equation in 1D

The standard, linear, 1D diffusion equation takes the form

$$\frac{\partial u}{\partial t} = \alpha \frac{\partial^2 u}{\partial x^2} + f(x,t), \quad x \in (0,L), \ t \in (0,T], \quad \text{(B.85)}$$

where $\alpha > 0$ is a constant diffusion coefficient. A more compact form of the diffusion equation is $u_t = \alpha u_{xx} + f$.

The spatial derivative in the diffusion equation, αu_{xx}, is commonly discretized as $[D_x D_x u]_i^n$. The time-derivative, however, can be treated by a variety of methods.

The Forward Euler scheme in time Let us start with the simple Forward Euler scheme:
$$[D_t^+ u = \alpha D_x D_x u + f]_i^n.$$

The truncation error arises as the residual R when inserting the exact solution u_e in the discrete equations:

$$[D_t^+ u_e = \alpha D_x D_x u_e + f + R]_i^n.$$

Now, using (B.11)–(B.12) and (B.17)–(B.18), we can transform the difference operators to derivatives:

$$u_{e,t}(x_i, t_n) + \frac{1}{2} u_{e,tt}(t_n) \Delta t + \mathcal{O}(\Delta t^2)$$
$$= \alpha u_{e,xx}(x_i, t_n) + \frac{\alpha}{12} u_{e,xxxx}(x_i, t_n) \Delta x^2 + \mathcal{O}(\Delta x^4) + f(x_i, t_n) + R_i^n.$$

The terms $u_{e,t}(x_i, t_n) - \alpha u_{e,xx}(x_i, t_n) - f(x_i, t_n)$ vanish because u_e solves the PDE. The truncation error then becomes

$$R_i^n = \frac{1}{2} u_{e,tt}(t_n) \Delta t + \mathcal{O}(\Delta t^2) - \frac{\alpha}{12} u_{e,xxxx}(x_i, t_n) \Delta x^2 + \mathcal{O}(\Delta x^4).$$

The Crank-Nicolson scheme in time The Crank-Nicolson method consists of using a centered difference for u_t and an arithmetic average of the u_{xx} term:

$$[D_t u]_i^{n+\frac{1}{2}} = \alpha \frac{1}{2}\left([D_x D_x u]_i^n + [D_x D_x u]_i^{n+1}\right) + f_i^{n+\frac{1}{2}}.$$

The equation for the truncation error is

$$[D_t u_e]_i^{n+\frac{1}{2}} = \alpha \frac{1}{2}\left([D_x D_x u_e]_i^n + [D_x D_x u_e]_i^{n+1}\right) + f_i^{n+\frac{1}{2}} + R_i^{n+\frac{1}{2}}.$$

To find the truncation error, we start by expressing the arithmetic average in terms of values at time $t_{n+\frac{1}{2}}$. According to (B.21)–(B.22),

$$\frac{1}{2}\left([D_x D_x u_e]_i^n + [D_x D_x u_e]_i^{n+1}\right) = [D_x D_x u_e]_i^{n+\frac{1}{2}} + \frac{1}{8}[D_x D_x u_{e,tt}]_i^{n+\frac{1}{2}}\Delta t^2$$
$$+ \mathcal{O}(\Delta t^4).$$

With (B.17)–(B.18) we can express the difference operator $D_x D_x u$ in terms of a derivative:

$$[D_x D_x u_e]_i^{n+\frac{1}{2}} = u_{e,xx}\left(x_i, t_{n+\frac{1}{2}}\right) + \frac{1}{12}u_{e,xxxx}\left(x_i, t_{n+\frac{1}{2}}\right)\Delta x^2 + \mathcal{O}(\Delta x^4).$$

The error term from the arithmetic mean is similarly expanded,

$$\frac{1}{8}[D_x D_x u_{e,tt}]_i^{n+\frac{1}{2}}\Delta t^2 = \frac{1}{8}u_{e,ttxx}\left(x_i, t_{n+\frac{1}{2}}\right)\Delta t^2 + \mathcal{O}(\Delta t^2 \Delta x^2).$$

The time derivative is analyzed using (B.5)–(B.6):

$$[D_t u]_i^{n+\frac{1}{2}} = u_{e,t}\left(x_i, t_{n+\frac{1}{2}}\right) + \frac{1}{24}u_{e,ttt}\left(x_i, t_{n+\frac{1}{2}}\right)\Delta t^2 + \mathcal{O}(\Delta t^4).$$

Summing up all the contributions and notifying that

$$u_{e,t}\left(x_i, t_{n+\frac{1}{2}}\right) = \alpha u_{e,xx}\left(x_i, t_{n+\frac{1}{2}}\right) + f\left(x_i, t_{n+\frac{1}{2}}\right),$$

the truncation error is given by

$$R_i^{n+\frac{1}{2}} = \frac{1}{8}u_{e,xx}\left(x_i, t_{n+\frac{1}{2}}\right)\Delta t^2 + \frac{1}{12}u_{e,xxxx}\left(x_i, t_{n+\frac{1}{2}}\right)\Delta x^2$$
$$+ \frac{1}{24}u_{e,ttt}\left(x_i, t_{n+\frac{1}{2}}\right)\Delta t^2 + \mathcal{O}(\Delta x^4) + \mathcal{O}(\Delta t^4) + \mathcal{O}(\Delta t^2 \Delta x^2).$$

B.6.2 Nonlinear Diffusion Equation in 1D

We address the PDE

$$\frac{\partial u}{\partial t} = \frac{\partial}{\partial x}\left(\alpha(u)\frac{\partial u}{\partial x}\right) + f(u),$$

with two potentially nonlinear coefficients $q(u)$ and $\alpha(u)$. We use a Backward Euler scheme with arithmetic mean for $\alpha(u)$,

$$\left[D^- u = D_x \overline{\alpha(u)}^x D_x u + f(u) \right]_i^n .$$

Inserting u_e defines the truncation error R:

$$\left[D^- u_e = D_x \overline{\alpha(u_e)}^x D_x u_e + f(u_e) + R \right]_i^n .$$

The most computationally challenging part is the variable coefficient with $\alpha(u)$, but we can use the same setup as in Sect. B.5.3 and arrive at a truncation error $\mathcal{O}(\Delta x^2)$ for the x-derivative term. The nonlinear term $[f(u_e)]_i^n = f(u_e(x_i, t_n))$ matches x and t derivatives of u_e in the PDE. We end up with

$$R_i^n = -\frac{1}{2} \frac{\partial^2}{\partial t^2} u_e(x_i, t_n) \Delta t + \mathcal{O}(\Delta x^2) .$$

B.7 Exercises

Exercise B.1: Truncation error of a weighted mean
Derive the truncation error of the weighted mean in (B.19)–(B.20).

Hint Expand u_e^{n+1} and u_e^n around $t_{n+\theta}$.
Filename: `trunc_weighted_mean`.

Exercise B.2: Simulate the error of a weighted mean
We consider the weighted mean

$$u_e(t_n) \approx \theta u_e^{n+1} + (1 - \theta) u_e^n .$$

Choose some specific function for $u_e(t)$ and compute the error in this approximation for a sequence of decreasing $\Delta t = t_{n+1} - t_n$ and for $\theta = 0, 0.25, 0.5, 0.75, 1$. Assuming that the error equals $C\Delta t^r$, for some constants C and r, compute r for the two smallest Δt values for each choice of θ and compare with the truncation error (B.19)–(B.20).
Filename: `trunc_theta_avg`.

Exercise B.3: Verify a truncation error formula
Set up a numerical experiment as explained in Sect. B.3.5 for verifying the formulas (B.15)–(B.16).
Filename: `trunc_backward_2level`.

Problem B.4: Truncation error of the Backward Euler scheme
Derive the truncation error of the Backward Euler scheme for the decay ODE $u' = -au$ with constant a. Extend the analysis to cover the variable-coefficient case $u' = -a(t)u + b(t)$.
Filename: `trunc_decay_BE`.

Exercise B.5: Empirical estimation of truncation errors

Use the ideas and tools from Sect. B.3.5 to estimate the rate of the truncation error of the Backward Euler and Crank-Nicolson schemes applied to the exponential decay model $u' = -au$, $u(0) = I$.

Hint In the Backward Euler scheme, the truncation error can be estimated at mesh points $n = 1, \ldots, N$, while the truncation error must be estimated at midpoints $t_{n+\frac{1}{2}}$, $n = 0, \ldots, N - 1$ for the Crank-Nicolson scheme. The `truncation_error(dt, N)` function to be supplied to the `estimate` function needs to carefully implement these details and return the right t array such that t[i] is the time point corresponding to the quantities R[i] and R_a[i].
Filename: `trunc_decay_BNCN`.

Exercise B.6: Correction term for a Backward Euler scheme

Consider the model $u' = -au$, $u(0) = I$. Use the ideas of Sect. B.3.6 to add a correction term to the ODE such that the Backward Euler scheme applied to the perturbed ODE problem is of second order in Δt. Find the amplification factor.
Filename: `trunc_decay_BE_corr`.

Problem B.7: Verify the effect of correction terms

Make a program that solves $u' = -au$, $u(0) = I$, by the θ-rule and computes convergence rates. Adjust a such that it incorporates correction terms. Run the program to verify that the error from the Forward and Backward Euler schemes with perturbed a is $\mathcal{O}(\Delta t^2)$, while the error arising from the Crank-Nicolson scheme with perturbed a is $\mathcal{O}(\Delta t^4)$.
Filename: `trunc_decay_corr_verify`.

Problem B.8: Truncation error of the Crank-Nicolson scheme

The variable-coefficient ODE $u' = -a(t)u + b(t)$ can be discretized in two different ways by the Crank-Nicolson scheme, depending on whether we use averages for a and b or compute them at the midpoint $t_{n+\frac{1}{2}}$:

$$[D_t u = -a\overline{u}^t + b]^{n+\frac{1}{2}}, \tag{B.86}$$

$$\left[D_t u = \overline{-au + b}^t\right]^{n+\frac{1}{2}}. \tag{B.87}$$

Compute the truncation error in both cases.
Filename: `trunc_decay_CN_vc`.

Problem B.9: Truncation error of $u' = f(u, t)$

Consider the general nonlinear first-order scalar ODE

$$u'(t) = f(u(t), t).$$

Show that the truncation error in the Forward Euler scheme,

$$[D_t^+ u = f(u, t)]^n,$$

and in the Backward Euler scheme,

$$[D_t^- u = f(u,t)]^n,$$

both are of first order, regardless of what f is.

Showing the order of the truncation error in the Crank-Nicolson scheme,

$$[D_t u = f(u,t)]^{n+\frac{1}{2}},$$

is somewhat more involved: Taylor expand u_e^n, u_e^{n+1}, $f(u_e^n, t_n)$, and $f(u_e^{n+1}, t_{n+1})$ around $t_{n+\frac{1}{2}}$, and use that

$$\frac{df}{dt} = \frac{\partial f}{\partial u} u' + \frac{\partial f}{\partial t}.$$

Check that the derived truncation error is consistent with previous results for the case $f(u,t) = -au$.

Filename: `trunc_nonlinear_ODE`.

Exercise B.10: Truncation error of $[D_t D_t u]^n$

Derive the truncation error of the finite difference approximation (B.17)–(B.18) to the second-order derivative.

Filename: `trunc_d2u`.

Exercise B.11: Investigate the impact of approximating $u'(0)$

Section B.4.1 describes two ways of discretizing the initial condition $u'(0) = V$ for a vibration model $u'' + \omega^2 u = 0$: a centered difference $[D_{2t} u = V]^0$ or a forward difference $[D_t^+ u = V]^0$. The program `vib_undamped.py` solves $u'' + \omega^2 u = 0$ with $[D_{2t} u = 0]^0$ and features a function `convergence_rates` for computing the order of the error in the numerical solution. Modify this program such that it applies the forward difference $[D_t^+ u = 0]^0$ and report how this simpler and more convenient approximation impacts the overall convergence rate of the scheme.

Filename: `trunc_vib_ic_fw`.

Problem B.12: Investigate the accuracy of a simplified scheme

Consider the ODE

$$mu'' + \beta |u'| u' + s(u) = F(t).$$

The term $|u'| u'$ quickly gives rise to nonlinearities and complicates the scheme. Why not simply apply a backward difference to this term such that it only involves known values? That is, we propose to solve

$$[m D_t D_t u + \beta |D_t^- u| D_t^- u + s(u) = F]^n.$$

Drop the absolute value for simplicity and find the truncation error of the scheme. Perform numerical experiments with the scheme and compared with the one based on centered differences. Can you illustrate the accuracy loss visually in real computations, or is the asymptotic analysis here mainly of theoretical interest?

Filename: `trunc_vib_bw_damping`.

C

Wave Equation Models for Software Engineering

C.1 A 1D Wave Equation Simulator

C.1.1 Mathematical Model

Let u_t, u_{tt}, u_x, u_{xx} denote derivatives of u with respect to the subscript, i.e., u_{tt} is a second-order time derivative and u_x is a first-order space derivative. The initial-boundary value problem implemented in the wave1D_dn_vc.py code is

$$u_{tt} = (q(x)u_x)_x + f(x,t), \qquad x \in (0,L), \ t \in (0,T] \qquad \text{(C.1)}$$
$$u(x,0) = I(x), \qquad x \in [0,L] \qquad \text{(C.2)}$$
$$u_t(x,0) = V(t), \qquad x \in [0,L] \qquad \text{(C.3)}$$
$$u(0,t) = U_0(t) \quad \text{or} \quad u_x(0,t) = 0, \qquad t \in (0,T] \qquad \text{(C.4)}$$
$$u(L,t) = U_L(t) \quad \text{or} \quad u_x(L,t) = 0, \qquad t \in (0,T]. \qquad \text{(C.5)}$$

We allow variable wave velocity $c^2(x) = q(x)$, and Dirichlet or homogeneous Neumann conditions at the boundaries.

C.1.2 Numerical Discretization

The PDE is discretized by second-order finite differences in time and space, with arithmetic mean for the variable coefficient

$$[D_t D_t u = D_x \overline{q}^x D_x u + f]_i^n. \qquad \text{(C.6)}$$

The Neumann boundary conditions are discretized by

$$[D_{2x}u]_i^n = 0,$$

at a boundary point i. The details of how the numerical scheme is worked out are described in Sect. 2.6 and 2.7.

C.1.3   A Solver Function

The general initial-boundary value problem (C.1)–(C.5) solved by finite difference methods can be implemented as shown in the following `solver` function (taken from the file `wave1D_dn_vc.py`). This function builds on simpler versions described in Sect. 2.3, 2.4 2.6, and 2.7. There are several quite advanced constructs that will be commented upon later. The code is lengthy, but that is because we provide a lot of flexibility with respect to input arguments, boundary conditions, and optimization (scalar versus vectorized loops).

```python
def solver(
    I, V, f, c, U_0, U_L, L, dt, C, T,
    user_action=None, version='scalar',
    stability_safety_factor=1.0):
    """Solve u_tt=(c^2*u_x)_x + f on (0,L)x(0,T)."""

    # --- Compute time and space mesh ---
    Nt = int(round(T/dt))
    t = np.linspace(0, Nt*dt, Nt+1)        # Mesh points in time

    # Find max(c) using a fake mesh and adapt dx to C and dt
    if isinstance(c, (float,int)):
        c_max = c
    elif callable(c):
        c_max = max([c(x_) for x_ in np.linspace(0, L, 101)])
    dx = dt*c_max/(stability_safety_factor*C)
    Nx = int(round(L/dx))
    x = np.linspace(0, L, Nx+1)            # Mesh points in space
    # Make sure dx and dt are compatible with x and t
    dx = x[1] - x[0]
    dt = t[1] - t[0]

    # Make c(x) available as array
    if isinstance(c, (float,int)):
        c = np.zeros(x.shape) + c
    elif callable(c):
        # Call c(x) and fill array c
        c_ = np.zeros(x.shape)
        for i in range(Nx+1):
            c_[i] = c(x[i])
        c = c_

    q = c**2
    C2 = (dt/dx)**2; dt2 = dt*dt    # Help variables in the scheme

    # --- Wrap user-given f, I, V, U_0, U_L if None or 0 ---
    if f is None or f == 0:
        f = (lambda x, t: 0) if version == 'scalar' else \
            lambda x, t: np.zeros(x.shape)
    if I is None or I == 0:
        I = (lambda x: 0) if version == 'scalar' else \
            lambda x: np.zeros(x.shape)
    if V is None or V == 0:
        V = (lambda x: 0) if version == 'scalar' else \
            lambda x: np.zeros(x.shape)
```

```
if U_0 is not None:
    if isinstance(U_0, (float,int)) and U_0 == 0:
        U_0 = lambda t: 0
if U_L is not None:
    if isinstance(U_L, (float,int)) and U_L == 0:
        U_L = lambda t: 0

# --- Make hash of all input data ---
import hashlib, inspect
data = inspect.getsource(I) + '_' + inspect.getsource(V) + \
       '_' + inspect.getsource(f) + '_' + str(c) + '_' + \
       ('None' if U_0 is None else inspect.getsource(U_0)) + \
       ('None' if U_L is None else inspect.getsource(U_L)) + \
       '_' + str(L) + str(dt) + '_' + str(C) + '_' + str(T) + \
       '_' + str(stability_safety_factor)
hashed_input = hashlib.sha1(data).hexdigest()
if os.path.isfile('.' + hashed_input + '_archive.npz'):
    # Simulation is already run
    return -1, hashed_input


# --- Allocate memomry for solutions ---
u    = np.zeros(Nx+1)   # Solution array at new time level
u_n  = np.zeros(Nx+1)   # Solution at 1 time level back
u_nm1 = np.zeros(Nx+1)  # Solution at 2 time levels back

import time;  t0 = time.clock()  # CPU time measurement

# --- Valid indices for space and time mesh ---
Ix = range(0, Nx+1)
It = range(0, Nt+1)

# --- Load initial condition into u_n ---
for i in range(0,Nx+1):
    u_n[i] = I(x[i])

if user_action is not None:
    user_action(u_n, x, t, 0)

# --- Special formula for the first step ---
for i in Ix[1:-1]:
    u[i] = u_n[i] + dt*V(x[i]) + \
    0.5*C2*(0.5*(q[i] + q[i+1])*(u_n[i+1] - u_n[i]) - \
            0.5*(q[i] + q[i-1])*(u_n[i] - u_n[i-1])) + \
    0.5*dt2*f(x[i], t[0])

i = Ix[0]
if U_0 is None:
    # Set boundary values (x=0: i-1 -> i+1 since u[i-1]=u[i+1]
    # when du/dn = 0, on x=L: i+1 -> i-1 since u[i+1]=u[i-1])
    ip1 = i+1
    im1 = ip1  # i-1 -> i+1
    u[i] = u_n[i] + dt*V(x[i]) + \
            0.5*C2*(0.5*(q[i] + q[ip1])*(u_n[ip1] - u_n[i])  - \
                    0.5*(q[i] + q[im1])*(u_n[i] - u_n[im1])) + \
    0.5*dt2*f(x[i], t[0])
else:
    u[i] = U_0(dt)
```

```
i = Ix[-1]
if U_L is None:
    im1 = i-1
    ip1 = im1  # i+1 -> i-1
    u[i] = u_n[i] + dt*V(x[i]) + \
           0.5*C2*(0.5*(q[i] + q[ip1])*(u_n[ip1] - u_n[i])  - \
                   0.5*(q[i] + q[im1])*(u_n[i] - u_n[im1])) + \
    0.5*dt2*f(x[i], t[0])
else:
    u[i] = U_L(dt)

if user_action is not None:
    user_action(u, x, t, 1)

# Update data structures for next step
#u_nm1[:] = u_n; u_n[:] = u  # safe, but slower
u_nm1, u_n, u = u_n, u, u_nm1


# --- Time loop ---
for n in It[1:-1]:
    # Update all inner points
    if version == 'scalar':
        for i in Ix[1:-1]:
            u[i] = -  u_nm1[i] + 2*u_n[i] + \
                C2*(0.5*(q[i] + q[i+1])*(u_n[i+1] - u_n[i])  - \
                    0.5*(q[i] + q[i-1])*(u_n[i] - u_n[i-1])) + \
            dt2*f(x[i], t[n])

    elif version == 'vectorized':
        u[1:-1] = -  u_nm1[1:-1] + 2*u_n[1:-1] + \
        C2*(0.5*(q[1:-1] + q[2:])*(u_n[2:] - u_n[1:-1]) -
            0.5*(q[1:-1] + q[:-2])*(u_n[1:-1] - u_n[:-2])) + \
        dt2*f(x[1:-1], t[n])
    else:
        raise ValueError('version=%s' % version)

    # Insert boundary conditions
    i = Ix[0]
    if U_0 is None:
        # Set boundary values
        # x=0: i-1 -> i+1 since u[i-1]=u[i+1] when du/dn=0
        # x=L: i+1 -> i-1 since u[i+1]=u[i-1] when du/dn=0
        ip1 = i+1
        im1 = ip1
        u[i] = -  u_nm1[i] + 2*u_n[i] + \
                C2*(0.5*(q[i] + q[ip1])*(u_n[ip1] - u_n[i])  - \
                    0.5*(q[i] + q[im1])*(u_n[i] - u_n[im1])) + \
        dt2*f(x[i], t[n])
    else:
        u[i] = U_0(t[n+1])

    i = Ix[-1]
    if U_L is None:
        im1 = i-1
        ip1 = im1
```

```
                    u[i] = - u_nm1[i] + 2*u_n[i] + \
                        C2*(0.5*(q[i] + q[ip1])*(u_n[ip1] - u_n[i])  - \
                            0.5*(q[i] + q[im1])*(u_n[i] - u_n[im1])) + \
                    dt2*f(x[i], t[n])
                else:
                    u[i] = U_L(t[n+1])

                if user_action is not None:
                    if user_action(u, x, t, n+1):
                        break

                # Update data structures for next step
                u_nm1, u_n, u = u_n, u, u_nm1

        cpu_time = time.clock() - t0
        return cpu_time, hashed_input
```

C.2   Saving Large Arrays in Files

Numerical simulations produce large arrays as results and the software needs to
store these arrays on disk. Several methods are available in Python. We recommend
to use tailored solutions for large arrays and not standard file storage tools such as
pickle (cPickle for speed in Python version 2) and shelve, because the tailored
solutions have been optimized for array data and are hence much faster than the
standard tools.

C.2.1   Using savez to Store Arrays in Files

Storing individual arrays The numpy.storez function can store a set of arrays
to a named file in a zip archive. An associated function numpy.load can be used
to read the file later. Basically, we call numpy.storez(filename, **kwargs),
where kwargs is a dictionary containing array names as keys and the corresponding
array objects as values. Very often, the solution at a time point is given a natural
name where the name of the variable and the time level counter are combined, e.g.,
u11 or v39. Suppose n is the time level counter and we have two solution arrays, u
and v, that we want to save to a zip archive. The appropriate code is

```
import numpy as np
u_name = 'u%04d' % n    # array name
v_name = 'v%04d' % n    # array name
kwargs = {u_name: u, v_name: v}    # keyword args for savez
fname = '.mydata%04d.dat' % n
np.savez(fname, **kwargs)
if n == 0:                  # store x once
    np.savez('.mydata_x.dat', x=x)
```

Since the name of the array must be given as a keyword argument to savez, and the
name must be constructed as shown, it becomes a little tricky to do the call, but with

a dictionary kwargs and **kwargs, which sends each key-value pair as individual keyword arguments, the task gets accomplished.

Merging zip archives Each separate call to np.savez creates a new file (zip archive) with extension .npz. It is very convenient to collect all results in one archive instead. This can be done by merging all the individual .npz files into a single zip archive:

```python
def merge_zip_archives(individual_archives, archive_name):
    """
    Merge individual zip archives made with numpy.savez into
    one archive with name archive_name.
    The individual archives can be given as a list of names
    or as a Unix wild chard filename expression for glob.glob.
    The result of this function is that all the individual
    archives are deleted and the new single archive made.
    """
    import zipfile
    archive = zipfile.ZipFile(

        archive_name, 'w', zipfile.ZIP_DEFLATED,
        allowZip64=True)
    if isinstance(individual_archives, (list,tuple)):
        filenames = individual_archives
    elif isinstance(individual_archives, str):
        filenames = glob.glob(individual_archives)

    # Open each archive and write to the common archive
    for filename in filenames:
        f = zipfile.ZipFile(filename, 'r',
                            zipfile.ZIP_DEFLATED)
        for name in f.namelist():
            data = f.open(name, 'r')
            # Save under name without .npy
            archive.writestr(name[:-4], data.read())
        f.close()
        os.remove(filename)
    archive.close()
```

Here we remark that savez automatically adds the .npz extension to the names of the arrays we store. We do not want this extension in the final archive.

Reading arrays from zip archives Archives created by savez or the merged archive we describe above with name of the form myarchive.npz, can be conveniently read by the numpy.load function:

```python
import numpy as np
array_names = np.load('myarchive.npz')
for array_name in array_names:
    # array_names[array_name] is the array itself
    # e.g. plot(array_names['t'], array_names[array_name])
```

C.2.2   Using `joblib` to Store Arrays in Files

The Python package `joblib` has nice functionality for efficient storage of arrays on disk. The following class applies this functionality so that one can save an array, or in fact any Python data structure (e.g., a dictionary of arrays), to disk under a certain name. Later, we can retrieve the object by use of its name. The name of the directory under which the arrays are stored by `joblib` can be given by the user.

```
class Storage(object):
    """
    Store large data structures (e.g. numpy arrays) efficiently
    using joblib.

    Use:

    >>> from Storage import Storage
    >>> storage = Storage(cachedir='tmp_u01', verbose=1)
    >>> import numpy as np
    >>> a = np.linspace(0, 1, 100000) # large array
    >>> b = np.linspace(0, 1, 100000) # large array
    >>> storage.save('a', a)
    >>> storage.save('b', b)
    >>> # later
    >>> a = storage.retrieve('a')
    >>> b = storage.retrieve('b')
    """
    def __init__(self, cachedir='tmp', verbose=1):
        """
        Parameters
        ----------
        cachedir: str
            Name of directory where objects are stored in files.
        verbose: bool, int
            Let joblib and this class speak when storing files
            to disk.
        """
        import joblib
        self.memory = joblib.Memory(cachedir=cachedir,
                                    verbose=verbose)
        self.verbose = verbose
        self.retrieve = self.memory.cache(
            self.retrieve, ignore=['data'])
        self.save = self.retrieve

    def retrieve(self, name, data=None):
        if self.verbose > 0:
            print 'joblib save of', name
        return data
```

The `retrive` and `save` functions, which do the work, seem quite magic. The idea is that `joblib` looks at the `name` parameter and saves the return value `data` to disk if the `name` parameter has not been used in a previous call. Otherwise, if `name` is already registered, `joblib` fetches the `data` object from file and returns it (this is an example of a memoize function, see Section 2.1.4 in [11] for a brief explanation]).

C.2.3 Using a Hash to Create a File or Directory Name

Array storage techniques like those outlined in Sect. C.2.2 and C.2.1 demand the user to assign a name for the file(s) or directory where the solution is to be stored. Ideally, this name should reflect parameters in the problem such that one can recognize an already run simulation. One technique is to make a hash string out of the input data. A hash string is a 40-character long hexadecimal string that uniquely reflects another potentially much longer string. (You may be used to hash strings from the Git version control system: every committed version of the files in Git is recognized by a hash string.)

Suppose you have some input data in the form of functions, numpy arrays, and other objects. To turn these input data into a string, we may grab the source code of the functions, use a very efficient hash method for potentially large arrays, and simply convert all other objects via `str` to a string representation. The final string, merging all input data, is then converted to an SHA1 hash string such that we represent the input with a 40-character long string.

```
def myfunction(func1, func2, array1, array2, obj1, obj2):
    # Convert arguments to hash
    import inspect, joblib, hashlib
    data = (inspect.getsource(func1),
            inspect.getsource(func2),
        joblib.hash(array1),
        joblib.hash(array2),
        str(obj1),
        str(obj2))
    hash_input = hashlib.sha1(data).hexdigest()
```

It is wise to use `joblib.hash` and not try to do a `str(array1)`, since that string can be *very* long, and `joblib.hash` is more efficient than `hashlib` when turning these data into a hash.

Remark: turning function objects into their source code is unreliable!

The idea of turning a function object into a string via its source code may look smart, but is not a completely reliable solution. Suppose we have some function

```
x0 = 0.1
f = lambda x: 0 if x <= x0 else 1
```

The source code will be `f = lambda x:  0 if x <= x0 else 1`, so if the calling code changes the value of x0 (which f remembers - it is a closure), the source remains unchanged, the hash is the same, and the change in input data is unnoticed. Consequently, the technique above must be used with care. The user can always just remove the stored files in disk and thereby force a recomputation (provided the software applies a hash to test if a zip archive or `joblib` subdirectory exists, and if so, avoids recomputation).

C.3   Software for the 1D Wave Equation

We use numpy.storez to store the solution at each time level on disk. Such actions must be taken care of outside the solver function, more precisely in the user_action function that is called at every time level.

We have, in the wave1D_dn_vc.py code, implemented the user_action callback function as a class PlotAndStoreSolution with a __call__(self, x, t, t, n) method for the user_action function. Basically, __call__ stores and plots the solution. The storage makes use of the numpy.savez function for saving a set of arrays to a zip archive. Here, in this callback function, we want to save one array, u. Since there will be many such arrays, we introduce the array names 'u%04d' % n and closely related filenames. The usage of numpy.savez in __call__ goes like this:

```
from numpy import savez
name = 'u%04d' % n    # array name
kwargs = {name: u}    # keyword args for savez
fname = '.' + self.filename + '_' + name + '.dat'
self.t.append(t[n])   # store corresponding time value
savez(fname, **kwargs)
if n == 0:            # store x once
    savez('.' + self.filename + '_x.dat', x=x)
```

For example, if n is 10 and self.filename is tmp, the above call to savez becomes savez('.tmp_u0010.dat', u0010=u). The actual filename becomes .tmp_u0010.dat.npz. The actual array name becomes u0010.npy.

Each savez call results in a file, so after the simulation we have one file per time level. Each file produced by savez is a zip archive. It makes sense to merge all the files into one. This is done in the close_file method in the PlotAndStoreSolution class. The code goes as follows.

```
class PlotAndStoreSolution:
    ...
    def close_file(self, hashed_input):
        """
        Merge all files from savez calls into one archive.
        hashed_input is a string reflecting input data
        for this simulation (made by solver).
        """
        if self.filename is not None:
            # Save all the time points where solutions are saved
            savez('.' + self.filename + '_t.dat',
                  t=array(self.t, dtype=float))
            # Merge all savez files to one zip archive
            archive_name = '.' + hashed_input + '_archive.npz'
            filenames = glob.glob('.' + self.filename + '*.dat.npz')
            merge_zip_archives(filenames, archive_name)
```

We use various ZipFile functionality to extract the content of the individual files (each with name filename) and write it to the merged archive (archive). There is only one array in each individual file (filename) so strictly speaking, there is no need for the loop `for name in f.namelist()` (as `f.namelist()` returns a list of length 1). However, in other applications where we compute more arrays at each time level, `savez` will store all these and then there is need for iterating over `f.namelist()`.

Instead of merging the archives written by `savez` we could make an alternative implementation that writes all our arrays into one archive. This is the subject of Exercise C.2.

C.3.1 Making Hash Strings from Input Data

The `hashed_input` argument, used to name the resulting archive file with all solutions, is supposed to be a hash reflecting all import parameters in the problem such that this simulation has a unique name. The `hashed_input` string is made in the `solver` function, using the `hashlib` and `inspect` modules, based on the arguments to `solver`:

```
# Make hash of all input data
import hashlib, inspect
data = inspect.getsource(I) + '_' + inspect.getsource(V) + \
       '_' + inspect.getsource(f) + '_' + str(c) + '_' + \
       ('None' if U_0 is None else inspect.getsource(U_0)) + \
       ('None' if U_L is None else inspect.getsource(U_L)) + \
       '_' + str(L) + str(dt) + '_' + str(C) + '_' + str(T) + \
       '_' + str(stability_safety_factor)
hashed_input = hashlib.sha1(data).hexdigest()
```

To get the source code of a function `f` as a string, we use `inspect.get-source(f)`. All input, functions as well as variables, is then merged to a string `data`, and then `hashlib.sha1` makes a unique, much shorter (40 characters long), fixed-length string out of `data` that we can use in the archive filename.

Remark

Note that the construction of the `data` string is not fool proof: if, e.g., `I` is a formula with parameters and the parameters change, the source code is still the same and `data` and hence the hash remains unaltered. The implementation must therefore be used with care!

C.3.2 Avoiding Rerunning Previously Run Cases

If the archive file whose name is based on `hashed_input` already exists, the simulation with the current set of parameters has been done before and one can avoid redoing the work. The `solver` function returns the CPU time and `hashed_input`, and a negative CPU time means that no simulation was run. In that case we should not call the `close_file` method above (otherwise we overwrite the archive with just the `self.t` array). The typical usage goes like

```
action = PlotAndStoreSolution(...)
dt = (L/Nx)/C  # choose the stability limit with given Nx
cpu, hashed_input = solver(
    I=lambda x: ...,
    V=0, f=0, c=1, U_0=lambda t: 0, U_L=None, L=1,
    dt=dt, C=C, T=T,
    user_action=action, version='vectorized',
    stability_safety_factor=1)
action.make_movie_file()
if cpu > 0:  # did we generate new data?
    action.close_file(hashed_input)
```

C.3.3 Verification

Vanishing approximation error Exact solutions of the numerical equations are always attractive for verification purposes since the software should reproduce such solutions to machine precision. With Dirichlet boundary conditions we can construct a function that is linear in t and quadratic in x that is also an exact solution of the scheme, while with Neumann conditions we are left with testing just a constant solution (see comments in Sect. 2.6.5).

Convergence rates A more general method for verification is to check the convergence rates. We must introduce one discretization parameter h and assume an error model $E = Ch^r$, where C and r are constants to be determine (i.e., r is the rate that we are interested in). Given two experiments with different resolutions h_i and h_i-1, we can estimate r by

$$r = \frac{\ln(E_i/E_{i-1})}{\ln(h_i/h_{i-1})},$$

where E_i is the error corresponding to h_i and E_{i-1} corresponds to h_{i-1}. Section 2.2.2 explains the details of this type of verification and how we introduce the single discretization parameter $h = \Delta t = \hat{c}\Delta t$, for some constant $\hat{c}$. To compute the error, we had to rely on a global variable in the user action function. Below is an implementation where we have a more elegant solution in terms of a class: the error variable is not a class attribute and there is no need for a global error (which is always considered an advantage).

```
def convergence_rates(
    u_exact,
    I, V, f, c, U_0, U_L, L,
    dt0, num_meshes,
    C, T, version='scalar',
    stability_safety_factor=1.0):
    """
    Half the time step and estimate convergence rates for
    for num_meshes simulations.
    """
    class ComputeError:
        def __init__(self, norm_type):
            self.error = 0
```

```
    def __call__(self, u, x, t, n):
        """Store norm of the error in self.E."""
        error = np.abs(u - u_exact(x, t[n])).max()
        self.error = max(self.error, error)

E = []
h = []  # dt, solver adjusts dx such that C=dt*c/dx
dt = dt0
for i in range(num_meshes):
    error_calculator = ComputeError('Linf')
    solver(I, V, f, c, U_0, U_L, L, dt, C, T,
           user_action=error_calculator,
           version='scalar',
           stability_safety_factor=1.0)
    E.append(error_calculator.error)
    h.append(dt)
    dt /= 2  # halve the time step for next simulation
print 'E:', E
print 'h:', h
r = [np.log(E[i]/E[i-1])/np.log(h[i]/h[i-1])
     for i in range(1,num_meshes)]
return r
```

The returned sequence r should converge to 2 since the error analysis in Sect. 2.10 predicts various error measures to behave like $\mathcal{O}(\Delta t^2) + \mathcal{O}(\Delta x^2)$. We can easily run the case with standing waves and the analytical solution $u(x,t) = \cos(\frac{2\pi}{L}t)\sin(\frac{2\pi}{L}x)$. The call will be very similar to the one provided in the test_convrate_sincos function in Sect. 2.3.4, see the file wave1D_dn_vc.py for details.

C.4   Programming the Solver with Classes

Many who know about class programming prefer to organize their software in terms of classes. This gives a richer application programming interface (API) since a function solver must have all its input data in terms of arguments, while a class-based solver naturally has a mix of method arguments and user-supplied methods. (Well, to be more precise, our solvers have demanded user_action to be a function provided by the user, so it is possible to mix variables and functions in the input also with a solver function.)

We will next illustrate how some of the functionality in wave1D_dn_vc.py may be implemented by using classes. Focusing on class implementation aspects, we restrict the example case to a simpler wave with constant wave speed c. Applying the method of manufactured solutions, we test whether the class based implementation is able to compute the known exact solution within machine precision.

We will create a class Problem to hold the physical parameters of the problem and a class Solver to hold the numerical solution parameters besides the solver function itself. As the number of parameters increases, so does the amount of repetitive code. We therefore take the opportunity to illustrate how this may be counteracted by introducing a super class Parameters that allows code to be parameterized. In addition, it is convenient to collect the arrays that describe the mesh in a special `Mesh` class and make a class `Function` for a mesh function (mesh point values and its mesh). All the following code is found in `wave1D_oo.py`.

C.4.1 Class Parameters

The classes `Problem` and `Solver` both inherit class `Parameters`, which handles reading of parameters from the command line and has methods for setting and getting parameter values. Since processing dictionaries is easier than processing a collection of individual attributes, the class `Parameters` requires each class `Problem` and `Solver` to represent their parameters by dictionaries, one compulsory and two optional ones. The compulsory dictionary, `self.prm`, contains all parameters, while a second and optional dictionary, `self.type`, holds the associated object types, and a third and optional dictionary, `self.help`, stores help strings. The `Parameters` class may be implemented as follows:

```python
class Parameters(object):
    def __init__(self):
        """
        Subclasses must initialize self.prm with
        parameters and default values, self.type with
        the corresponding types, and self.help with
        the corresponding descriptions of parameters.
        self.type and self.help are optional, but
        self.prms must be complete and contain all parameters.
        """
        pass


    def ok(self):
        """Check if attr. prm, type, and help are defined."""
        if hasattr(self, 'prm') and \
            isinstance(self.prm, dict) and \
            hasattr(self, 'type') and \
            isinstance(self.type, dict) and \
            hasattr(self, 'help') and \
            isinstance(self.help, dict):

            return True
        else:
            raise ValueError(
                'The constructor in class %s does not '\
                'initialize the\ndictionaries '\
                'self.prm, self.type, self.help!' %
                self.__class__.__name__)

    def _illegal_parameter(self, name):
        """Raise exception about illegal parameter name."""
        raise ValueError(
            'parameter "%s" is not registered.\nLegal '\
            'parameters are\n%s' %
            (name, ' '.join(list(self.prm.keys()))))
```

```python
    def set(self, **parameters):
        """Set one or more parameters."""
        for name in parameters:
            if name in self.prm:
                self.prm[name] = parameters[name]
            else:
                self._illegal_parameter(name)

    def get(self, name):
        """Get one or more parameter values."""
        if isinstance(name, (list,tuple)):    # get many?
            for n in name:
                if n not in self.prm:
                    self._illegal_parameter(name)
            return [self.prm[n] for n in name]
        else:
            if name not in self.prm:
                self._illegal_parameter(name)
            return self.prm[name]

    def __getitem__(self, name):
        """Allow obj[name] indexing to look up a parameter."""
        return self.get(name)

    def __setitem__(self, name, value):
        """
        Allow obj[name] = value syntax to assign a parameter's value.
        """
        return self.set(name=value)


    def define_command_line_options(self, parser=None):
        self.ok()
        if parser is None:
            import argparse
            parser = argparse.ArgumentParser()

        for name in self.prm:
            tp = self.type[name] if name in self.type else str
            help = self.help[name] if name in self.help else None
            parser.add_argument(
                '--' + name, default=self.get(name), metavar=name,
                type=tp, help=help)

        return parser

    def init_from_command_line(self, args):
        for name in self.prm:
            self.prm[name] = getattr(args, name)
```

C.4.2   Class Problem

Inheriting the Parameters class, our class Problem is defined as:

```
class Problem(Parameters):
    """
    Physical parameters for the wave equation
    u_tt = (c**2*u_x)_x + f(x,t) with t in [0,T] and
    x in (0,L). The problem definition is implied by
    the method of manufactured solution, choosing
    u(x,t)=x(L-x)(1+t/2) as our solution. This solution
    should be exactly reproduced when c is const.
    """

    def __init__(self):
        self.prm  = dict(L=2.5, c=1.5, T=18)
        self.type = dict(L=float, c=float, T=float)
        self.help = dict(L='1D domain',

                         c='coefficient (wave velocity) in PDE',
                         T='end time of simulation')
    def u_exact(self, x, t):
        L = self['L']
        return x*(L-x)*(1+0.5*t)
    def I(self, x):
        return self.u_exact(x, 0)
    def V(self, x):
        return 0.5*self.u_exact(x, 0)
    def f(self, x, t):
        c = self['c']
        return 2*(1+0.5*t)*c**2
    def U_0(self, t):
        return self.u_exact(0, t)
    U_L = None
```

C.4.3   Class Mesh

The Mesh class can be made valid for a space-time mesh in any number of space dimensions. To make the class versatile, the constructor accepts either a tuple/list of number of cells in each spatial dimension or a tuple/list of cell spacings. In addition, we need the size of the hypercube mesh as a tuple/list of 2-tuples with lower and upper limits of the mesh coordinates in each direction. For 1D meshes it is more natural to just write the number of cells or the cell size and not wrap it in a list. We also need the time interval from t0 to T. Giving no spatial discretization information implies a time mesh only, and vice versa. The Mesh class with documentation and a doc test should now be self-explanatory:

```
import numpy as np

class Mesh(object):
    """
    Holds data structures for a uniform mesh on a hypercube in
    space, plus a uniform mesh in time.

    ======== =====================================================
    Argument               Explanation
    ======== =====================================================
    L        List of 2-lists of min and max coordinates
             in each spatial direction.
    T        Final time in time mesh.
    Nt       Number of cells in time mesh.
    dt       Time step. Either Nt or dt must be given.
    N        List of number of cells in the spatial directions.
    d        List of cell sizes in the spatial directions.
             Either N or d must be given.
    ======== =====================================================

    Users can access all the parameters mentioned above, plus
    ``x[i]`` and ``t`` for the coordinates in direction ``i``
    and the time coordinates, respectively.

    Examples:

    >>> from UniformFDMesh import Mesh
    >>>
    >>> # Simple space mesh
    >>> m = Mesh(L=[0,1], N=4)
    >>> print m.dump()
    space: [0,1] N=4 d=0.25
    >>>
    >>> # Simple time mesh
    >>> m = Mesh(T=4, dt=0.5)
    >>> print m.dump()
    time: [0,4] Nt=8 dt=0.5
    >>>
    >>> # 2D space mesh
    >>> m = Mesh(L=[[0,1], [-1,1]], d=[0.5, 1])
    >>> print m.dump()
    space: [0,1]x[-1,1] N=2x2 d=0.5,1
    >>>
    >>> # 2D space mesh and time mesh
    >>> m = Mesh(L=[[0,1], [-1,1]], d=[0.5, 1], Nt=10, T=3)
    >>> print m.dump()

    """
    def __init__(self,
                 L=None, T=None, t0=0,
                 N=None, d=None,
                 Nt=None, dt=None):
        if N is None and d is None:
            # No spatial mesh
            if Nt is None and dt is None:
                raise ValueError(
                'Mesh constructor: either Nt or dt must be given')
            if T is None:
                raise ValueError(
                'Mesh constructor: T must be given')
```

```
if Nt is None and dt is None:
    if N is None and d is None:
        raise ValueError(
        'Mesh constructor: either N or d must be given')
    if L is None:
        raise ValueError(
        'Mesh constructor: L must be given')

# Allow 1D interface without nested lists with one element
if L is not None and isinstance(L[0], (float,int)):
    # Only an interval was given
    L = [L]
if N is not None and isinstance(N, (float,int)):
    N = [N]
if d is not None and isinstance(d, (float,int)):
    d = [d]

# Set all attributes to None
self.x = None
self.t = None
self.Nt = None
self.dt = None
self.N = None
self.d = None
self.t0 = t0

if N is None and d is not None and L is not None:
    self.L = L
    if len(d) != len(L):
        raise ValueError(
            'd has different size (no of space dim.) from '
            'L: %d vs %d', len(d), len(L))
    self.d = d
    self.N = [int(round(float(self.L[i][1] -
                              self.L[i][0])/d[i]))
              for i in range(len(d))]
if d is None and N is not None and L is not None:
    self.L = L
    if len(N) != len(L):
        raise ValueError(
            'N has different size (no of space dim.) from '
            'L: %d vs %d', len(N), len(L))
    self.N = N
    self.d = [float(self.L[i][1] - self.L[i][0])/N[i]
              for i in range(len(N))]

if Nt is None and dt is not None and T is not None:
    self.T = T
    self.dt = dt
    self.Nt = int(round(T/dt))
if dt is None and Nt is not None and T is not None:
    self.T = T
    self.Nt = Nt
    self.dt = T/float(Nt)

if self.N is not None:
    self.x = [np.linspace(
                self.L[i][0], self.L[i][1], self.N[i]+1)
                for i in range(len(self.L))]
if Nt is not None:
    self.t = np.linspace(self.t0, self.T, self.Nt+1)
```

```
def get_num_space_dim(self):
    return len(self.d) if self.d is not None else 0

def has_space(self):
    return self.d is not None

def has_time(self):
    return self.dt is not None

def dump(self):
    s = ''
    if self.has_space():
        s += 'space: ' + \
            'x'.join(['[%g,%g]' % (self.L[i][0], self.L[i][1])
                     for i in range(len(self.L))]) + ' N='
        s += 'x'.join([str(Ni) for Ni in self.N]) + ' d='
        s += ','.join([str(di) for di in self.d])
    if self.has_space() and self.has_time():
        s += ' '
    if self.has_time():
        s += 'time: ' + '[%g,%g]' % (self.t0, self.T) + \
            ' Nt=%g' % self.Nt + ' dt=%g' % self.dt
    return s
```

We rely on attribute access – not get/set functions!
Java programmers, in particular, are used to get/set functions in classes to access internal data. In Python, we usually apply direct access of the attribute, such as m.N[i] if m is a Mesh object. A widely used convention is to do this as long as access to an attribute does not require additional code. In that case, one applies a property construction. The original interface remains the same after a property is introduced (in contrast to Java), so user will not notice a change to properties.

The only argument against direct attribute access in class Mesh is that the attributes are read-only so we could avoid offering a set function. Instead, we rely on the user that she does not assign new values to the attributes.

C.4.4   Class Function

A class Function is handy to hold a mesh and corresponding values for a scalar or vector function over the mesh. Since we may have a time or space mesh, or a combined time and space mesh, with one or more components in the function, some if tests are needed for allocating the right array sizes. To help the user, an indices attribute with the name of the indices in the final array u for the function values is made. The examples in the doc string should explain the functionality.

```
class Function(object):
    """
    A scalar or vector function over a mesh (of class Mesh).

    ========== ============================================================
    Argument                        Explanation
    ========== ============================================================
    mesh       Class Mesh object: spatial and/or temporal mesh.
    num_comp   Number of components in function (1 for scalar).
    space_only True if the function is defined on the space mesh
               only (to save space). False if function has values
               in space and time.
    ========== ============================================================

    The indexing of ``u``, which holds the mesh point values of the
    function, depends on whether we have a space and/or time mesh.

    Examples:

    >>> from UniformFDMesh import Mesh, Function
    >>>
    >>> # Simple space mesh
    >>> m = Mesh(L=[0,1], N=4)
    >>> print m.dump()
    space: [0,1] N=4 d=0.25
    >>> f = Function(m)
    >>> f.indices
    ['x0']
    >>> f.u.shape
    (5,)
    >>> f.u[4]   # space point 4
    0.0
    >>>
    >>> # Simple time mesh for two components
    >>> m = Mesh(T=4, dt=0.5)
    >>> print m.dump()
    time: [0,4] Nt=8 dt=0.5
    >>> f = Function(m, num_comp=2)
    >>> f.indices
    ['time', 'component']
    >>> f.u.shape
    (9, 2)
    >>> f.u[3,1]  # time point 3, comp=1 (2nd comp.)
    0.0
    >>>
    >>> # 2D space mesh
    >>> m = Mesh(L=[[0,1], [-1,1]], d=[0.5, 1])
    >>> print m.dump()
    space: [0,1]x[-1,1] N=2x2 d=0.5,1
    >>> f = Function(m)
    >>> f.indices
    ['x0', 'x1']
    >>> f.u.shape
    (3, 3)
    >>> f.u[1,2]  # space point (1,2)
    0.0
```

```
>>>
>>> # 2D space mesh and time mesh
>>> m = Mesh(L=[[0,1],[-1,1]], d=[0.5,1], Nt=10, T=3)
>>> print m.dump()
space: [0,1]x[-1,1] N=2x2 d=0.5,1 time: [0,3] Nt=10 dt=0.3
>>> f = Function(m, num_comp=2, space_only=False)
>>> f.indices
['time', 'x0', 'x1', 'component']
>>> f.u.shape
(11, 3, 3, 2)
>>> f.u[2,1,2,0]  # time step 2, space point (1,2), comp=0
0.0
>>> # Function with space data only
>>> f = Function(m, num_comp=1, space_only=True)
>>> f.indices
['x0', 'x1']
>>> f.u.shape
(3, 3)
>>> f.u[1,2]  # space point (1,2)
0.0
"""

def __init__(self, mesh, num_comp=1, space_only=True):
    self.mesh = mesh
    self.num_comp = num_comp
    self.indices = []

    # Create array(s) to store mesh point values
    if (self.mesh.has_space() and not self.mesh.has_time()) or \
       (self.mesh.has_space() and self.mesh.has_time() and \
        space_only):
        # Space mesh only
        if num_comp == 1:
            self.u = np.zeros(
                [self.mesh.N[i] + 1
                 for i in range(len(self.mesh.N))])
            self.indices = [
                'x'+str(i) for i in range(len(self.mesh.N))]
        else:
            self.u = np.zeros(
                [self.mesh.N[i] + 1
                 for i in range(len(self.mesh.N))] +
                [num_comp])
            self.indices = [
                'x'+str(i)
                for i in range(len(self.mesh.N))] +\
                ['component']
    if not self.mesh.has_space() and self.mesh.has_time():
        # Time mesh only
        if num_comp == 1:
            self.u = np.zeros(self.mesh.Nt+1)
            self.indices = ['time']
        else:
            # Need num_comp entries per time step
            self.u = np.zeros((self.mesh.Nt+1, num_comp))
            self.indices = ['time', 'component']
```

```
    if self.mesh.has_space() and self.mesh.has_time() \
        and not space_only:
        # Space-time mesh
        size = [self.mesh.Nt+1] + \
                [self.mesh.N[i]+1
                 for i in range(len(self.mesh.N))]
        if num_comp > 1:
            self.indices = ['time'] + \
                            ['x'+str(i)
                             for i in range(len(self.mesh.N))] +\
                            ['component']
            size += [num_comp]
        else:
            self.indices = ['time'] + ['x'+str(i)
                             for i in range(len(self.mesh.N))]
        self.u = np.zeros(size)
```

C.4.5   Class Solver

With the Mesh and Function classes in place, we can rewrite the solver function, but we make it a method in class Solver:

```
class Solver(Parameters):
    """
    Numerical parameters for solving the wave equation
    u_tt = (c**2*u_x)_x + f(x,t) with t in [0,T] and
    x in (0,L). The problem definition is implied by
    the method of manufactured solution, choosing
    u(x,t)=x(L-x)(1+t/2) as our solution. This solution
    should be exactly reproduced, provided c is const.
    We simulate in [0, L/2] and apply a symmetry condition
    at the end x=L/2.
    """

    def __init__(self, problem):
        self.problem = problem
        self.prm  = dict(C = 0.75, Nx=3, stability_safety_factor=1.0)
        self.type = dict(C=float, Nx=int, stability_safety_factor=float)
        self.help = dict(C='Courant number',
                         Nx='No of spatial mesh points',
                         stability_safety_factor='stability factor')

        from UniformFDMesh import Mesh, Function
        # introduce some local help variables to ease reading
        L_end = self.problem['L']
        dx = (L_end/2)/float(self['Nx'])
        t_interval = self.problem['T']
        dt = dx*self['stability_safety_factor']*self['C']/ \
                                float(self.problem['c'])
        self.m = Mesh(L=[0,L_end/2],
                      d=[dx],
                      Nt = int(round(t_interval/float(dt))),
                      T=t_interval)
        # The mesh function f will, after solving, contain
        # the solution for the whole domain and all time steps.
        self.f = Function(self.m, num_comp=1, space_only=False)
```

```python
def solve(self, user_action=None, version='scalar'):
    # ...use local variables to ease reading
    L, c, T = self.problem['L c T'.split()]
    L = L/2     # compute with half the domain only (symmetry)
    C, Nx, stability_safety_factor = self[
                        'C Nx stability_safety_factor'.split()]
    dx = self.m.d[0]
    I = self.problem.I
    V = self.problem.V
    f = self.problem.f
    U_0 = self.problem.U_0
    U_L = self.problem.U_L
    Nt = self.m.Nt
    t = np.linspace(0, T, Nt+1)      # Mesh points in time
    x = np.linspace(0, L, Nx+1)      # Mesh points in space

    # Make sure dx and dt are compatible with x and t
    dx = x[1] - x[0]
    dt = t[1] - t[0]


    # Treat c(x) as array
    if isinstance(c, (float,int)):
        c = np.zeros(x.shape) + c
    elif callable(c):
        # Call c(x) and fill array c
        c_ = np.zeros(x.shape)
        for i in range(Nx+1):
            c_[i] = c(x[i])
        c = c_

    q = c**2
    C2 = (dt/dx)**2; dt2 = dt*dt     # Help variables in the scheme

    # Wrap user-given f, I, V, U_0, U_L if None or 0
    if f is None or f == 0:
        f = (lambda x, t: 0) if version == 'scalar' else \
            lambda x, t: np.zeros(x.shape)
    if I is None or I == 0:
        I = (lambda x: 0) if version == 'scalar' else \
            lambda x: np.zeros(x.shape)
    if V is None or V == 0:
        V = (lambda x: 0) if version == 'scalar' else \
            lambda x: np.zeros(x.shape)
    if U_0 is not None:
        if isinstance(U_0, (float,int)) and U_0 == 0:
            U_0 = lambda t: 0
    if U_L is not None:
        if isinstance(U_L, (float,int)) and U_L == 0:
            U_L = lambda t: 0

    # Make hash of all input data
    import hashlib, inspect
    data = inspect.getsource(I) + '_' + inspect.getsource(V) + \
           '_' + inspect.getsource(f) + '_' + str(c) + '_' + \
           ('None' if U_0 is None else inspect.getsource(U_0)) + \
           ('None' if U_L is None else inspect.getsource(U_L)) + \
           '_' + str(L) + str(dt) + '_' + str(C) + '_' + str(T) + \
           '_' + str(stability_safety_factor)
```

```
                              y        y
hashed_input = hashlib.sha1(data).hexdigest()
if os.path.isfile('.' + hashed_input + '_archive.npz'):
    # Simulation is already run
    return -1, hashed_input

# use local variables to make code closer to mathematical
# notation in computational scheme
u_1 = self.f.u[0,:]
u   = self.f.u[1,:]

import time;  t0 = time.clock()  # CPU time measurement

Ix = range(0, Nx+1)
It = range(0, Nt+1)

# Load initial condition into u_1
for i in range(0,Nx+1):
    u_1[i] = I(x[i])

if user_action is not None:
    user_action(u_1, x, t, 0)


# Special formula for the first step
for i in Ix[1:-1]:
    u[i] = u_1[i] + dt*V(x[i]) + \
    0.5*C2*(0.5*(q[i] + q[i+1])*(u_1[i+1] - u_1[i]) - \
            0.5*(q[i] + q[i-1])*(u_1[i] - u_1[i-1])) + \
    0.5*dt2*f(x[i], t[0])

i = Ix[0]
if U_0 is None:
    # Set boundary values (x=0: i-1 -> i+1 since u[i-1]=u[i+1]
    # when du/dn = 0, on x=L: i+1 -> i-1 since u[i+1]=u[i-1])
    ip1 = i+1
    im1 = ip1  # i-1 -> i+1
    u[i] = u_1[i] + dt*V(x[i]) + \
            0.5*C2*(0.5*(q[i] + q[ip1])*(u_1[ip1] - u_1[i])  - \
                    0.5*(q[i] + q[im1])*(u_1[i] - u_1[im1])) + \
    0.5*dt2*f(x[i], t[0])
else:
    u[i] = U_0(dt)

i = Ix[-1]
if U_L is None:
    im1 = i-1
    ip1 = im1  # i+1 -> i-1
    u[i] = u_1[i] + dt*V(x[i]) + \
            0.5*C2*(0.5*(q[i] + q[ip1])*(u_1[ip1] - u_1[i])  - \
                    0.5*(q[i] + q[im1])*(u_1[i] - u_1[im1])) + \
    0.5*dt2*f(x[i], t[0])
else:
    u[i] = U_L(dt)

if user_action is not None:
    user_action(u, x, t, 1)
```

```
for n in It[1:-1]:
    # u corresponds to u^{n+1} in the mathematical scheme
    u_2 = self.f.u[n-1,:]
    u_1 = self.f.u[n,:]
    u   = self.f.u[n+1,:]

    # Update all inner points
    if version == 'scalar':
        for i in Ix[1:-1]:
            u[i] = -  u_2[i] + 2*u_1[i] + \
                C2*(0.5*(q[i] + q[i+1])*(u_1[i+1] - u_1[i])  - \
                    0.5*(q[i] + q[i-1])*(u_1[i] - u_1[i-1])) + \
                dt2*f(x[i], t[n])

    elif version == 'vectorized':
        u[1:-1] = - u_2[1:-1] + 2*u_1[1:-1] + \
        C2*(0.5*(q[1:-1] + q[2:])*(u_1[2:] - u_1[1:-1]) -
            0.5*(q[1:-1] + q[:-2])*(u_1[1:-1] - u_1[:-2])) + \
        dt2*f(x[1:-1], t[n])
    else:
        raise ValueError('version=%s' % version)


    # Insert boundary conditions
    i = Ix[0]
    if U_0 is None:
        # Set boundary values
        # x=0: i-1 -> i+1 since u[i-1]=u[i+1] when du/dn=0
        # x=L: i+1 -> i-1 since u[i+1]=u[i-1] when du/dn=0
        ip1 = i+1
        im1 = ip1
        u[i] = -  u_2[i] + 2*u_1[i] + \
                C2*(0.5*(q[i] + q[ip1])*(u_1[ip1] - u_1[i])  - \
                    0.5*(q[i] + q[im1])*(u_1[i] - u_1[im1])) + \
                dt2*f(x[i], t[n])
    else:
        u[i] = U_0(t[n+1])

    i = Ix[-1]
    if U_L is None:
        im1 = i-1
        ip1 = im1
        u[i] = -  u_2[i] + 2*u_1[i] + \
                C2*(0.5*(q[i] + q[ip1])*(u_1[ip1] - u_1[i])  - \
                    0.5*(q[i] + q[im1])*(u_1[i] - u_1[im1])) + \
                dt2*f(x[i], t[n])
    else:
        u[i] = U_L(t[n+1])

    if user_action is not None:
        if user_action(u, x, t, n+1):
            break

cpu_time = time.clock() - t0
return cpu_time, hashed_input
```

```
def assert_no_error(self):
    """Run through mesh and check error"""
    Nx = self['Nx']
    Nt = self.m.Nt
    L, T = self.problem['L T'].split()
    L = L/2        # only half the domain used (symmetry)
    x = np.linspace(0, L, Nx+1)      # Mesh points in space
    t = np.linspace(0, T, Nt+1)      # Mesh points in time

    for n in range(len(t)):
        u_e = self.problem.u_exact(x, t[n])
        diff = np.abs(self.f.u[n,:] - u_e).max()
        print 'diff:', diff
        tol = 1E-13
        assert diff < tol
```

Observe that the solutions from all time steps are stored in the mesh function, which allows error assessment (in `assert_no_error`) to take place after all solutions have been found. Of course, in 2D or 3D, such a strategy may place too high demands on available computer memory, in which case intermediate results could be stored on file.

Running `wave1D_oo.py` gives a printout showing that the class-based implementation performs as expected, i.e. that the known exact solution is reproduced (within machine precision).

C.5   Migrating Loops to Cython

We now consider the `wave2D_u0.py` code for solving the 2D linear wave equation with constant wave velocity and homogeneous Dirichlet boundary conditions $u = 0$. We shall in the present chapter extend this code with computational modules written in other languages than Python. This extended version is called `wave2D_u0_adv.py`.

The `wave2D_u0.py` file contains a `solver` function, which calls an `advance_*` function to advance the numerical scheme one level forward in time. The function `advance_scalar` applies standard Python loops to implement the scheme, while `advance_vectorized` performs corresponding vectorized arithmetics with array slices. The statements of this solver are explained in Sect. 2.12, in particular Sect. 2.12.1 and 2.12.2.

Although vectorization can bring down the CPU time dramatically compared with scalar code, there is still some factor 5-10 to win in these types of applications by implementing the finite difference scheme in compiled code, typically in Fortran, C, or C++. This can quite easily be done by adding a little extra code to our program. Cython is an extension of Python that offers the easiest way to nail our Python loops in the scalar code down to machine code and achieve the efficiency of C.

Cython can be viewed as an extended Python language where variables are declared with types and where functions are marked to be implemented in C. Migrating Python code to Cython is done by copying the desired code segments to functions (or classes) and placing them in one or more separate files with extension .pyx.

C.5.1 Declaring Variables and Annotating the Code

Our starting point is the plain advance_scalar function for a scalar implementation of the updating algorithm for new values $u_{i,j}^{n+1}$:

```
def advance_scalar(u, u_n, u_nm1, f, x, y, t, n, Cx2, Cy2, dt2,
                   V=None, step1=False):
    Ix = range(0, u.shape[0]);  Iy = range(0, u.shape[1])
    if step1:
        dt = sqrt(dt2)  # save
        Cx2 = 0.5*Cx2;  Cy2 = 0.5*Cy2; dt2 = 0.5*dt2  # redefine
        D1 = 1;  D2 = 0
    else:
        D1 = 2;  D2 = 1
    for i in Ix[1:-1]:
        for j in Iy[1:-1]:
            u_xx = u_n[i-1,j] - 2*u_n[i,j] + u_n[i+1,j]
            u_yy = u_n[i,j-1] - 2*u_n[i,j] + u_n[i,j+1]
            u[i,j] = D1*u_n[i,j] - D2*u_nm1[i,j] + \
                     Cx2*u_xx + Cy2*u_yy + dt2*f(x[i], y[j], t[n])
            if step1:
                u[i,j] += dt*V(x[i], y[j])
    # Boundary condition u=0
    j = Iy[0]
    for i in Ix: u[i,j] = 0
    j = Iy[-1]
    for i in Ix: u[i,j] = 0
    i = Ix[0]
    for j in Iy: u[i,j] = 0
    i = Ix[-1]
    for j in Iy: u[i,j] = 0
    return u
```

We simply take a copy of this function and put it in a file wave2D_u0_loop_cy.pyx. The relevant Cython implementation arises from declaring variables with types and adding some important annotations to speed up array computing in Cython. Let us first list the complete code in the .pyx file:

```
import numpy as np
cimport numpy as np
cimport cython
ctypedef np.float64_t DT    # data type

@cython.boundscheck(False)  # turn off array bounds check
@cython.wraparound(False)   # turn off negative indices (u[-1,-1])
cpdef advance(
    np.ndarray[DT, ndim=2, mode='c'] u,
    np.ndarray[DT, ndim=2, mode='c'] u_n,
    np.ndarray[DT, ndim=2, mode='c'] u_nm1,
    np.ndarray[DT, ndim=2, mode='c'] f,
    double Cx2, double Cy2, double dt2):
```

```
cdef:
    int Ix_start = 0
    int Iy_start = 0
    int Ix_end = u.shape[0]-1
    int Iy_end = u.shape[1]-1
    int i, j
    double u_xx, u_yy

for i in range(Ix_start+1, Ix_end):
    for j in range(Iy_start+1, Iy_end):
        u_xx = u_n[i-1,j] - 2*u_n[i,j] + u_n[i+1,j]
        u_yy = u_n[i,j-1] - 2*u_n[i,j] + u_n[i,j+1]
        u[i,j] = 2*u_n[i,j] - u_nm1[i,j] + \
                 Cx2*u_xx + Cy2*u_yy + dt2*f[i,j]
# Boundary condition u=0
j = Iy_start
for i in range(Ix_start, Ix_end+1): u[i,j] = 0
j = Iy_end
for i in range(Ix_start, Ix_end+1): u[i,j] = 0
i = Ix_start
for j in range(Iy_start, Iy_end+1): u[i,j] = 0
i = Ix_end
for j in range(Iy_start, Iy_end+1): u[i,j] = 0
return u
```

This example may act as a recipe on how to transform array-intensive code with loops into Cython.

1. Variables are declared with types: for example, `double v` in the argument list instead of just `v`, and `cdef double v` for a variable `v` in the body of the function. A Python `float` object is declared as `double` for translation to C by Cython, while an `int` object is declared by `int`.

2. Arrays need a comprehensive type declaration involving
   - the type `np.ndarray`,
   - the data type of the elements, here 64-bit floats, abbreviated as DT through `ctypedef np.float64_t DT` (instead of DT we could use the full name of the data type: `np.float64_t`, which is a Cython-defined type),
   - the dimensions of the array, here `ndim=2` and `ndim=1`,
   - specification of contiguous memory for the array (`mode='c'`).

3. Functions declared with `cpdef` are translated to C but are also accessible from Python.

4. In addition to the standard numpy import we also need a special Cython import of numpy: `cimport numpy as np`, to appear *after* the standard import.

5. By default, array indices are checked to be within their legal limits. To speed up the code one should turn off this feature for a specific function by placing `@cython.boundscheck(False)` above the function header.

6. Also by default, array indices can be negative (counting from the end), but this feature has a performance penalty and is therefore here turned off by writing `@cython.wraparound(False)` right above the function header.

7. The use of index sets `Ix` and `Iy` in the scalar code cannot be successfully translated to C. One reason is that constructions like `Ix[1:-1]` involve negative

indices, and these are now turned off. Another reason is that Cython loops must take the form `for i in xrange` or `for i in range` for being translated into efficient C loops. We have therefore introduced `Ix_start` as `Ix[0]` and `Ix_end` as `Ix[-1]` to hold the start and end of the values of index i. Similar variables are introduced for the j index. A loop `for i in Ix` is with these new variables written as `for i in range(Ix_start, Ix_end+1)`.

Array declaration syntax in Cython

We have used the syntax `np.ndarray[DT, ndim=2, mode='c']` to declare numpy arrays in Cython. There is a simpler, alternative syntax, employing typed memory views[1], where the declaration looks like `double [:,:]`. However, the full support for this functionality is not yet ready, and in this text we use the full array declaration syntax.

C.5.2 Visual Inspection of the C Translation

Cython can visually explain how successfully it translated a code from Python to C. The command

---
Terminal

```
Terminal> cython -a wave2D_u0_loop_cy.pyx
```
---

produces an HTML file `wave2D_u0_loop_cy.html`, which can be loaded into a web browser to illustrate which lines of the code that have been translated to C. Figure C.1 shows the illustrated code. Yellow lines indicate the lines that Cython

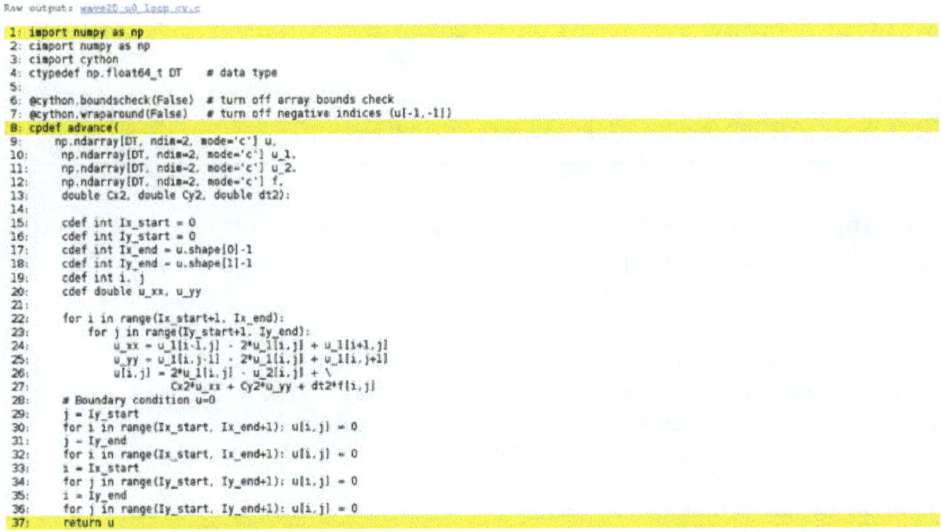

```
Raw output: wave2D_u0_loop_cy.c
1: import numpy as np
2: cimport numpy as np
3: cimport cython
4: ctypedef np.float64_t DT    # data type
5:
6: @cython.boundscheck(False)  # turn off array bounds check
7: @cython.wraparound(False)   # turn off negative indices (u[-1,-1])
8: cpdef advance(
9:     np.ndarray[DT, ndim=2, mode='c'] u,
10:    np.ndarray[DT, ndim=2, mode='c'] u_1,
11:    np.ndarray[DT, ndim=2, mode='c'] u_2,
12:    np.ndarray[DT, ndim=2, mode='c'] f,
13:    double Cx2, double Cy2, double dt2):
14:
15:    cdef int Ix_start = 0
16:    cdef int Iy_start = 0
17:    cdef int Ix_end = u.shape[0]-1
18:    cdef int Iy_end = u.shape[1]-1
19:    cdef int i, j
20:    cdef double u_xx, u_yy
21:
22:    for i in range(Ix_start+1, Ix_end):
23:        for j in range(Iy_start+1, Iy_end):
24:            u_xx = u_1[i-1,j] - 2*u_1[i,j] + u_1[i+1,j]
25:            u_yy = u_1[i,j-1] - 2*u_1[i,j] + u_1[i,j+1]
26:            u[i,j] = 2*u_1[i,j] - u_2[i,j] + \
27:                     Cx2*u_xx + Cy2*u_yy + dt2*f[i,j]
28:    # Boundary condition u=0
29:    j = Iy_start
30:    for i in range(Ix_start, Ix_end+1): u[i,j] = 0
31:    j = Iy_end
32:    for i in range(Ix_start, Ix_end+1): u[i,j] = 0
33:    i = Ix_start
34:    for j in range(Iy_start, Iy_end+1): u[i,j] = 0
35:    i = Iy_end
36:    for j in range(Iy_start, Iy_end+1): u[i,j] = 0
37:    return u
```

Fig. C.1 Visual illustration of Cython's ability to translate Python to C

---
[1] http://docs.cython.org/src/userguide/memoryviews.html

did not manage to translate to efficient C code and that remain in Python. For the present code we see that Cython is able to translate all the loops with array computing to C, which is our primary goal.

You can also inspect the generated C code directly, as it appears in the file `wave2D_u0_loop_cy.c`. Nevertheless, understanding this C code requires some familiarity with writing Python extension modules in C by hand. Deep down in the file we can see in detail how the compute-intensive statements have been translated into some complex C code that is quite different from what a human would write (at least if a direct correspondence to the mathematical notation was intended).

C.5.3   Building the Extension Module

Cython code must be translated to C, compiled, and linked to form what is known in the Python world as a *C extension module*. This is usually done by making a `setup.py` script, which is the standard way of building and installing Python software. For an extension module arising from Cython code, the following `setup.py` script is all we need to build and install the module:

```
from distutils.core import setup
from distutils.extension import Extension
from Cython.Distutils import build_ext

cymodule = 'wave2D_u0_loop_cy'
setup(
  name=cymodule
  ext_modules=[Extension(cymodule, [cymodule + '.pyx'],)],
  cmdclass={'build_ext': build_ext},
)
```

We run the script by

```
—————————————————————— Terminal ——————————————————————
Terminal> python setup.py build_ext --inplace
```

The `-inplace` option makes the extension module available in the current directory as the file `wave2D_u0_loop_cy.so`. This file acts as a normal Python module that can be imported and inspected:

```
>>> import wave2D_u0_loop_cy
>>> dir(wave2D_u0_loop_cy)
['__builtins__', '__doc__', '__file__', '__name__',
 '__package__', '__test__', 'advance', 'np']
```

The important output from the `dir` function is our Cython function `advance` (the module also features the imported numpy module under the name `np` as well as many standard Python objects with double underscores in their names).

The `setup.py` file makes use of the `distutils` package in Python and Cython's extension of this package. These tools know how Python was built on the computer and will use compatible compiler(s) and options when building other code in Cython, C, or C++. Quite some experience with building large program systems is needed to do the build process manually, so using a `setup.py` script is strongly recommended.

Simplified build of a Cython module

When there is no need to link the C code with special libraries, Cython offers a shortcut for generating and importing the extension module:

```
import pyximport; pyximport.install()
```

This makes the `setup.py` script redundant. However, in the `wave2D_u0_adv.py` code we do not use `pyximport` and require an explicit build process of this and many other modules.

C.5.4 Calling the Cython Function from Python

The `wave2D_u0_loop_cy` module contains our `advance` function, which we now may call from the Python program for the wave equation:

```
import wave2D_u0_loop_cy
advance = wave2D_u0_loop_cy.advance
...
for n in It[1:-1]:                     # time loop
    f_a[:,:] = f(xv, yv, t[n])         # precompute, size as u
    u = advance(u, u_n, u_nm1, f_a, x, y, t, Cx2, Cy2, dt2)
```

Efficiency For a mesh consisting of 120×120 cells, the scalar Python code requires 1370 CPU time units, the vectorized version requires 5.5, while the Cython version requires only 1! For a smaller mesh with 60×60 cells Cython is about 1000 times faster than the scalar Python code, and the vectorized version is about 6 times slower than the Cython version.

C.6 Migrating Loops to Fortran

Instead of relying on Cython's (excellent) ability to translate Python to C, we can invoke a compiled language directly and write the loops ourselves. Let us start with Fortran 77, because this is a language with more convenient array handling than C (or plain C++), because we can use the same multi-dimensional indices in the Fortran code as in the numpy arrays in the Python code, while in C these arrays are one-dimensional and require us to reduce multi-dimensional indices to a single index.

C.6.1   The Fortran Subroutine

We write a Fortran subroutine advance in a file wave2D_u0_loop_f77.f for implementing the updating formula (2.117) and setting the solution to zero at the boundaries:

```fortran
      subroutine advance(u, u_n, u_nm1, f, Cx2, Cy2, dt2, Nx, Ny)
      integer Nx, Ny
      real*8 u(0:Nx,0:Ny), u_n(0:Nx,0:Ny), u_nm1(0:Nx,0:Ny)
      real*8 f(0:Nx,0:Ny), Cx2, Cy2, dt2
      integer i, j
      real*8 u_xx, u_yy
Cf2py intent(in, out) u

C     Scheme at interior points
      do j = 1, Ny-1
         do i = 1, Nx-1
            u_xx = u_n(i-1,j) - 2*u_n(i,j) + u_n(i+1,j)
            u_yy = u_n(i,j-1) - 2*u_n(i,j) + u_n(i,j+1)
            u(i,j) = 2*u_n(i,j) - u_nm1(i,j) + Cx2*u_xx + Cy2*u_yy +
     &                dt2*f(i,j)
         end do
      end do

C     Boundary conditions
      j = 0
      do i = 0, Nx
         u(i,j) = 0
      end do
      j = Ny
      do i = 0, Nx
         u(i,j) = 0
      end do
      i = 0
      do j = 0, Ny
         u(i,j) = 0
      end do
      i = Nx
      do j = 0, Ny
         u(i,j) = 0
      end do
      return
      end
```

This code is plain Fortran 77, except for the special Cf2py comment line, which here specifies that u is both an input argument *and* an object to be returned from the advance routine. Or more precisely, Fortran is not able return an array from a function, but we need a *wrapper code* in C for the Fortran subroutine to enable calling it from Python, and from this wrapper code one can return u to the calling Python code.

Tip: Return all computed objects to the calling code

It is not strictly necessary to return u to the calling Python code since the advance function will modify the elements of u, but the convention in Python is to get all output from a function as returned values. That is, the right way of calling the above Fortran subroutine from Python is

```
u = advance(u, u_n, u_nm1, f, Cx2, Cy2, dt2)
```

The less encouraged style, which works and resembles the way the Fortran subroutine is called from Fortran, reads

```
advance(u, u_n, u_nm1, f, Cx2, Cy2, dt2)
```

C.6.2 Building the Fortran Module with f2py

The nice feature of writing loops in Fortran is that, without much effort, the tool f2py can produce a C extension module such that we can call the Fortran version of advance from Python. The necessary commands to run are

```
Terminal
Terminal> f2py -m wave2D_u0_loop_f77 -h wave2D_u0_loop_f77.pyf \
          --overwrite-signature wave2D_u0_loop_f77.f
Terminal> f2py -c wave2D_u0_loop_f77.pyf --build-dir build_f77 \
          -DF2PY_REPORT_ON_ARRAY_COPY=1 wave2D_u0_loop_f77.f
```

The first command asks f2py to interpret the Fortran code and make a Fortran 90 specification of the extension module in the file wave2D_u0_loop_f77.pyf. The second command makes f2py generate all necessary wrapper code, compile our Fortran file and the wrapper code, and finally build the module. The build process takes place in the specified subdirectory build_f77 so that files can be inspected if something goes wrong. The option -DF2PY_REPORT_ON_ARRAY_COPY=1 makes f2py write a message for every array that is copied in the communication between Fortran and Python, which is very useful for avoiding unnecessary array copying (see below). The name of the module file is wave2D_u0_loop_f77.so, and this file can be imported and inspected as any other Python module:

```
>>> import wave2D_u0_loop_f77
>>> dir(wave2D_u0_loop_f77)
['__doc__', '__file__', '__name__', '__package__',
 '__version__', 'advance']
>>> print wave2D_u0_loop_f77.__doc__
This module 'wave2D_u0_loop_f77' is auto-generated with f2py....
Functions:
  u = advance(u,u_n,u_nm1,f,cx2,cy2,dt2,
      nx=(shape(u,0)-1),ny=(shape(u,1)-1))
```

Examine the doc strings!

Printing the doc strings of the module and its functions is extremely important after having created a module with f2py. The reason is that f2py makes Python interfaces to the Fortran functions that are different from how the functions are declared in the Fortran code (!). The rationale for this behavior is that f2py creates *Pythonic* interfaces such that Fortran routines can be called in the same way as one calls Python functions. Output data from Python functions is always returned to the calling code, but this is technically impossible in Fortran. Also, arrays in Python are passed to Python functions without their dimensions because that information is packed with the array data in the array objects. This is not possible in Fortran, however. Therefore, f2py removes array dimensions from the argument list, and f2py makes it possible to return objects back to Python.

Let us follow the advice of examining the doc strings and take a close look at the documentation f2py has generated for our Fortran advance subroutine:

```
>>> print wave2D_u0_loop_f77.advance.__doc__
This module 'wave2D_u0_loop_f77' is auto-generated with f2py
Functions:
  u = advance(u,u_n,u_nm1,f,cx2,cy2,dt2,
              nx=(shape(u,0)-1),ny=(shape(u,1)-1))
.
advance - Function signature:
  u = advance(u,u_n,u_nm1,f,cx2,cy2,dt2,[nx,ny])
Required arguments:
  u : input rank-2 array('d') with bounds (nx + 1,ny + 1)
  u_n : input rank-2 array('d') with bounds (nx + 1,ny + 1)
  u_nm1 : input rank-2 array('d') with bounds (nx + 1,ny + 1)
  f : input rank-2 array('d') with bounds (nx + 1,ny + 1)
  cx2 : input float
  cy2 : input float
  dt2 : input float
Optional arguments:
  nx := (shape(u,0)-1) input int
  ny := (shape(u,1)-1) input int
Return objects:
  u : rank-2 array('d') with bounds (nx + 1,ny + 1)
```

Here we see that the nx and ny parameters declared in Fortran are optional arguments that can be omitted when calling advance from Python.

We strongly recommend to print out the documentation of *every* Fortran function to be called from Python and make sure the call syntax is exactly as listed in the documentation.

C.6.3 How to Avoid Array Copying

Multi-dimensional arrays are stored as a stream of numbers in memory. For a two-dimensional array consisting of rows and columns there are two ways of creating such a stream: *row-major ordering*, which means that rows are stored consecutively in memory, or *column-major ordering*, which means that the columns are stored one after each other. All programming languages inherited from C, including Python, apply the row-major ordering, but Fortran uses column-major storage. Thinking of a two-dimensional array in Python or C as a matrix, it means that Fortran works with the transposed matrix.

Fortunately, f2py creates extra code so that accessing u(i,j) in the Fortran sub-routine corresponds to the element u[i,j] in the underlying numpy array (without the extra code, u(i,j) in Fortran would access u[j,i] in the numpy array). Technically, f2py takes a copy of our numpy array and reorders the data before sending the array to Fortran. Such copying can be costly. For 2D wave simulations on a 60×60 grid the overhead of copying is a factor of 5, which means that almost the whole performance gain of Fortran over vectorized numpy code is lost!

To avoid having f2py to copy arrays with C storage to the corresponding Fortran storage, we declare the arrays with Fortran storage:

```
order = 'Fortran' if version == 'f77' else 'C'
u    = zeros((Nx+1,Ny+1), order=order)   # solution array
u_n  = zeros((Nx+1,Ny+1), order=order)   # solution at t-dt
u_nm1 = zeros((Nx+1,Ny+1), order=order)   # solution at t-2*dt
```

In the compile and build step of using f2py, it is recommended to add an extra option for making f2py report on array copying:

```
                          ┌──────────┐
──────────────────────────│ Terminal │──────────────────────────
                          └──────────┘
Terminal> f2py -c wave2D_u0_loop_f77.pyf --build-dir build_f77 \
          -DF2PY_REPORT_ON_ARRAY_COPY=1 wave2D_u0_loop_f77.f
─────────────────────────────────────────────────────────────────
```

It can sometimes be a challenge to track down which array that causes a copying. There are two principal reasons for copying array data: either the array does not have Fortran storage or the element types do not match those declared in the Fortran code. The latter cause is usually effectively eliminated by using real*8 data in the Fortran code and float64 (the default float type in numpy) in the arrays on the Python side. The former reason is more common, and to check whether an array before a Fortran call has the right storage one can print the result of isfortran(a), which is True if the array a has Fortran storage.

Let us look at an example where we face problems with array storage. A typical problem in the wave2D_u0.py code is to set

```
f_a = f(xv, yv, t[n])
```

before the call to the Fortran advance routine. This computation creates a new array with C storage. An undesired copy of f_a will be produced when sending f_a to a Fortran routine. There are two remedies, either direct insertion of data in an array with Fortran storage,

```
f_a = zeros((Nx+1, Ny+1), order='Fortran')
...
f_a[:,:] = f(xv, yv, t[n])
```

or remaking the f(xv, yv, t[n]) array,

```
f_a = asarray(f(xv, yv, t[n]), order='Fortran')
```

The former remedy is most efficient if the asarray operation is to be performed a large number of times.

Efficiency The efficiency of this Fortran code is very similar to the Cython code. There is usually nothing more to gain, from a computational efficiency point of view, by implementing the *complete* Python program in Fortran or C. That will just be a lot more code for all administering work that is needed in scientific software, especially if we extend our sample program wave2D_u0.py to handle a real scientific problem. Then only a small portion will consist of loops with intensive array calculations. These can be migrated to Cython or Fortran as explained, while the rest of the programming can be more conveniently done in Python.

C.7 Migrating Loops to C via Cython

The computationally intensive loops can alternatively be implemented in C code. Just as Fortran calls for care regarding the storage of two-dimensional arrays, working with two-dimensional arrays in C is a bit tricky. The reason is that numpy arrays are viewed as one-dimensional arrays when transferred to C, while C programmers will think of u, u_n, and u_nm1 as two dimensional arrays and index them like u[i][j]. The C code must declare u as double* u and translate an index pair [i][j] to a corresponding single index when u is viewed as one-dimensional. This translation requires knowledge of how the numbers in u are stored in memory.

C.7.1 Translating Index Pairs to Single Indices

Two-dimensional numpy arrays with the default C storage are stored row by row. In general, multi-dimensional arrays with C storage are stored such that the last index has the fastest variation, then the next last index, and so on, ending up with the slowest variation in the first index. For a two-dimensional u declared as zeros((Nx+1,Ny+1)) in Python, the individual elements are stored in the following order:

```
u[0,0], u[0,1], u[0,2], ..., u[0,Ny], u[1,0], u[1,1], ...,
u[1,Ny], u[2,0], ..., u[Nx,0], u[Nx,1], ..., u[Nx, Ny]
```

Viewing u as one-dimensional, the index pair (i, j) translates to $i(N_y + 1) + j$. So, where a C programmer would naturally write an index u[i][j], the indexing must read u[i*(Ny+1) + j]. This is tedious to write, so it can be handy to define a C macro,

```
#define idx(i,j) (i)*(Ny+1) + j
```

so that we can write u[idx(i,j)], which reads much better and is easier to debug.

> **Be careful with macro definitions**
>
> Macros just perform simple text substitutions: `idx(hello,world)` is expanded
> to `(hello)*(Ny+1) + world`. The parentheses in `(i)` are essential – us-
> ing the natural mathematical formula `i*(Ny+1) + j` in the macro definition,
> `idx(i-1,j)` would expand to `i-1*(Ny+1) + j`, which is the wrong formula.
> Macros are handy, but require careful use. In C++, inline functions are safer and
> replace the need for macros.

C.7.2 The Complete C Code

The C version of our function `advance` can be coded as follows.

```c
#define idx(i,j) (i)*(Ny+1) + j

void advance(double* u, double* u_n, double* u_nm1, double* f,
        double Cx2, double Cy2, double dt2, int Nx, int Ny)
{
  int i, j;
  double u_xx, u_yy;
  /* Scheme at interior points */
  for (i=1; i<=Nx-1; i++) {
    for (j=1; j<=Ny-1; j++) {
      u_xx = u_n[idx(i-1,j)] - 2*u_n[idx(i,j)] + u_n[idx(i+1,j)];
      u_yy = u_n[idx(i,j-1)] - 2*u_n[idx(i,j)] + u_n[idx(i,j+1)];
      u[idx(i,j)] = 2*u_n[idx(i,j)] - u_nm1[idx(i,j)] +
      Cx2*u_xx + Cy2*u_yy + dt2*f[idx(i,j)];
    }
  }
  /* Boundary conditions */
  j = 0;  for (i=0; i<=Nx; i++) u[idx(i,j)] = 0;
  j = Ny; for (i=0; i<=Nx; i++) u[idx(i,j)] = 0;
  i = 0;  for (j=0; j<=Ny; j++) u[idx(i,j)] = 0;
  i = Nx; for (j=0; j<=Ny; j++) u[idx(i,j)] = 0;
}
```

C.7.3 The Cython Interface File

All the code above appears in the file `wave2D_u0_loop_c.c`. We need to compile
this file together with C wrapper code such that `advance` can be called from Python.
Cython can be used to generate appropriate wrapper code. The relevant Cython
code for interfacing C is placed in a file with extension `.pyx`. This file, called
`wave2D_u0_loop_c_cy.pyx`[2], looks like

```python
import numpy as np
cimport numpy as np
cimport cython
```

---

[2] http://tinyurl.com/nu656p2/softeng2/wave2D_u0_loop_c_cy.pyx

```
cdef extern from "wave2D_u0_loop_c.h":
    void advance(double* u, double* u_n, double* u_nm1, double* f,
                 double Cx2, double Cy2, double dt2,
                 int Nx, int Ny)

@cython.boundscheck(False)
@cython.wraparound(False)
def advance_cwrap(
    np.ndarray[double, ndim=2, mode='c'] u,
    np.ndarray[double, ndim=2, mode='c'] u_n,
    np.ndarray[double, ndim=2, mode='c'] u_nm1,
    np.ndarray[double, ndim=2, mode='c'] f,
    double Cx2, double Cy2, double dt2):
    advance(&u[0,0], &u_n[0,0], &u_nm1[0,0], &f[0,0],
            Cx2, Cy2, dt2,
            u.shape[0]-1, u.shape[1]-1)
    return u
```

We first declare the C functions to be interfaced. These must also appear in a C header file, `wave2D_u0_loop_c.h`,

```
extern void advance(double* u, double* u_n, double* u_nm1, double* f,
            double Cx2, double Cy2, double dt2,
            int Nx, int Ny);
```

The next step is to write a Cython function with Python objects as arguments. The name advance is already used for the C function so the function to be called from Python is named `advance_cwrap`. The contents of this function is simply a call to the advance version in C. To this end, the right information from the Python objects must be passed on as arguments to advance. Arrays are sent with their C pointers to the first element, obtained in Cython as `&u[0,0]` (the & takes the address of a C variable). The Nx and Ny arguments in advance are easily obtained from the shape of the numpy array u. Finally, u must be returned such that we can set u = advance(...) in Python.

C.7.4   Building the Extension Module

It remains to build the extension module. An appropriate `setup.py` file is

```
from distutils.core import setup
from distutils.extension import Extension
from Cython.Distutils import build_ext

sources = ['wave2D_u0_loop_c.c', 'wave2D_u0_loop_c_cy.pyx']
module = 'wave2D_u0_loop_c_cy'
setup(
  name=module,
  ext_modules=[Extension(module, sources,
                         libraries=[], # C libs to link with
                         )],
  cmdclass={'build_ext': build_ext},
)
```

All we need to specify is the `.c` file(s) and the `.pyx` interface file. Cython is automatically run to generate the necessary wrapper code. Files are then compiled and linked to an extension module residing in the file `wave2D_u0_loop_c_cy.so`. Here is a session with running `setup.py` and examining the resulting module in Python

```
Terminal
Terminal> python setup.py build_ext --inplace
Terminal> python
>>> import wave2D_u0_loop_c_cy as m
>>> dir(m)
['__builtins__', '__doc__', '__file__', '__name__', '__package__',
 '__test__', 'advance_cwrap', 'np']
```

The call to the C version of advance can go like this in Python:

```
import wave2D_u0_loop_c_cy
advance = wave2D_u0_loop_c_cy.advance_cwrap
...
f_a[:,:] = f(xv, yv, t[n])
u = advance(u, u_n, u_nm1, f_a, Cx2, Cy2, dt2)
```

Efficiency In this example, the C and Fortran code runs at the same speed, and there are no significant differences in the efficiency of the wrapper code. The overhead implied by the wrapper code is negligible as long as there is little numerical work in the advance function, or in other words, that we work with small meshes.

C.8   Migrating Loops to C via f2py

An alternative to using Cython for interfacing C code is to apply `f2py`. The C code is the same, just the details of specifying how it is to be called from Python differ. The `f2py` tool requires the call specification to be a Fortran 90 module defined in a `.pyf` file. This file was automatically generated when we interfaced a Fortran subroutine. With a C function we need to write this module ourselves, or we can use a trick and let `f2py` generate it for us. The trick consists in writing the signature of the C function with Fortran syntax and place it in a Fortran file, here `wave2D_u0_loop_c_f2py_signature.f`:

```
      subroutine advance(u, u_n, u_nm1, f, Cx2, Cy2, dt2, Nx, Ny)
Cf2py intent(c) advance
      integer Nx, Ny, N
      real*8 u(0:Nx,0:Ny), u_n(0:Nx,0:Ny), u_nm1(0:Nx,0:Ny)
      real*8 f(0:Nx, 0:Ny), Cx2, Cy2, dt2
Cf2py intent(in, out) u
Cf2py intent(c) u, u_n, u_nm1, f, Cx2, Cy2, dt2, Nx, Ny
      return
      end
```

Note that we need a special f2py instruction, through a Cf2py comment line, to specify that all the function arguments are C variables. We also need to tell that the function is actually in C: intent(c) advance.

Since f2py is just concerned with the function signature and not the complete contents of the function body, it can easily generate the Fortran 90 module specification based solely on the signature above:

```
Terminal> f2py -m wave2D_u0_loop_c_f2py \
          -h wave2D_u0_loop_c_f2py.pyf --overwrite-signature \
          wave2D_u0_loop_c_f2py_signature.f
```

The compile and build step is as for the Fortran code, except that we list C files instead of Fortran files:

```
Terminal> f2py -c wave2D_u0_loop_c_f2py.pyf \
          --build-dir tmp_build_c \
          -DF2PY_REPORT_ON_ARRAY_COPY=1 wave2D_u0_loop_c.c
```

As when interfacing Fortran code with f2py, we need to print out the doc string to see the exact call syntax from the Python side. This doc string is identical for the C and Fortran versions of advance.

C.8.1 Migrating Loops to C++ via f2py

C++ is a much more versatile language than C or Fortran and has over the last two decades become very popular for numerical computing. Many will therefore prefer to migrate compute-intensive Python code to C++. This is, in principle, easy: just write the desired C++ code and use some tool for interfacing it from Python. A tool like SWIG[3] can interpret the C++ code and generate interfaces for a wide range of languages, including Python, Perl, Ruby, and Java. However, SWIG is a comprehensive tool with a correspondingly steep learning curve. Alternative tools, such as Boost Python[4], SIP[5], and Shiboken[6] are similarly comprehensive. Simpler tools include PyBindGen[7].

A technically much easier way of interfacing C++ code is to drop the possibility to use C++ classes directly from Python, but instead make a C interface to the C++ code. The C interface can be handled by f2py as shown in the example with pure C code. Such a solution means that classes in Python and C++ cannot be mixed and that only primitive data types like numbers, strings, and arrays can be transferred between Python and C++. Actually, this is often a very good solution because it

---

[3] http://swig.org/

[4] http://www.boost.org/doc/libs/1_51_0/libs/python/doc/index.html

[5] http://riverbankcomputing.co.uk/software/sip/intro

[6] http://qt-project.org/wiki/Category:LanguageBindings::PySide::Shiboken

[7] http://code.google.com/p/pybindgen/

forces the C++ code to work on array data, which usually gives faster code than if fancy data structures with classes are used. The arrays coming from Python, and looking like plain C/C++ arrays, can be efficiently wrapped in more user-friendly C++ array classes in the C++ code, if desired.

C.9 Exercises

Exercise C.1: Explore computational efficiency of numpy.sum versus built-in sum

Using the task of computing the sum of the first n integers, we want to compare the efficiency of numpy.sum versus Python's built-in function sum. Use IPython's %timeit functionality to time these two functions applied to three different arguments: range(n), xrange(n), and arange(n).
Filename: sumn.

Exercise C.2: Make an improved numpy.savez function

The numpy.savez function can save multiple arrays to a zip archive. Unfortunately, if we want to use savez in time-dependent problems and call it multiple times (once per time level), each call leads to a separate zip archive. It is more convenient to have all arrays in one archive, which can be read by numpy.load. Section C.2 provides a recipe for merging all the individual zip archives into one archive. An alternative is to write a new savez function that allows multiple calls and storage into the same archive prior to a final close method to close the archive and make it ready for reading. Implement such an improved savez function as a class Savez.

The class should pass the following unit test:

```python
def test_Savez():
    import tempfile, os
    tmp = 'tmp_testarchive'
    database = Savez(tmp)
    for i in range(4):
        array = np.linspace(0, 5+i, 3)
        kwargs = {'myarray_%02d' % i: array}
        database.savez(**kwargs)
    database.close()

    database = np.load(tmp+'.npz')

    expected = {
        'myarray_00': np.array([ 0. ,   2.5,  5. ]),
        'myarray_01': np.array([ 0.,   3.,   6.])
        'myarray_02': np.array([ 0. ,   3.5,  7. ]),
        'myarray_03': np.array([ 0.,   4.,   8.]),
        }
    for name in database:
        computed = database[name]
        diff = np.abs(expected[name] - computed).max()
        assert diff < 1E-13
    database.close
    os.remove(tmp+'.npz')
```

Hint Study the source code[8] for function `savez` (or more precisely, function `_savez`).
Filename: Savez.

Exercise C.3: Visualize the impact of the Courant number

Use the `pulse` function in the `wave1D_dn_vc.py` to simulate a pulse through two media with different wave velocities. The aim is to visualize the impact of the Courant number C on the quality of the solution. Set `slowness_factor=4` and `Nx=100`.

Simulate for $C = 1, 0.9, 0.75$ and make an animation comparing the three curves (use the `animate_archives.py` program to combine the curves and make animations on the screen and video files). Perform the investigations for different types of initial profiles: a Gaussian pulse, a "cosine hat" pulse, half a "cosine hat" pulse, and a plug pulse.
Filename: `pulse1D_Courant`.

Exercise C.4: Visualize the impact of the resolution

We solve the same set of problems as in Exercise C.3, except that we now fix $C = 1$ and instead study the impact of Δt and Δx by varying the Nx parameter: 20, 40, 160. Make animations comparing three such curves.
Filename: `pulse1D_Nx`.

---

[8] https://github.com/numpy/numpy/blob/master/numpy/lib/npyio.py

References

1. O. Axelsson. *Iterative Solution Methods*. Cambridge University Press, 1996.
2. R. Barrett, M. Berry, T. F. Chan, J. Demmel, J. Donato, J. Dongarra, V. Eijkhout, R. Pozo, C. Romine, and H. Van der Vorst. *Templates for the Solution of Linear Systems: Building Blocks for Iterative Methods*. SIAM, second edition, 1994. http://www.netlib.org/linalg/html_templates/Templates.html.
3. D. Duran. *Numerical Methods for Fluid Dynamics - With Applications to Geophysics*. Springer, second edition, 2010.
4. C. A. J. Fletcher. *Computational Techniques for Fluid Dynamics, Vol. 1: Fundamental and General Techniques*. Springer, second edition, 2013.
5. C. Greif and U. M. Ascher. *A First Course in Numerical Methods*. Computational Science and Engineering. SIAM, 2011.
6. E. Hairer, S. P. Nørsett, and G. Wanner. *Solving Ordinary Differential Equations I. Nonstiff Problems*. Springer, 1993.
7. M. Hjorth-Jensen. *Computational Physics*. Institute of Physics Publishing, 2016. https://github.com/CompPhysics/ComputationalPhysics1/raw/gh-pages/doc/L%ectures/lectures2015.pdf.
8. C. T. Kelley. *Iterative Methods for Linear and Nonlinear Equations*. SIAM, 1995.
9. H. P. Langtangen. *Finite Difference Computing with Exponential Decay Models*. Lecture Notes in Computational Science and Engineering. Springer, 2016. http://hplgit.github.io/decay-book/doc/web/.
10. H. P. Langtangen. *A Primer on Scientific Programming with Python*. Texts in Computational Science and Engineering. Springer, fifth edition, 2016.
11. H. P. Langtangen and G. K. Pedersen. *Scaling of Differential Equations*. Simula Springer Brief Series. Springer, 2016. http://hplgit.github.io/scaling-book/doc/web/.
12. L. Lapidus and G. F. Pinder. *Numerical Solution of Partial Differential Equations in Science and Engineering*. Wiley, 1982.
13. R. LeVeque. *Finite Difference Methods for Ordinary and Partial Differential Equations: Steady-State and Time-Dependent Problems*. SIAM, 2007.
14. I. P. Omelyan, I. M. Mryglod, and R. Folk. Optimized forest-ruth- and suzuki-like algorithms for integration of motion in many-body systems. *Computer Physics Communication*, 146(2):188–202, 2002.
15. R. Rannacher. Finite element solution of diffusion problems with irregular data. *Numerische Mathematik*, 43:309–327, 1984.
16. Y. Saad. *Iterative Methods for Sparse Linear Systems*. SIAM, second edition, 2003. http://www-users.cs.umn.edu/~saad/IterMethBook_2ndEd.pdf.
17. J. Strikwerda. *Numerical Solution of Partial Differential Equations in Science and Engineering*. SIAM, second edition, 2007.
18. L. N. Trefethen. *Trefethen's index cards - Forty years of notes about People, Words and Mathematics*. World Scientific, 2011.

Series Editors

Timothy J. Barth
NASA Ames Research Center
NAS Division
Moffett Field, CA 94035, USA
barth@nas.nasa.gov

Michael Griebel
Institut für Numerische Simulation
der Universität Bonn
Wegelerstr. 6
53115 Bonn, Germany
griebel@ins.uni-bonn.de

David E. Keyes
Mathematical and Computer Sciences
and Engineering
King Abdullah University of Science
and Technology
P.O. Box 55455
Jeddah 21534, Saudi Arabia
david.keyes@kaust.edu.sa

and

Department of Applied Physics
and Applied Mathematics
Columbia University
500 W. 120 th Street
New York, NY 10027, USA
kd2112@columbia.edu

Risto M. Nieminen
Department of Applied Physics
Aalto University School of Science
and Technology
00076 Aalto, Finland
risto.nieminen@aalto.fi

Dirk Roose
Department of Computer Science
Katholieke Universiteit Leuven
Celestijnenlaan 200A
3001 Leuven-Heverlee, Belgium
dirk.roose@cs.kuleuven.be

Tamar Schlick
Department of Chemistry
and Courant Institute
of Mathematical Sciences
New York University
251 Mercer Street
New York, NY 10012, USA
schlick@nyu.edu

Editor for Computational Science
and Engineering at Springer:
Martin Peters
Springer-Verlag
Mathematics Editorial IV
Tiergartenstrasse 17
69121 Heidelberg, Germany
martin.peters@springer.com

Permissions

Index

N

Nonlinear Diffusion Equation, 166, 236

Numerical Artifacts, 35

Numerical Solution, 11-12, 28, 30-31, 34, 36, 58, 121, 133, 140, 142, 194, 206-207, 223, 239, 251, 281

O

Operator Notation, 30, 141, 153, 177, 200, 226

Operator Splitting, 180

P

Periodic Boundary Condition, 127

Picard Iteration, 148, 150-154, 156-159, 162-165, 167, 169, 171, 173-174, 176-179, 194-195, 197-199

Preconditioning, 79, 112

Pulse Function, 280

Q

Quadratic Solution, 58, 223

S

Solver Function, 115, 119, 128-129, 241, 248-249, 251, 260, 264

Space Dimension, 2

Space Scheme, 125-126, 134, 143

Stability Criterion, 34-35, 167

Stability Limit, 9, 33, 36, 39, 187, 250

Staggered Mesh, 229-230

Stationary Solution, 1, 41, 70

Strang Splitting, 181-189, 191-192

Symbolic Software, 205-206, 214

Symmetry Boundary Condition, 36-37

T

Test Function, 7, 9, 58, 86, 92-93, 114

Truncation Error, 28, 30-31, 59, 117, 137, 153, 195, 205-209, 212-231, 233-239

V

Variable Coefficient, 46, 171, 232, 234, 237, 240

Vectorization, 5, 57, 63, 72, 84, 89, 91, 101, 264

Vibration Problems, 195, 224

Volume Flux, 109

W

Wave Component, 27-30, 32, 34-35, 121, 193

Wave Equation, 1-2, 8, 230, 232, 234, 240, 254, 260, 264, 269

Wave Equations, 117, 230

Wave Velocity, 134-135, 231, 234, 240, 254, 264